COAL

THE AUSTRALIAN

STORY

VOLUME 2

1970-2020

Published in 2020 by Connor Court Publishing Pty Ltd

Connor Court Publishing Pty Ltd
PO Box 7257
Redland Bay QLD 4165

sales@connorcourt.com
www.connorcourtpublishing.com.au

ISBN: 9781922449313

Front Cover Design: Maria Giordano
Cover photo: Dragline at BHP Mitsubishi Alliance's Cavil Ridge mine, Bowen Basin, Queensland. Courtesy BHP Group Ltd

Printed in Australia

COAL

THE AUSTRALIAN

STORY

VOLUME 2

1970-2020

Denis Porter

Connor Court Publishing

Previous Volume

COAL

THE AUSTRALIAN STORY

From convict mining to the birth of a world leader

ISBN: 9781925826609

List of shortened forms

ABARE	Australian Bureau of Agricultural and Resource Economics
ABS	Australian Bureau of Statistics
ACA	Australian Coal Association
ACIRRT	Australian Centre for Industrial Relations Research and Training
ACSA	Australian Collieries Staff Association (now part of Professionals Australia)
AGPS	Australian Government Publishing Service
AIRC	Australian Industrial Relations Commission
ALP	Australian Labor Party
AMIC	Australian Mining Industry Council; now the Minerals Coucil of Australia
AMWU	Amalgamated Metal Workers Union
ANU	Australian National University
APESMA	Association of Professional Engineers, Scientists and Managers Australia; now Professionals Australia
ATO	Australian Taxation Office
AusIMM	Australasian Institute of Mining and Metallurgy
AWU	Australian Workers' Union
BCA	Business Council of Australia
BHP	Broken Hill Proprietary Company Limited; now BHP Group Limited
BHPB	BHP Billiton Limited
BIE	Bureau of Industry Economics
BMA	BHP Mitsubishi Alliance
BREE	Bureau of Resource Economics and Energy (now Office of Chief Economist in Department of Industry, Science, Energy and Resources)
CFMEU	Construction Forestry Mining and Energy Union
CFMMEU	Construction Forestry Maritime Mining and Energy Union

CIT	Coal Industry Tribunal
CQCA	Central Queensland Coal Associates
CRA	Conzinc Rio Tinto of Australia Limited
DBCT	Dalrymple Bay Coal Terminal
EIA	Energy Information Administration; part of US Department of Energy
EIS	Environmental impact statement
EPDC	Electric Power Development Company of Japan (also known as J-Power)
ETS	Emissions trading scheme
ETU	Electrical Trades Union
Federation	Miners' Federation; now the Mining and Energy Division of the CFMMEU
FEDFA	Federated Engine Drivers' and Firemen's Association
FIRB	Foreign Investment Review Board
IEA	International Energy Agency
JCB	Joint Coal Board
JFY	Japanese financial year; commences 1 April
JPU	Japanese power utilities
JSM	Japanese steel mills
LCA	Local Coal Authority
MIM	Mount Isa Mines
MITI	Japanese Ministry of International Trade and Industry; now Ministry of Economy, Trade and Industry
NCIG	Newcastle Coal Infrastructure Group
NSW	State of New South Wales, Australia
NSWCCPA	NSW Combined Colliery Proprietors' Association
NSWMC	NSW Minerals Council
OECD	Organisation for Economic Cooperation and Development
PKCT	Port Kembla Coal Terminal
PWCS	Port Waratah Coal Services
QCA	Queensland Coal Association
QCB	Queensland Coal Board

QMC	Queensland Minerals Council; now Queensland Resources Council
QRC	Queensland Resources Council
Qld	State of Queensland, Australia
RBA	Reserve Bank of Australia
SA	State of South Australia, Australia
SMCA	State Mines Control Authority of NSW
TDM	Thiess Dampier Mitsui Coal Pty Limited
TPM	Thiess Peabody Mitsui Coal Pty Limited
WA	State of Western Australia, Australia

Foreword

The first volume of "Coal – the Australian Story", published in 2019, serviced a wide readership of people with economic, technical, historical, industrial and regional interests in coal. This second volume has even wider appeal.

Denis Porter not only completes the history of coal to the present time for those same readers but also will attract new ones. Those who worked in and lived through the exciting growth of coal of the last 50 years and those interested in the place of coal in the decades ahead will find this book compelling reading. Denis himself is such a person, having held senior positions in the mining industry during that time, including CEO of the New South Wales Minerals Council.

This book chronicles coal mining in Queensland and New South Wales from 1970. Denis details the development of the industry, the changes in commercial ownership, foreign and local investment, employment, the impact on established coal communities and the growth of new ones, industrial relations issues, the involvement of governments, and the huge expansion of coal exports.

These years comprise the third and most sustained of the three periods over the last two centuries in which mining has lifted Australia to the forefront of global living standards. Exports of coal, together with iron ore, bauxite and natural gas combined with gold and other metals and minerals to deliver unprecedented benefits to Australians.

Coal also provided the crucial internationally competitive advantage of cheap electricity that enabled so much value to be added to Australian minerals in Australia, particularly by the power-hungry refining and smelting of bauxite into alumina and aluminium.

While this book provides a comprehensive resource on coal, more broadly it challenges Australians on their attitude to the mining industry.

What do Australians want of the coal industry in the future, both as a generator of export revenue and domestically?

Denis concludes with a review of forecasts of global demand for both metallurgical coal in steel making and thermal coal in power generation, forecasts that point to continuing large-scale use of coal for decades to come, forecasts that vary as governments address the carbon emissions issue.

Coal is currently a major part of an Australian mining industry that contributes more than half of the nation's export revenue. Australia clearly has the capacity to supply high quality coals to meet international demand.

Australians have the luxury of deciding whether they wish to do so.

Whether they wish to keep Australia as a supplier of high quality coals in the forthcoming decades of high coal usage around the world, particularly in Asia.

Whether they wish to maintain the high skills advantage that Australia possesses in mining (the number of Australian mining engineering graduates has fallen from about 300 to 50 a year over the past 50 years).

Whether they wish to regain Australia's low-cost electricity advantage at home.

This book not only records the Australian story of coal, but also stimulates profound consideration of coal – and mining - in Australia's future.

Collin Myers AM

Table of Contents

NSW royalty hike in 1989; company takeovers, comings and goings; BHP buys Utah; CSR bursts on to the energy scene; the oil companies invest, then sell out; Peko Wallsend rocks the status quo; CRA, the Blair Athol saga and foreign investment policy; CRA and CSR fight for Hail Creek; demands for a coal marketing authority; coal industrial relations in the 1980s; Australian Industrial relations – the precursor to a new era; pit ponies become union members; Japan's coal industry death spiral continues

Queensland development resumes; NSW mining developments also get underway; Billiton enters Australian coal industry; coal prices, strikes and the Taylor Review; the 1994 coal price settlements; the Taylor inquiry; Japan and Australian coal trade; steaming coal market becomes more competitive; export controls on coal abolished; coal market growth in the 1990s; price discrimination and excess supply; the Joint Coal Board's powers reduced; the end of the JCB and QCB era; the QCB is abolished; the end of the Coal Industry Tribunal; the coal and mineral associations unite; the industrial battles of the 1990s; Gordonstone – a fight to the death; Rio Tinto's Vickery and Hunter Valley No.1 disputes; coal mines restructure in both States; BHP acknowledges the role of management; Shell Coal's IR experience was mixed; Camberwell was innovative, but still frustrated; old habits die hard; BHP says goodbye to the Hunter Valley and Newcastle; State Governments give up de facto rail taxes; coal research – a success story for the industry.

Appin 1979; Box Flat 1972; Moura and Kianga 1975, 1986, 1994; Gretley 1996; NSW Mine Safety Review; the industry post Moura No.2 and Gretley; black lung disease reappears; new manslaughter provisions in Queensland.

BHP and Billiton merge; BHPBilliton's urge to merge with Rio Tinto; Rio Tinto's joint venture with Chinalco implodes; the rise of Glencore – and the demise of MIM; Gauci fights a losing battle; Glasenberg accuses other mining CEOs of creating excess supply; Rio Tinto exits coal; Peabody in, out, in again; Centennial – from small beginnings to a major producer; NSW Government exits coal mining after 90 years; China and India become major coal importers and investors; China's investments in Australian coal; India now a major investor and importer; Japan still a dominant overseas investor; coal pricing is no longer controversial; Queensland privatises QR – but not without a fight; NSW rail privatisation goes smoothly; carbon taxes and emissions trading; the Rudd and Gillard super profits taxes; the golden years for mining profits; Treasury gets it wrong; the productivity puzzle; from an industry in turmoil to a world leader.

Introduction

This book is the second in a series of two which tell the story of the development of the coal mining industry in the States of New South Wales (NSW) and Queensland. The first book – *Coal: the Australian Story, from convict mining to the birth of a world leader* – covers the industry from the early days after colonisation by the British up to and including the 1960s, the decade when the industry began its development to supply the growing market in Japan for metallurgical coal. This second volume takes up the story from the 1970s through to the present day, with some brief thoughts on the future.

The first century following colonisation by the British saw the coal industry grow in both states to become a key component of everyday life. By the time of Federation in 1901, the industry had established itself as an integral and critical part of the economy of the new Australian nation, supplying coal for the factories, railways and gas plants in NSW and Queensland and in other states. There was a also a substantial interstate trade in coal, with NSW supplying major users including gas companies, railways and factories in Melbourne, Adelaide, Perth and other centres. There was also a significant export trade, with NSW coal mines supplying coal to New Zealand, the Sandwich Islands (modern day Hawaii), Java (Indonesia) and other markets. The path was then a rocky one in the following three decades, interrupted by World War One and devastated by the Great Depression.

During World War Two the industry in NSW was wracked by industrial disputes and other problems and failed to produce sufficient coal for the nation's needs. At the end of the war the industry in both states was in a poor condition, unprepared to meet the challenges and demands of a growing economy. The NSW and Commonwealth Governments then established the Joint Coal Board to reform the NSW coal industry.

The Coal Industry Tribunal was established under the same Acts of the Commonwealth and NSW Parliaments which established the Joint Coal Board; its powers also extended into the Queensland coal industry. The Queensland Government set up its own coal board, also with strong powers to restructure that state's industry and control its development. The 1950s then saw major programs to modernise and mechanise the industry, with consequent restructuring and job losses. Major changes also took place in the local market for coal, with railways converting to diesel powered locomotives, oil replacing coal as the fuel used by many factories, and the decline in the use of coal for producing town gas. On the positive side, new coal fired power stations were being planned and developed, stimulating the development and expansion of a number of coal mines. And with the growth of industries such as aluminium, the demand for electricity continued to surge.

Following the visit of the first major Japanese coal industry group in 1958, the first longer term export contracts for supply of metallurgical coal were signed by several NSW south coast producers and the Japanese steel mills (JSM), followed shortly after by the development of the Thiess Peabody Mitsui Moura and Kianga mines in Queensland. The big contracts won by Thiess to supply the JSM were then eclipsed by huge contracts won by US mining company Utah Development for its new Goonyella and Peak Downs mines. The massive development of the coal industry in the Bowen Basin was underway. At the end of the 1960s, the Australian coal industry was well positioned to become a major force in the international coal market, perhaps even the dominant player. This book picks up the story from the 1970s, a decade which would see economic upheavals around the world, primarily due to the oil shocks of 1973 and 1979, the war in Vietnam and also the major currency realignments involving the US dollar, Japanese yen and other currencies. These upheavals would see Japanese steel production peak in 1973, but also herald the emergence of a major new world trade in thermal coal in the late 1970s as countries reacted to the hikes in oil prices with plans for new power stations fired by coal, gas and nuclear energy. The 1970s were also the decade which saw the Labor Government elected in Australia in

1972, with the new government introducing controls over the export of minerals, major restrictions on foreign investment and an early de facto super profits tax, the coal export duty.

The decades of the 1970s, 1980s and 1990s also proved to be turbulent and challenging, but saw Australia develop into the leading global coal exporter, overtaking the USA as the dominant metallurgical supplier and also becoming the leading thermal coal exporter. The coal industry and the international coal trade were often centre stage in political and public debates, with Japan accused of paying too little for Australia's coal and of encouraging an excess supply by investing in coal mines in a number of countries or underwriting their development with long term contracts. These controversies about the market also saw the dominant coal industry union, the Miners' Federation (later the CFMEU) demanding that the Australian Government establish a national authority to market coal or at least actively oversee its marketing. Export controls on the industry, introduced in early 1973, continued to be applied until their final abolition in 1997. The union's demands and the controversies continued into the 1990s, but have now faded away with the emergence of other major markets including China and India and with consequent lessening in the power of the JSM and the Japanese power utilities.

The rapid development of the industry in Australia from the 1960s to supply new and growing markets initially in Japan and later in a range of other countries including Korea, Taiwan and countries in western Europe brought with it pressures on the industry to become more efficient. Pressures also grew on state governments to invest more in rail and port infrastructure and to lift the efficiency of those rail and port facilities, as did pressures on both state and federal governments to rationalise the regulation of the coal industry, in particular by bringing that regulation more into the mainstream of Australian industry. With the Queensland export coal industry not constrained by the Queensland Coal Board, the end of the Joint Coal Board in NSW was arguably only a matter of time, although that would come in two stages – a restructured and much less powerful board in the early 1990s and its transformation into an industry-owned company in the early 2000s. Who would have thought that the CFMEU and the NSW Minerals Council would ever have become shareholders in such an organisation? The Queensland Coal Board also exited in the 1990s.

The mainstreaming of the industry also involved the abolition of the Coal Industry Tribunal in 1995, ending a period of almost 50 years during which it was effectively separated from the mainstream industrial relations system and headed by just two chairmen. The Tribunal was finally absorbed into the Australian Industrial Relations Commission in 1995 by a Labor Government, although not without massive resistance from the CFMEU and politicians from coal electorates. That historic change was then followed by the election of the Coalition Government in 1996, with a new industrial relations Act becoming law by the start of 1997. The mid to late 1990s also saw some major industrial disputes, including those involving the Gordonstone mine in Queensland and the Hunter Valley No.1 and Vickery mines in NSW which would change the industrial relations landscape.

The 1990s would also see tragic loss of life at the Moura mine in Queensland and the Gretley mine in NSW, as well as a number of other fatal accidents, which would prove to be the catalyst for reforms to coal mine safety legislation and regulation. The coal industry's tragic safety record of the past had been improving with new mining technologies such as longwall mining and with greater company and management focus. However, the two tragedies, both involving multiple fatalities, forced regulators and the industry to confront the industry's unacceptable safety performance.

One of the features of the industry over the last 50 years has been the rapid change in the ownership structure of the major producers. The late 1970s, the 1980s and the 1990s saw the big oil producers – BP, ARCO, Exxon and Shell – investing heavily in acquiring coal mines and developing new mines in Australia, only to exit after struggling to make reasonable returns. They were not alone, with companies like CSR and US miner Phelps Dodge's Cyprus Mining also making substantial investments only to exit the industry within a fairly short time. The

2000s have continued to see some major ownership changes, with Anglo becoming a major producer, Peabody selling out only to re-enter the industry a few years later, and BHP merging with Billiton in 2001, only to spin off a significant number of its assets in 2015 to the new South32. Rio Tinto, formed in 1995 through the merger of CRA and RTZ, expanded its coal business in 2001 with the purchase of the Peabody assets in Australia, but has now sold out of the coal industry in Australia. In contrast, Glencore has become a major force in the industry since it began its spree of acquisitions in the 1990s and 2000s which included the takeover of MIM in 2003. Glencore and Chinese company Yancoal are now the dominant thermal coal producers in Australia.

In contrast to the passing parade of mining companies, Japanese trading companies Mitsui and Mitsubishi, pioneer investors in the Queensland metallurgical sector in the 1960s and 1970s, have been a constant presence over recent decades, although Mitsubishi has sold out of its interests in thermal coal producers in recent years. Other Japanese companies including Sojitz, Marubeni, Nippon Steel and Sumitomo Metal Corporation, EPDC and JCoal have also been long term participants in the industry. These Japanese companies, often criticised in the past, have provided valuable funds for the development of new mines and in many cases have assisted coal producers to secure vital contracts with the Japanese steel mills and power utilities.

The rise of China and India has also had major impact on our coal industry. China, previously a significant coal exporter, became a significant importer in the early 2000s and has also made significant investments in the Australian industry, with Yancoal now one of the leading thermal coal producers. China now produces over half the world's crude steel, and its rise as a steel producer has seen it become the largest metallurgical coal importer, with Australia its major supplier. India has also become a major market for Australian metallurgical coal and may become a more significant market for thermal coal in the years ahead.

The Australian coal industry, a basket case at the end of World War Two, has developed and matured into a world leading industry. It however faces an uncertain future, with community and political pressures opposing the development of new thermal coal mines. Australia's coal fired electricity generation sector has seen widespread closures of generating capacity in the last decade, with the next major closure to be Liddell in the Hunter Valley in 2022 and 2023. What the future holds for the remaining coal fired stations, some of which have a potential life extending into the 2040s, remains to be seen. The international market outlook for thermal coal is a mixed one, with coal demand on the decline in Europe and USA, but potentially increasing strongly in a number of countries in south and southeast Asia. Australian thermal coal producers, with a generally high quality product, have the potential to capture a significant share of these Asian markets in competition with competitors including Indonesia, Russia and South Africa. The outlook for metallurgical coal, an essential ingredient in steel production, is more promising, and Australia is well placed to continue to be the dominant supplier to the major markets in Asia which account for the bulk of global steel capacity.

I hope that readers of this book will gain a greater understanding of the path of development of the Australian coal industry and of the many challenges it has faced over a long period. While any book such as this can only cover a limited range of issues and stories. I also hope that it covers many of the key issues, including those which have been central to the way the industry has evolved and matured into the industry of today. While not attempting to ignore the many problems of the past and the current challenges, Australians can be proud to recognise what is a world leading industry.

Acknowledgements

While taking full responsibility for all the content of this book, I would like to acknowledge the assistance of a number of people, not least my wife Lesley for her patience and understanding over the long period during which I researched and wrote the two volumes. To Collin Myers, who was kind enough to write the foreword for this volume and to carefully check draft chapters for grammatical and other errors, my sincere thanks. Thanks also to Tony Haraldson, Bob Humphris, Kevin Dixon, Ross Taylor and Ross Graham for their valuable comments on the industry and aspects of its development. And thanks to Anthony Cappello and his colleagues at Connor Court for publishing both volumes of Coal: the Australian Story.

1

Oil shocks, price controversies and the new Labor Government

Japan, the yen and US economic woes

During the 1950s and 1960s, Japan's economy grew on average by close to 10% per year in real terms, a phenomenal performance for any country, let alone one devastated by the war. Its recovery was due to many factors, not least the reforms to its political system imposed by the Allies led by the USA. But Japan was already an industrialised country before the war, and given the work ethic of its people, and its know-how, and support from other countries, it would only be a matter of time before it recovered and regained its place as a major economic power. The early post war growth of the Japanese economy also owed much to the growth of its export industries, including industries such as steel and textiles. That growth in turn was assisted increasingly during the 1960s by the low value of the yen which had been fixed at a constant 360 to the US dollar since 1950. Under the 1944 Bretton Woods Agreement, countries had agreed to a stable system of exchange rates, with currencies tied to the US dollar which in turn was tied to gold at US$35 per ounce. But this agreement could hardly have anticipated the way in which the post war world would develop, and the strains which would be placed on it by differing economic rates of growth and balance of payments movements.

By the late 1960s, with exports booming and the country's international currency reserves growing to unprecedented levels, Japan came under increasing pressure from other leading countries to revalue its currency. But the Japanese were not prepared to countenance a revaluation, fearing of course the impact that might have on its booming export industries. In July 1971, despite growing international pressure, the Japanese government, led by its senior bureaucrats, fought on and announced its Eight Point Plan to Avoid Revaluation of the Yen. The plan proposed liberalising imports, speeding up the reduction in tariffs, expanding foreign aid, a more orderly expansion of exports and more flexible monetary and fiscal policies. The plan was doomed to fail.

However it was the economic problems in the USA which brought into play a new era for the international economic system. Burdened by the costs of the war in Vietnam, President Johnson's Great Society programs, higher inflation, an over-valued dollar, competition from imports, and balance of payments and debt problems, in August 1971 President Nixon made the momentous unilateral decision to temporarily suspend the ability of holders of gold to be able to convert it into US dollars. Nixon also imposed as a temporary measure a 10% surcharge on imports. His New Economic Policy was aimed at ensuring that American industry was protected against what his administration saw as "unfair exchange rates."[1] Nixon's policy included a range of measures designed to stimulate the US economy, restrain inflation and boost employment. He also attacked "international speculators" who he said had been "waging an all-out war on the American economy." He went out of his way to say that the import tax was "not directed against any other country", and so while Japan was not mentioned once during the President's statement, or any other country for that matter, Japan was one of the primary targets.

Then in December 1971, the ten major economies, meeting as the Group of Ten, came together to consider how to fix the global exchange rate system that was clearly breaking down. The Smithsonian Agreement which emerged from this meeting had the US agreeing to fix its dollar at US$38 per ounce of gold, with 2.25% bands allowed for trading, and other countries agreeing to revalue their currencies. Japan revalued the yen by 16.88% immediately after the Agreement. However the Agreement did not have the desired effect, and the USA devalued in February 1973 by 10%, with Japan and European countries soon deciding to let their currencies float. The Bretton Woods system was close to dead, although it took around another ten years before all key western economies, including Australia, allowed their currencies to float.

The Vietnam War (or as the Vietnamese called it, the American War) ended in 1975, but one legacy of the war was the US government deficit. By 1980, and with the added impact of the oil shocks, that deficit had ballooned to almost US$60 billion, or around 2% of US GDP, a level not matched to that time since the end of World War Two. Inflation in the USA hit over 6% in 1973, and surged to 11% in 1974. With recession in the USA in 1974 and 1975, inflation eased back to just under 6% in 1976, before accelerating again in 1977, and climbing to over 11% in 1979 and over 13% in 1980. Unemployment was also worsening in the 1970s, reaching 7.5% in 1980. The combination of poor economic growth, high inflation and increasing unemployment in the 1970s ushered a new word into the economic lexicon – stagflation.

With export coal contracts generally written in US dollars, any changes in the relative values of currencies vis a vis the US dollar were bound to have an impact on the parties involved. Japan's revaluation in December 1971 of almost 17% meant that it would have to pay less for its coal and other raw material imports from Australia and other countries, but it would also be receiving less in yen terms for its exports of steel to the USA and for exports of products such as automobiles which used a lot of steel. The net impact on the Japanese steel mills in the short term, however, would be significant and would be negative. The currency shocks of the early 1970s would not be the only massive shocks to hit major countries in the 1970s. The oil shocks in 1974 and 1979 would see oil prices rise from $US3.29 per barrel in 1973 to almost $US37 by the end of the decade, causing major economic shocks and a fundamental reassessment by many countries of their energy policies. We will return to the oil shocks later in this chapter.

NSW industry visits to Japan see meagre results

Australian coal exports to Japan surged by around 25% to almost 18 million tonnes in 1969-70, with NSW seeing an increase of almost 19% (to 12 million tonnes) and Queensland an increase of 40% (to almost 6 million tonnes). Export prospects were promising, but a major problem confronting the industry in NSW was the capacity of its coal ports and rail system. In October 1970, Australian Coal Association chairman, Sir Edward Warren, said that the industry in NSW was about to reach absolute handling capacity in Newcastle and Port Kembla: "I cannot stress too strongly the need for a thorough investigation by the Government of sites along the coast near Newcastle and Port Kembla suitable for building of offshore loading installations. These should be capable of accommodating vessels of at least 150,000 tons capacity ...compared with ships of around 50,000 tons, which are the biggest we can handle in existing ports. "[2] Warren also drew attention to the developments in Canada where the new Roberts Bank export terminal had been commissioned to handle ships of up to 150,000 tons capacity, and with a loading rate three times what NSW ports could achieve. Canada also was using 100-wagon trains to move its coal from mines some 800-1100 kilometres by rail, thereby reducing its rail freight costs. The NSW industry was clearly facing a major challenge to upgrade and modernise its rail and port capacity if it was to take advantage of the potential for growth in the export market.

We will come back to the port problems in NSW, but in 1971, it became clear that the problems facing the industry in Australia, and in particular NSW, were more serious than just port and rail capacity constraints. A delegation from the NSW coal industry visited Japan in October and November 1971 for meetings with the Japanese Steel Mills (JSM). Led by Sir Edward Warren in his capacity as chairman of the NSW producers (the NSW Combined Colliery Proprietors' Association or NSWCCPA), the JSM were told that the NSW industry was in a state of emergency, and that an urgent review of coal prices was required.[3] The Australian industry had seen wages increase, hours of work fall (miners had won a 35-hour week in July 1970 in two steps – 37.5 hours from August 1970 and 34 from July 1971), and rail and other costs increase. "The impact of these factors on the NSW industry has been such that a situation of emergency has been reached," Sir Edward told the Japanese. "We need emergency relief as a matter of urgency," he said, and asked for the JSM's support. The Japanese were asked to look at the way coal contracts had been structured on essentially a cost-plus basis, as this was damaging the viability of NSW producers.[4]

The giant of the Japanese steel industry, Nippon Steel Corporation, was formed in March 1970 with the merger of two large steel companies – Fuji Iron and Steel and Yawata Iron and Steel. The marriage of the two steel companies came 20 years after the breakup of Japan Iron & Steel under the Allied occupation of Japan. To encourage competition in the steel industry, Japan Iron & Steel was broken up into four companies which included Yawata and Fuji. The new Nippon Steel opened its Australian office in Sydney in November 1970, and in May 1972 its senior executive in charge of raw material procurement, Saburo Tanabe, paid Australia a visit. There was a follow-up mission to Japan by senior executives from NSW coal companies in October 1972, led again by Sir Edward Warren. The report of that mission provides important insight into the visit by Tanabe to Australia in May 1972, as well as the October 1972 mission itself, and also visits to Japan by Sir Reginald Swartz, Australia's National Development Minister, in June, and NSW Mines Minster, Wal Fife, in July of that year.[5]

The mission's first meetings were with senior executives of the JSM, led by Nippon Steel's Tanabe. Tanabe briefed the NSW producers about recent major events including Nixon's policy announcements in August of 1971, the revaluation of the yen, the stagnation of the Japanese economy since late 1970, the formation of a cartel by the JSM to force up the price of steel, the drop in earnings by steel companies and the slump in the steel market.[6] Tanabe also sounded a positive note, saying that the Japanese economy was finally recovering after 20 months of recession, and that steel production was also looking better, with production for that financial year (the year ending April 1973) expected to be around 98 million tonnes, compared with earlier estimates of 95 million tonnes.

Tanabe's reference to recession in Japan was in fact incorrect, although the slowdown in growth may well have felt like a recession to many. Japan's GDP growth in 1970 had slumped to only 0.4%, compared with an average of around 10% per year in the 1960s. However growth in GDP recovered in 1971, up 4.7% and surged again in 1972 and 1973 with growth of around 8% each year. The steel industry, however, was hard hit by the domestic slowdown and by world markets, with production slumping in 1970 by over 10%.

The key message to the JSM in October 1972 was delivered by Edward (Ted) Ryan, General Manager of South Coast company Bellambi Coal. In the first round of talks on October 2, Ryan referred to Tanabe's visit in May and said that he and other Nippon Steel executives had been given a blunt message. He said that a crisis was approaching in the NSW coal industry and that there was a mutual problem: the possibility of government interference in the traditional private relationship between buyers and sellers of Australian coal. Ryan said that this problem had arisen because of a lack of orderly development and marketing of Australian coal resources, evidence of which was the large contracts for open cut mines in Queensland at a time when NSW stockpiles were growing and when there were threats of closure of mines.

The phrase "orderly development and marketing" was meant to convey to the JSM the need to look after the NSW underground sector which was struggling to compete with the massive new open cut mines developed in Queensland, particularly by Utah Development. Swartz used similar words when he met senior Japanese in June 1972, telling them that there were three key issues: first, that open cut mining was cheaper than underground and this was causing problems; second, that there should be balanced development between open cut and underground; and third, that Australian coals according to quality should command prices comparable with those from other countries. Swartz added that to fix prices on the basis of cost of production plus a margin for profit was not the best arrangement for the long term association between Japan and Australia, although the Australian Government felt that matters of this type could best be solved by negotiation between buyer and seller. Swartz did, however, note that he was pleased that the latest major Australian coal contract agreed was at a price much closer to what the Government considered to be the going market price for similar coal from other sources. However, he warned the Japanese that while he hoped good sense would prevail, the Government would take action if necessary (as it had on a couple of occasions, including once with iron ore).[7] NSW Mines Minister Wal Fife in July also told the Japanese that a "proper balance of development" was not being maintained and unless some action was taken to correct it, the NSW Government would have no alternative but to make representations to the Commonwealth about this matter.

On behalf of the NSWCCPA group, Ryan told the JSM that the practical solution was for the mills to agree to new long term contracts for NSW producers to commence on 1 April 1973 (existing contracts expired at the end of JFY 1972 or JFY 1973). He said that while it was a matter for individual negotiation, tonnages in new contracts needed to be more appropriate to the existing capacity and future potential capacity of each mine; and that while each company had to be competitive in the Japanese steel market against underground mines elsewhere in the world, he asked the mills to accept the principle that new prices should allow mines to operate economically. Ryan also said that new contracts should include escalation provisions which recognised the higher rate of inflation in Australia. The question of tonnages allocated to NSW coking coal producers was a particularly sore point with the NSW companies. With the downturn in the Japanese steel industry in 1971, the JSM effectively reneged on their contracts, and the NSW exporters had been selling less than the contracted tonnes. But the media statement by Tanabe at the close of the meetings revealed that the lower tonnages for NSW producers were the result of more complex factors than simply lower Japanese steel production.

On October 4, the NSWCCPA mission had further talks with Nippon Steel's Mr Iimura, who provided some details of the possible levels of Japanese steel production for the period between 1975 and 1980. These numbers provided more hope for the embattled NSW producers, although extra tonnages, of course, would depend not only on the level of Japanese steel production, but also on what share of the coal requirement the JSM were prepared to assign to NSW. Iimura said that the JSM now estimated that annual crude steel production was likely to be in the range of 110 to 130 million tonnes in the period between 1975 and 1980, with coal requirements between 58 and 73 million tonnes, depending on the ratio of coal to steel, as well as the level of steel production. Of that coal requirement, the NSW share could vary between 9.5 million tonnes and 14 million tonnes.[8] Ryan suggested to Iimura that the JSM's figures would be unlikely to encourage the NSW Government to invest in upgrading the state's rail and port facilities. Iimura countered, urging the NSW Government to go ahead with upgrading these facilities.[9]

On the last day of the NSW mission, statements were issued to the media by Sir Edward on behalf of the NSW mission, and by Tanabe on behalf of the JSM. Sir Edward's statement said that the NSWCCPA has not achieved all it was hoping, and that the NSW coal industry had no prospects of growth in the next few years. He noted that the official JSM forecast of steel production in 1975 of 110 million tonnes would require only 9.5 to 11.5 million tonnes of NSW coal, which was less than the tonnage the JSM had already contracted to take in both 1972 and 1973. On a positive note, Sir Edward said that the mission did receive an assurance that the JSM would negotiate with individual companies for new contracts for 5-year terms from 1 April 1973, and that there was an understanding that the JSM would re-consider price escalation provisions in individual contracts.

Tanabe's statement to the media agreed with Sir Edward that new 5-year contracts would be written to commence from April 1973. He also acknowledged that current escalation clauses were not covering cost increases in Australia, and said that they should be revised. Tanabe committed the JSM to buying more coal from NSW, but warned that some mines may actually see reduced tonnages. Tanabe also made some interesting comments which threw more light on the negotiations and the JSM's attitudes. He said that with this agreement, "damage which the depression has caused for the NSW coal industry and the antipathy towards Japan from that damage will be calmed down for the time being." He went on to acknowledge that NSW coking coals were an important source of supply, but had played a secondary role in recent times to coal from Queensland, as the JSM had been dramatically increasing their purchases of Queensland coal. The reasons for favouring Queensland coal included the fact that NSW underground coal was more expensive, and the fact that the JSM had been able to be more selective about where they chose to buy coal; before the economic downturn in Japan they bought whatever was available because of the growth in their industry. Tanabe also said that the steel mills had made concessions on the basis of a political judgement that they did not want to cause any further deterioration in the relationship between Japan and its major supplier of coal and iron ore.[10]

While Sir Edward was clearly guarded in his media comments, there were other coal executives who were not so diplomatic. In an article in the Sydney Morning Herald headlined *NSW dismay at coal talks*, the writer said that the colliery proprietors had failed to get guarantees of price rises, bigger shipments and new contracts. It quoted one company director as saying: "You know what we got? We got nothing", while another member of the mission said that the Japanese steel companies had made "the usual promises" that they would do all they could in the future to help the NSW industry, but had made no commitments. And another company executive was even more forthright, saying: "It is obvious what we face now. They will play us off one by one against the cheaper coal from Queensland."[11]

The gap between the costs of production in NSW and Queensland, and the resulting lack of competitiveness of the NSW producers, had become major issues for the producer associations, the Joint Coal Board, and the State and Commonwealth Governments, and the pressure was on the Commonwealth in particular to do something about the prices the Japanese were paying for Australian coal. The desirability of a more coordinated or united approach to selling to Japan that was being advocated by the JCB was also winning support in the media, for example in The Sydney Morning Herald. The SMH put its view forward in an

editorial in October 1972, just after the Miners' Federation had proposed
to Minister Swartz that a National Fuel Board should be established:
"Utah Mining will make tremendous profits while paying a paltry few
million dollars in royalties to the Queensland Government. But you can't
blame Utah for getting on to a good thing. What had really happened
is that the Queensland Government had given Utah lower rail and
port charges to enable it to undercut almost everybody in the business.
Even the Queensland producers like Thiess and TPM have distinct
cost production advantages over NSW producers. This is all because
the Government allows exporters to go their own way. There must be
a united front for all our natural resources if we're not to be completely
exploited by foreign interests."[12] Senator Swartz in a Ministerial
statement to Parliament in October 1972 effectively threatened Japan
that the Government would use its export control powers to discourage
the development of new projects if "goodwill and good sense on both
sides" did not see problems in the Australian industry resolved. With
the election of the Labor Government in December 1972, it would be
the new resources and energy Minister, Rex Connor, who would soon
announce sweeping changes relating to regulation of mineral exports,
and in particular coal.[13]

Sir Edward Warren resigns industry roles

Less than two weeks after the 1972 NSW mission returned to Australia,
Sir Edward Warren resigned as chairman of the NSW Colliery
Proprietors' Association and the Australian Coal Association. Officially,
his resignation was due to pressure of his other commitments, but one
wonders whether there was some other reason for the resignation coming
as it did so soon after the mission to Japan, and with the fundamental
dilemma over the major issue of the competitiveness of the NSW
underground sector. Sir Edward's contribution to the coal industry had
been enormous. He commenced work with Abermain Seaham Collieries
as a messenger and served with distinction in World War One. He re-
joined Abermain Seaham after the war, becoming accountant in 1921,

and was acting company secretary for a time in 1928. With the merger in 1931 between Abermain Seaham and J&A Brown to form J&A Brown and Abermain Seaham Collieries (JABAS), he was accountant for the new group. Warren was then appointed assistant manager in 1939 and general manager in 1944. JABAS and Caledonian Collieries merged in 1960 to form Coal & Allied Industries and Warren became the new managing director of Coal & Allied, retiring from the company at the end of 1975 at the age of 80.

Warren became chairman of the Northern Colliery Proprietors' Association and the NSW Combined Colliery Proprietors' Association in 1949, a critical time for the industry with the recent formation of the JCB and the reform agenda it was forcing the industry to adopt. He was also the inaugural chairman of the Australian Coal Association from 1955, the ACA having been established as a federation of the NSW and Queensland coal employer associations to provide a national voice for the coal industry. He led a number of coal industry delegations overseas to promote export opportunities, including the 1972 mission to Japan just before he stepped down from his NSW and Australian industry roles. He was a member of a delegation led by JCB chairman Sam Cochran in 1952 which travelled to the USA and Europe to assess mechanisation and other industry trends. Warren also travelled extensively for the company, seeking out new markets for JABAS and Coal & Allied. In 1971, he said that he had been to Japan 57 times since the end of World War Two, and had travelled to Europe and the USA once per year since 1948.[14] Among many honours, Warren was knighted in 1956, with higher British honours in 1959 and 1969, and was the first Australian to receive the First Order of Merit of the Order of the Sacred Treasure of Japan, one of that country's most senior honours, for his services to relations between Australia and Japan. In 1954, Warren became a member of the NSW Legislative Council, a position also held by his old boss, Thomas Armstrong. He served two terms in the Council, retiring from that position in 1978.

One of the more puzzling aspects of Sir Edward's career was his attitude to Queensland. Apart from small interests in the Ipswich area which arose out of the merger with Caledonian Collieries, Coal & Allied was not involved in the massive developments in Queensland from the 1960s. In around 1964, Sir Edward sent a young geologist, Brian Vitnell, to look into the possibility of a joint venture with the Mount Morgan company which had a small anthracite mine. Vitnell did not see any future in such a project, but spent some time investigating the potential for new developments in the Bowen Basin, which was already seeing the Thiess Peabody Mitsui consortium pushing ahead with developing Moura and Utah about to be granted the exploration rights over almost 6500 square kilometres of the Basin. However despite the geologist's findings, Warren was not keen to expand into Queensland and in fact was not at all happy that Vitnell had spent valuable time departing from his assigned task.[15]

How different Coal & Allied's future may have been is difficult to say, but as its strength in the 1960s was in underground mining, and the Bowen Basin developments were open cut operations, Sir Edward's stance is understandable. Not long after Sir Edward retired from Coal & Allied, the company moved into the open cut sector in a big way, with the Hunter Valley No. 1 mine its first major new surface development. And during the 1980s the company would further rationalise its underground mines, closing a number and finally selling off the last couple to Coal Operations Australia Limited (COAL), a company led by its ex CEO Tony Haraldson.

JCB tries to balance open cut and underground sectors

The long standing JCB policy which required most open cut mines in NSW to be operated by companies which integrated their surface operations with underground operations continued into the 1970s. There were a number of open cut mines operating in the early 1970s apart from Ravensworth and Swamp Creek which supplied the Liddell power station. These open cuts were operated as part of integrated open cut/ underground operations which marketed their coal as a common product. The JCB policy meant that at times - such as in 1970-71 - open cut production at some collieries was reduced in order to maintain production and employment in the undergrounds. In November 1972, with export demand weak and coal stock levels high, the JCB ordered those producers affected to reduce their output from the open cuts, with the result that underground production was maintained or increased.[16] RW Miller was one company affected, with the JCB directing it to cut production from the open cut section of its Wallsend Borehole colliery. RW Miller immediately issued retrenchment notices to over 300 miners at Wallsend Borehole and two other mines, giving them one week's notice. Miners' Federation general president Evan Phillips said the company was engaging in a lockout and accused Miller of using its employees as pawns in order to force the JCB to withdraw its directive. NSW Mines Minister Wal Fife called the parties together and R W Miller was informed that they had not been singled out by the JCB for a cut in production, with Fife assuring Miller that all open cuts had had production limits imposed; Miller quickly withdrew the dismissal notices.[17]

The problems in 1972, however, did precede a relaxation in thinking by the JCB. The Board's balanced development policy remained, but its annual report for 1972-73 indicated the need for more flexibility in the coming years. If existing commitments for local and overseas markets were to be met during the 1970s, the Board said, "and the ratio between open cut and underground is to be modified, the process (of change) must necessarily be a gradual one."[18] The JCB's powers clearly gave it control over the NSW industry and the development of new mines, but in looking to attempt to apply the balanced development policy to the export market it was never likely to be successful. That same annual report had the JCB saying that "If coal is sold at a price based on cost of production plus a margin rather than a price based at full value of the coal in the market place, then a buyer has a real interest in obtaining the cheaper open cut coal…A prerequisite therefore of balanced development is that all Australian coals supplying the export market should be sold at prices comparable on a quality basis, both as between themselves and in relation to the coals supplied by other countries. Furthermore, such prices should provide an adequate return on coal either wholly or predominantly won by underground methods."[19] Japanese buyers certainly had an interest in buying the cheapest coal provided their quality requirements were met. However, it is doubtful whether they would have agreed with the Board that the market price should reflect the "full value" of the coal, whatever that may have meant. If the "full value" of underground coal reflected the price needed to give the higher cost underground producers an adequate return on investment, how was that value going to be achieved when, for example, low cost open cut production in Queensland was increasing and other suppliers in Canada were entering the market?

The balanced development policy in NSW was soon to change more rapidly than the JCB had anticipated. By its 1974-75 annual report, the JCB was stating that the policy still applied, and that it had been necessary for the Board to maintain restrictions on the production of open cut coal and that there would be "no departure from the Board's long established policy that the continued operation of existing underground mines and the security of employment of the skilled and experienced underground mineworker is a matter of first concern." However, the benefits of open cut mining in terms of coal recovery rates were now being better understood, the JCB acknowledging that "a high recovery factor associated with open cut mining is so important that, in general, no coal deposits should be worked by underground methods if it is practical to mine such deposits by open cut methods, using equipment and techniques either currently available or expected to become available."[20]

The evolution in the JCB's thinking did not come too soon as interest from local producers and overseas buyers in new open cut developments would soon emerge. By 1976-77 Thiess' Drayton project was beginning to be progressed and the JCB gave approval for the company to mine coal for a bulk sample. In 1977-78 R W Miller's Mt Thorley project and Coal & Allied's Hunter Valley No.1 project were also progressing and received JCB approval for mining of bulk samples. There were also nine major underground projects under development by 1978, an indication that the underground sector was showing new life. These underground projects were Peko Wallsend's Ellalong, Buchanan Borehole Collieries' Buchanan Borehole, BHP's Macquarie, Bloomfield Collieries' Rathluba,

Newcom's Angus Place, Coalex's Clarence, Clutha's Tahmoor and AI&S' Cordeaux and Tower.

Japan's coal pricing favours USA and Canada

It was clear to everyone in the industry in early 1972 that there was a fundamental dilemma facing coal industry stakeholders in Australia: the divide between NSW and Queensland in terms of the cost structure of the industry. This divide was exacerbated by the pricing policies of the JSM which saw Australian coal receive significantly lower prices than coal from Canada and the USA. A submission from the Minister for National Development, Sir Reginald Swartz, to Federal Cabinet in April 1972 (which was withdrawn and so not formally considered by Cabinet) throws some interesting light on the state of the industry and the attitudes of key players.[21] One of the most contentious issues in the 1970s, and one which continued into the 1980s and 1990s, was the question of the price Australian exporters received for their coking coal compared with US exporters (which at that time were the established and dominant coking coal suppliers to the Japanese steel mills). The Japanese import statistics clearly showed that Australian exports were priced well below the prices being received by the USA and Canada, the other major suppliers.

The submission detailed average A$ values per long ton (c.i.f. Japan) for hard coking coal with less than 8% ash imported into Japan in 1970 and 1971:

	1970	1971
Australia	13.16	13.79
USA	22.30	24.76
Canada	16.06	17.54
All suppliers	20.59	22.07

The submission also gave data for higher ash hard coking coal and other coking coals. Again Australian coals were significantly cheaper than the US coals. While Australia had a distinct advantage over the USA in terms of the shipping freight component of the landed cost into Japan, with a difference in the landed cost of hard coking coals of around A$11 per ton in 1971, it was clear that the Japanese were using their market power to maintain the higher cost producers in the US Appalachian region as a major supply source. The JSM no doubt had a good idea of the cost structure of the Australian industry and had set prices for Australian coking coal exports so that at least some of the high cost NSW exporters were able to cover their costs, while Utah was rewarded as the major new supplier and the source of much lower cost coal over the long term.

While it could be argued that there were other quality factors which meant that US coals merited a higher price, the evidence points to the fact that the Japanese were able to buy high quality coking coals from Australia at a significant discount to the US coals. The Minister's submission stated that there was "strong evidence to suggest that our coals are being undersold on the world market." The submission is also important in clarifying the Government's position on the coal industry in the period leading up to the historic election in December 1972, or at least the position of the Minister, his department and the Joint Coal Board. There was a concern about the potential for Australia to be denuded of its readily mined surface coals in the Bowen Basin, and concerns about the competitiveness of underground coal reserves. With those readily accessible Bowen Basin coals also being the cheapest and best quality, the Minister believed that it was in the national interest to seek a balanced production of open cut and underground. The ability of NSW mines to compete was another key concern, with some NSW producers "forced to export at prices which barely cover costs" and continued development of open cut production in Queensland was seen as likely to endanger underground production in NSW and also in Queensland.

The Japanese steel industry was acknowledged as a monolithic buying cartel, adept at playing one supplier against another. However the Minister cautioned that if Australian prices were forced up, the Japanese may consider turning to other sources of supply, adding that the JSM above all valued reliability and quality, and were anxious to maintain a diversification in their sources of supply. On this last point, the Minister seemed to be implying that perhaps Australia could force up the prices paid by Japan to some extent without causing the JSM to react too negatively.

The Minister stated that there had been a meeting the previous month of coal exporters to attempt to reach an agreement on coordinating coal exports, but that with Utah and Clutha being US owned and subject to US anti-trust laws, no agreement was possible. However, he believed that there was a growing feeling in NSW that some action should be taken, perhaps by the Commonwealth Government, to ensure higher prices and prevent hardship in the industry. The Minister also said that he had just met the NSW and Queensland Ministers for Mines, both of whom had previously publicly indicated their support for a more united front by the export industry. The Queensland Minister, however, said that he had not had any official approaches from the industry regarding any controls, and was not happy about the prospect of Commonwealth intervention in the market. NSW, on the other hand, was believed to be likely to support Commonwealth intervention, and the NSW Minister agreed to consult the NSW producers to suggest that they could take the issue up with the three Governments. The Miners' Federation had also written to Swartz to register its concerns, advocating a coordinated approach to the development of the industry, including a better balance between the underground and open cut sectors. The Federation was concerned about the closure of NSW mines which had not been able to compete with the Queensland open cuts, and about the potential for further damage to the underground sector and the jobs of members.

One of Swartz's key sources of advice was the Joint Coal Board, which was in favour of Commonwealth intervention in the market to achieve the much hoped for balance between the underground and open cut sectors and improved prices. However, given the structure of the industry in NSW and Queensland and the competitive advantage of the Queensland producers, a simple resolution was not likely. Schwartz concluded that pressures were mounting for action to be taken by the

Commonwealth, and that the introduction of export controls may be desirable, although such controls would not solve all the industry's problems. He saw export controls as requiring the Commonwealth to insist on customers paying a fair price for comparable grades of coal, and possibly also seeing the Government expecting an export producer to mine and export a certain tonnage of underground coal as part of the conditions for gaining approval to export its open cut coal. How such an arrangement might have worked for a producer such as Utah, which at that time was only operating open cut mines, was not addressed by the Minister and probably indicates the problems his departmental advisers were having in attempting to navigate the politics and economic realities of the industry's problems at that time. But the Coalition Government's days were numbered, and when the Whitlam Labor Government gained power in December 1972 the issue of export prices and controls soon came to a head.

From the Japanese perspective, of course, the situation was a little different. Its steel industry had relied on US coking coal, supplemented by supplies from Poland and the Soviet Union. The US mines were underground mines, primarily located in the Appalachian region on the eastern side of the continent, with shipping charges a significant factor in the landed cost into Japan. With the development of new open cut mines in the Bowen Basin beginning in the 1960s, Japan now had a new low cost player in the market. Japan also understood that the Australian producers, or at least the Bowen Basin producers, could make reasonable profits even with prices below those available to the USA and Canada. It was not about to be generous to Australia by offering our producers the higher prices paid to our competitors.

A Senate Select Committee inquiry in 1972 provided an opportunity for Utah (UDC) to place its marketing and price negotiating policies on the public record. Utah in effect said that it was not seeking to maximise its prices and profits in the short term; it had a longer term strategy: "UDC is prepared to forgo available short term high profits which may be non-recurring for the stability provided by long term sales contracts with cost escalation protection." Utah said that its pricing policy recognised that "a. Long term contract prices will, in most circumstances, be below prices which might be obtained on spot sales or under short term supply arrangements; b. Prices negotiated are also influenced by aggregate levels of tonnages for which buyers are committed and the length of long term contracts; c. Long term coal sales contracts negotiated by UDC contain cost escalation protections (reflected in increased prices per ton) and the base price to some degree reflects the future value to the producer of cost protection."[22] However by 1974, the intervention by Resources and Energy Minister Rex Connor and his success in forcing the JSM to significantly increase coal prices indicated that Utah may have been too cautious in its approach to pricing.

Japan promotes Canada as a competitor to Australia

In the 1950s, the USA was the dominant source of coking coal supply for the JSM. As the Japanese steel industry grew and with it its need for ever increasing volumes of coking coal, Australia was fortunate to be in a position to take a significant share of the Japanese market. However, the Japanese were keen to further diversify, and looked to Canada to become a third force in the international coking coal trade. The coal industry in the western provinces of Canada dates back to the late 1800s, when coal was discovered in 1885 in the Elk Valley of British Columbia, near the borders with the USA and the province of Alberta; mining there began in the 1890s.

In the mid-1960s the JSM contracted with a local Canadian company, Crowsnest Industries, to supply 1 million tonnes of coal per year for 3 years. The mine was then bought by Kaiser Resources, part of the Kaiser Steel group, in the late 1960s, and Kaiser Resources signed a contract with the JSM to supply 45 million tonnes of coking coal over 15 years, and proceeded to develop the Balmer mine. Next to enter the scene were three other new mines, Luscar, Smoky Hills and Fording River, the latter developed by Canadian Pacific Railways subsidiary Fording Coal. These mines commenced producing between 1969 and 1972 and were underwritten by export contracts with the JSM.

The other key parts of the development picture were also put in place during this period, with Canadian Pacific and the Government-owned Canadian National Railroad building rail links to the mines, and two new export coal terminals built by the mining companies, one in Vancouver and another at Roberts Bank, south of Vancouver. With the export terminals open for business in 1970, Canada was set to become another significant source of coking coal for Japan, receiving higher prices for its coal than Australia. The JSM were clearly prepared to pay what was required to ensure that its major new supplier, Australia, was not too dominant in the market.

The JSM also were instrumental in the development of the South African export coal industry in the 1970s, agreeing in 1972 with local coal mining companies to take 27 million tonnes of metallurgical coal over 10 years. This contract then saw the development of the Richards Bay coal terminal which opened in 1976 to supply the Japanese market, later developing into one of the world's largest export terminals, mainly supplying thermal coal to markets in Asia and Europe. The JSM were now successful in bringing a third new supplier into the metallurgical coal market.

Whitlam Government's new export controls

Following Robert Menzies' retirement on Australia Day 1966 after 18 years as Prime Minister, Harold Holt became Liberal Leader and Prime Minister. Holt served for almost two years, before losing his life in December 1967, apparently drowned in the ocean near Portsea in Victoria. Holt was followed by John McEwen, the Country Party leader, as a stopgap Prime Minister. McEwen handed the job over to John Gorton in January 1968, with Gorton serving for a little over three years, before he was replaced by William McMahon in March 1971. McMahon took the Coalition to the historic "It's Time" election in December 1972 which saw the Whitlam Labor Government elected, ending the Coalition's 23 year hold on power. Gough Whitlam and his deputy, Lance Barnard, ran the new Government as a ministry of two for the initial couple of weeks after taking the reins. The full Ministry was appointed late in December, with Rex Connor the new Resources and Energy Minister. The old Department of National Development was renamed Resources and Energy, and Sir Lennox Hewitt became Connor's new departmental head. The minerals and energy industries were never to be quite the same again. One of the new Cabinet's first agenda items related to the minerals industry, and Connor put a submission to his colleagues in January 1973 to extend export controls to all minerals. Connor also submitted an urgent and more detailed submission to Cabinet on the coal industry, much of the wording of which was similar to the March 1972 submission by the Coalition Minister, Sir Reginald Swartz, indicating that the same senior officials in the department prepared both submissions. Connor outlined the problems in the industry, with Australian coal prices well below the prices the JSM were prepared to pay competitors from other countries, and with NSW producers suffering while new open cut producers in Queensland were winning an increasing share of the Japanese market for coking coal.[23] The urgency in Connor's coal industry submission was attributed to the problems which Clutha had been having with its NSW mines. Connor said that in discussions between Barnard and Clutha in December 1972, the company had agreed to take no action on closing mines and dismissing employees in NSW until early in 1973, with Barnard agreeing to the Government looking into the industry's problems as a matter of urgency. He also stated that the JCB had been advised that some Hunter Valley producers were planning to reduce their output, with implications for more dismissals.

With little increase in exports from NSW over the previous 2-3 years, but a significant growth in exports from Queensland, the major concern of the Commonwealth Government was the situation in NSW, where 56 of the 81 mines were in the export business, employing around 8,000 workers. Of these 8,000, around 6,000 were stated to work in mines highly dependent on exports, with exports accounting for at least 75% of their production.[24] There was also concern about the underground sector in the Ipswich area in Queensland, where open cut mining in the same area was said to be seriously affecting underground employment.

With the support of the NSW producers and the Miners' Federation (both organisations having made representations to the Government that it intervene in the market), and also the support of the Joint Coal Board, Connor recommended a series of policy changes for an initial six month period to bring what he saw as being some sense into the market. At the centre of the new suite of policies were export controls, with the Government to use its constitutional trade powers to place stringent conditions on the approval of new export contracts. However as a short term measure, Connor proposed that as the responsible Minister, he would use these powers to refuse to approve any new export commitments from open cut mines until longer term policies were developed. Export prices would need to be reasonable when compared with prices being received by other overseas suppliers, and export contracts for new projects would be approved only if Connor was satisfied that comparable coals were not available at reasonable prices from existing mines.

These two Cabinet submissions were referred to the Economic Committee (EC) of Cabinet, rather than going straight to all the members of Cabinet. That Committee considered Connor's proposals and recommended a more realistic stance, including the introduction of export controls on all minerals as sought by Connor, and the "balanced development" of Australia's mineral resources to ensure that export production was consistent with the national interest (although what that balanced development meant in policy terms was not clear). However, the EC stopped short of giving Connor the power to ban any new open cut projects. The EC also recommended that a special committee of Cabinet, to be chaired by the Treasurer, be established to develop detailed proposals for the minerals industry which would achieve the Government's objectives. Connor now had been granted a massive extension of his powers, but would also have to contend with the new special Cabinet committee, a potential restraint on the way in which he may wish to re-shape the minerals industry. The full Cabinet approved the EC recommendations on 30 January, and the new export control regime was announced. Export controls now extended to all semi-processed and unprocessed minerals, and the new policy meant that all export contracts for all mineral products would now need government approval on a shipment by shipment basis.

In March 1973 Whitlam went into the lion's den and addressed a meeting of the Australian Mining Industry Council (now the Minerals Council of Australia). Whitlam referred to the new export control regime, and also put the industry on notice regarding the level of foreign investment in mining: "The notion of unrestricted mining and exploitation is one we cannot accept. We must be concerned at the effects of unbridled competition on the stability of international trade... We must be concerned to see that our minerals are sold at reasonable prices in world markets ...The (mining) industry is already 62% foreign owned. We do not need that figure to go higher...judicious use of the export controls will add to our foreign exchange earnings...we believe it would be in the interests of both Japan and ourselves to work together more closely to secure for Japan a more reliable source of supply and to secure for Australian exporters more reasonable prices...The steps we have already taken will ensure that the Government is kept fully aware of marketing arrangements covering export prices and supply commitments. This is a prime purpose of our export controls."[25]

The Commonwealth Government has power to control exports under Section 51 of the Constitution which refers to the regulation of trade and commerce with other countries. Prior to the election of the Whitlam Government, the Commonwealth controlled the exports of uranium, zircon, manganese, copper, tin and iron ore. Until the 1960s, iron ore exports were controlled because of a belief that Australia lacked sufficient resources for our own domestic steel industry. The massive iron ore resources of the Pilbara had only begun to be discovered in the 1950s, with the scale becoming increasingly evident as the 1960s unfolded. Now coal exports were drawn into the Government's regulatory net.

Connor in 1973 was concerned about the prices being received by Australian exporters and was also surprised by the lack of information on the international coal market which was available to the Government when it came to power. In October 1973 he took pleasure in criticising the previous government over this lack of information and also the mining companies for signing export contracts written in US dollars: "Our major customer is Japan. For the first time we are dealing with the Japanese in a proper, businesslike way. The so-called government of extreme socialists in fact is capable of doing a better job than the so-called businessman's government. For the first time we have adequate information. Export controls were imposed... for the purpose of obtaining information, and we have information on prices, quantities, times of delivery and, above all, the denomination of currency in which contracts are written. I have had major companies come to me cringing because they were too stupid and the previous Government also was too stupid not to have their contracts written in Australian currency. This applies to some of the major companies. I certainly will not commit a breach of faith by disclosing their names, but they do exist. What have been the results of our activities? In the minds of members of the Opposition, we have committed an unpardonable crime. We have been a commercial success."[26]

In 1975 Connor's Department[27] said that the aims of the controls were threefold: "to ensure export prices are at a reasonable level in relation to export prices from other countries; to ensure that the Australian Government has a full knowledge of all details of export contracts, particularly the currency denomination and protection clauses, if any; and to see a balanced development of Australian minerals resources so that production for export should be consistent with the best interests of Australia." A later ABARE report on export controls noted that the extension of controls to all minerals in 1973 "was intended to address a number of concerns related to the international marketing of minerals".[28] These concerns included the level of returns on Australian mineral exports to Japan; the extent of foreign ownership of Australian mines, and hence possible offshore control of their export strategies; the possibility that some long term contract conditions, still in force, might no longer reflect current market conditions; the need for a balanced development of Australian mineral resources; and the possibility that some companies were avoiding taxation obligations through the use of 'transfer pricing'. In relation to coal, the ABARE report noted that the government's main objective was "to counter the possible effect of collective buying practices in international coal trade. It was believed that large buyers might be inducing individual Australian coal exporters to accept needlessly low prices. The government stated at that time that its aim in applying ... export controls was to ensure that prices for Australian coal exports (were) at reasonable levels in relation to export prices from other countries."[29]

Export controls on coal continued to be applied through the 1980s and into the 1990s, although in a more light-handed manner in the latter years. It would not be until the 1990s that export controls on coal were finally abolished

Fitzgerald Report slams mining companies

Rex Connor was one of the most memorable of the Whitlam government ministers, nicknamed, among other things, "the strangler". He was from Wollongong, with a keen interest in minerals and resources generally, but a low opinion of the previous government's policies. "He dressed characteristically in a dark suit, white shirt, dark tie, braces and a wide-brimmed felt hat. A massive yet subtle and complex man, he could be chillingly abrupt and sometimes bellicose. Suspicious and strongly inclined to be secretive, he could also be courtly, patient, persuasive, and enthusiastic, especially when speculating about advanced technology or expatiating on a prized project".[30] According to Whitlam, Connor had established himself by 1974 as his Government's most popular Minister: "His prompt action in extending Australia's export controls in January 1973 gave him an early and immense authority in the Cabinet. Connor did not clutter the Cabinet agenda with trivial submissions….At the election for the Ministry after the 1974 double dissolution Connor topped the poll. At the 1975 National Conference of the ALP at Terrigal the National President, Hawke, praised Connor as his favourite Minister. Moreover, the Conference resoundingly carried a resolution endorsing the minerals and energy policy embodied in the 1972 and 1975 Labor policy speeches and now being implemented so ably by Connor, and (urging) that no departure from the said policy be tolerated." [31]

Connor and his Departmental Head, Sir Lennox Hewitt, ran the minerals portfolio and brooked no interference. John Menadue was in no doubt as to who was in charge of minerals when he was appointed to head up the Department of Prime Minister and Cabinet in 1974. "On arrival in Canberra I had learned what I suspected: minerals and energy exploration and development was practically a 'no go' area for the Prime Minister and most of the Government. It was the exclusive domain of the Minister, Rex Connor, and the Head of the Department, Sir Lennox Hewitt. Connor was from Wollongong, an Australian nationalist par excellence, suspicious of foreigners and with a great love of the mining industry. He spoke about coal with knowledge and passion. A great resource was being plundered by the Japanese. …. Hewitt gave Connor unswerving support. He intensified Connor's suspicions. Hewitt told me when I went to visit him after I became Head of PM&C, 'I won't allow interdepartmental committees to get their fingers into this department.' It was a very clear warning to keep out of his department's territory, but at least I knew where I stood."[32]

However, Connor's appointment as minister responsible for the minerals and resources sectors was to cause major concerns in the boardrooms and senior executive offices of Australian companies, and was ultimately a major factor in the downfall of the Labor Government in 1975. The imposition of export controls in 1973 was the first major decision of the Whitlam Government which affected the mining industry and brought it under the firm control of Connor and his department. That decision was soon followed by the commissioning by Connor's department of a special review of the mining industry undertaken by Tom Fitzgerald, previously a senior journalist and finance editor of the Sydney Morning Herald. The Fitzgerald report looked at the contribution of the minerals industry to Australia, focusing on its profits, payments to governments and its subsidies and tax concessions. The review was commenced in 1973 and the final report was released by Connor in April 1974. Connor and Whitlam then had a field day with the report which accused the mining industry of not paying its way.

Fitzgerald's report contained data on the six years to 1973 for the mining industry which showed that the industry had earned $2072 million in profits before company tax. However the Government's net take from the industry was minus $55 million, with government assistance to the industry of $341 million and Federal government receipts of only $286 million. The cost of the assistance to the industry included the major item, income tax concessions, and also subsidies and the costs of the Federal Government's Bureau of Mineral Resources (BMR). Royalties paid to state governments of $263 million were shown separately.[33] Connor and Whitlam were now able to accuse the mining industry of being subsidised by the Federal Government to the tune of $55 million.

CRA's annual general meeting was held on the day of the official release of the report, and its chairman Sir Maurice Mawby made some pertinent comments. CRA's Hamersley iron ore subsidiary had been mentioned in the report, having paid little tax over the period in question. Mawby noted that the tax arrangements only involved a deferment of tax due to the ability of mining companies to write off capital expenditure against earnings, with tax payable increasing to above the normal rate when the benefits of accelerated depreciation cut out. He said that the viability of Hamersley was initially questioned and financing the project including its social infrastructure had been difficult, but that Hamersley had invested in townships, hospitals, schools, water and electricity supplies and amenities for a population of 11,000 people in a previously uninhabited area.[34]

The Australian Mining Industry Council (AMIC) held back from a quick response to the report, taking its time to analyse it in more detail and waiting until after the May election, won by Labor. AMIC said that rather than the industry receiving $55 million more from the Government than it had contributed, mining and smelting had contributed a net $700 million to State and Federal Governments over the 6 year period. AMIC accused the Government of lending its authority to a report which was "too narrowly conceived, distorted by the particular statistics selected and written without any consultation with the mining industry either collectively or on an individual company basis."[35]   While AMIC's response raised some important questions about the validity of the Fitzgerald report, the mining industry's reputation had been damaged. The major problem for the mining industry in the context of the debate on the Fitzgerald report was the taxation law which allowed companies to write off the value of fixed assets and exploration at a rapid rate. These arrangements, as Fitzgerald himself noted, were not unique to Australia, but he said that Australia was unusual in the range and scale of the write-off provisions for capital expenditure.[36]

Connor had been quick to signal that there would be changes to the taxation laws for the mining industry. In a statement to the media the day before the official release of the report, Connor said that tougher company tax arrangements would be introduced when the government was re-elected in the election to be held the following month.[37] With Labor back in power, the Budget in September 1974 saw major changes to the taxation provisions for mining. Capital expenditure was now to be written off over the estimated life of the mine, or if the mine life was over 25 years, 4% of the value of undeducted expenditure was able to be written off each year; exploration expenditure was now able to be written off, but only to offset income from mining. The tax changes, of course, did not please the mining companies. Coal & Allied, for example, complained in its annual report in 1975 that it would take many years to write off capital expenditure: "...Consequently, there will be a severe effect on cash flow and, coupled with the burden of the new export duty, mining operations requiring heavy capital expenditure may not be viable in the future. Underground mining is very different from open cut operations, and the Federal Government has failed to give proper recognition to this fact."[38]

The Government also directed the Industries Assistance Commission (the forerunner to today's Productivity Commission) to hold an inquiry into the mining and petroleum industries in 1975. Connor made a submission to the Commission in July 1975 in which he referred to the desirability of a super profits tax on some sectors of the industry, a tax he had flagged earlier in the year. A de facto super profits tax did come that year in the Federal Budget, but only on coal and in the form of an export duty.

Labor's new foreign investment controls

By the early 1970s, the level of foreign investment in, and control of, Australian mining, energy and manufacturing industries had begun to take a prominent place in political debates. Foreign investment and the significant role it played in these Australian industries, was a somewhat touchy subject with the public. A survey of over 2,000 people in June 1972, only a few months before the election of the Whitlam Government, showed that almost 90% of those surveyed supported placing a limit on the number of shares which overseas investors could buy in Australian companies. About two in three people were in favour of limiting overseas shareholdings to less than 50% of the shares in Australian companies. A similar survey by the same company in 1968 had produced very similar results. The survey company said that most people in favour of limiting foreign investment wanted to "keep control in Australia" or "keep profits in Australia".[39]

Whitlam as Opposition Leader issued a statement on foreign investment in May 1972 referring to the level of ownership and control of a wide range of industries, attacking the Coalition Government for doing little, and signalling that a Labor Government would act to introduce guidelines to control foreign investment and use the Australian Industry Development Corporation to "ensure Australian equity in large-scale projects."[40] Whitlam's statement was short on detail, but was a clear sign that change would occur when Labor won government later that year. In September Prime Minister McMahon announced changes to the Coalition's policy, the centrepiece of which was the creation of a new authority to review foreign takeovers, together with measures to ban short term borrowings overseas by Australian companies, allow foreign companies to borrow funds in Australia and allow Australians to purchase shares in foreign companies. The power to review and if necessary reject foreign takeovers would be legislated, and the new authority, in its assessment of whether a takeover was in the national interest, would be subject to detailed guidelines laid down by the government. The new policy was to take effect immediately, but with the defeat of the Coalition in the December election, it barely saw the light of day, and it would not be until 1976 that it would be largely resurrected. A few weeks after McMahon's new policy was announced, a Senate Committee on Foreign Ownership and Control of Australian Resources chaired by Liberal Reg Withers released a preliminary report which urged the Government to consider nominating key strategic or sensitive areas where it believed foreign investment should be limited or excluded. The committee had found that there was a lack of any national economic policy objectives which could be used as a framework to measure the benefits of foreign investment.[41]

The Whitlam Government did not move immediately on foreign investment, waiting until October 1973, when Whitlam travelled to Japan as the new Prime Minister to meet the Japanese leaders. Whitlam was accompanied by Trade Minister Jim Cairns, Treasurer Frank Crean, Minerals and Energy Minister Rex Connor and Primary Industry Minister Ken Wriedt. Whitlam announced details of the government's new policy on foreign investment at talks between Australian and Japanese ministers in Tokyo. While stressing that there was no prohibition on foreign investment in mining, Whitlam in effect said that there would be a ban on new investment in uranium, oil, natural gas and black coal, but that in most mineral and industrial development projects foreign investment would be welcome. All the Australian ministers present were reported to have participated in developing the new policy.[42] Whitlam was clear that 100% Australian ownership was his government's policy in relation to uranium projects, and that it was also a "desirable objective" in relation to the oil, natural gas and black coal industries. He said that with Australia's limited resources of capital and technology and the large size of many projects, Australia would "need to call on overseas expertise, technology and capital to contribute to the proper development of these vital energy resources." While equity in new projects in these four industries would be restricted to local investors, Whitlam said that his government was looking for overseas companies to contribute through "access to technology, loans and especially long-term contracts."[43]

The wording used by Whitlam seemed to indicate that the Government would be more flexible in relation to overseas investment in oil, natural gas and coal. The capital requirements for major new projects were significant and beyond the ability and preparedness of Australian investors to fulfil. But Whitlam and Connor were looking to the Australian Industry Development Corporation as a key financier, and the soon to be created Petroleum and Minerals Authority would also be seen as playing a major role. The Japanese were reported to have had concerns over Whitlam's foreign investment policy, but also to have accepted that the policy was the "bad news they had been expecting."[44] The Japanese ministers involved in the talks were reported to have made no attempts to disguise their dislike of the new policy, but also to have made it clear that they understood why it was being imposed. Whitlam was apparently pleasantly surprised that in his private talks with Japanese Prime Minister Tanaka, Tanaka readily accepted the new policy. Just days after returning from Japan, Whitlam issued new foreign investment guidelines which appeared to indicate a more flexible approach than was contained in the policy presented to the Japanese.

By late 1974, the Whitlam government was coming under increasing pressure to ease its foreign investment guidelines and encourage greater investment from foreign companies wishing to be involved in development of minerals and energy projects. In November the Government released guidelines relating to foreign participation through equity investment in the minerals industry. While the guidelines were quite specific in saying that the objective was to maximise Australian ownership and control, they also recognised that with many companies having major foreign ownership, "for this reason, the promotion of Australian equity ... must be viewed as a longer-term objective."[45]

Connor's Petroleum and Minerals Authority

Rex Connor's plans for the minerals and energy sectors in Australia extended far beyond export controls on minerals and restrictions on foreign investment; his new Petroleum and Minerals Authority (PMA) would be the government's major vehicle for developing Australia's resources and for promoting Australian ownership and control of Australian resources and resource industries. The PMA Bill was introduced into Parliament in December 1973 and gave the Government wide ranging powers, and in particular the power to explore for and develop petroleum and minerals. Connor said the Bill required the PMA to assist mining developments to "get off the ground", thereby helping them to remain Australian and save them from falling into foreign hands.[46] "Finance for the development of minerals and petroleum in future would flow through the authority" and "Australia would move from being primarily an exporter of raw materials to becoming a substantial exporter of semi-processed and processed materials." Connor said, however, that the PMA's powers would not affect supplies under existing arrangements, which would "be honoured in full" and would also not affect future supplies of minerals from Australia. He stressed that the Government would, "of course, insist upon fair and reasonable returns based on world market prices for our exports." The PMA Bill was rejected in the Senate, but following a double dissolution election, was finally passed at the first ever joint sitting of the House of Representatives and Senate in August 1974, along with several other Bills. Opposition resources spokesman Doug Anthony described the Bill as "a monstrous piece of legislation, tainted with bureaucracy and with socialism."[47]

Connor was now able to proceed with his plans for the PMA to become a major resources industry player and he received Executive Council approval to borrow $4 billion overseas to invest in oil and minerals projects in Australia. This approval was reduced to $2 billion in January 1975, although it was subsequently revoked in May 1975. The list of projects on the Government's agenda to be funded through these borrowings was extensive and included natural gas pipelines, uranium projects in the Northern Territory, electrification of the heavy freight rail system in NSW, upgrading of coal export ports (Hay Creek, Gladstone, Newcastle and Port Kembla), coal hydrogenation, the Cooper Basin gas project and a petrochemical plant in South Australia.[48]

The PMA Bill was subject to legal challenge by several States and was ruled invalid by the High Court in June 1975. Connor's PMA was now dead, but Connor claimed that the decision by the High Court had been anticipated and revealed that in February the Government had established a company registered in the ACT, Petroleum and Minerals Company of Australia Pty Ltd. He also said that the commitments entered into by the PMA would be honoured by the Government through this company, and that the Government was considering a number of other possible investments. One of the interesting but often forgotten decisions of the Whitlam Government was its intention to acquire a major shareholding in a new mine in the Hunter Valley. Wambo Mining Corporation, whose major shareholder was the Hartogen group, held mining leases in the Upper Hunter Valley near Singleton and, as it had been unsuccessful in finding local investors to take up equity in the proposed Wambo mine, it was considering accessing foreign investment. In 1974 the company approached the Federal Government and was successful in securing its agreement to take up a 49% shareholding. In October1974, Connor announced that the vehicle for the Government's equity would be his Petroleum and Minerals Authority. Connor said that Wambo, not finding local funding, was forced to look overseas, and identified the Anglo American Corporation as the proposed new partner in the project. Anglo later denied that there was any agreement to take over Wambo.[49]

There was some coal industry support for Connor's investment in Wambo, at least from NSW, with Sir Edward Warren supporting the move by the Government on the basis of retaining Australian control over our coal resources in "preference to further domination of natural resources by overseas capital and interests". Warren said that there was too much overseas control of our coal resources, particularly in Queensland, and that this also had the "effect of weakening the international competitive aspect" of the industry.[50] Warren was no doubt referring to Utah Development in Queensland, and possibly also the Japanese interests involved in Utah and TPM and implying that these companies were selling coal too cheaply and to the detriment of NSW exporters.

The PMA did advance some loan funding to Wambo in 1975, announcing that it was negotiating to take a shareholding in the company. However with the election of the Fraser Government in December 1975, and the winding up of the PMA, Connor's plan to become a major shareholder in a coal mining company was overturned. The PMA's other interests included a loan to Mareeba Mining NL, a small exploration company operating in the north Queensland, and equity in the Cooper Basin oil and gas industry. The loan to Mareeba was subsequently repaid and the Coalition Government negotiated with the government of South Australia to divest its interest in the Cooper Basin. Before its closure, the PMA also had a number of other projects in the pipeline; it was involved in coal exploration in NSW and planning to extend into Queensland and other States, and was studying the procedures for off-shore petroleum exploration and development. During the PMA's final weeks, it took decisions to become involved in a number of other projects and companies including oil and gas exploration in the Northern Territory, coal mining in NSW, mineral exploration in WA and gold and antimony production in WA. At the time of the High Court's decision, the PMA was considering "some 69 proposals for petroleum and mineral development, large and small, covering all States and the Northern Territory".[51]

With coal export ports one of the priorities for the Commonwealth, Newcastle MP and Transport Minister, Charlie Jones, made a submission to Cabinet in December 1974, noting that the NSW Government had approved the construction of a new privately owned coal loader which would double the capacity of the port of Newcastle. A subsidiary of Gollin & Company had negotiated a loan from Nippon Steel to finance the construction and was seeking approval from the Federal Government under the foreign exchange control system. Jones said that he "thought it would be a retrograde step to allow control of these facilities to fall into private hands", noting that the main facilities at that time were owned by NSW Government authorities. Jones recommended to Cabinet that it agree in principle to Commonwealth Government participation in a scheme for upgrading certain coal ports in NSW and Queensland; that State Governments be advised that the Commonwealth was willing to provide financial assistance on the understanding that it would share in the ownership and control of port facilities in proportion to its financial contribution; that the Commonwealth offer to provide the necessary funds for the new Newcastle loader on the understanding that the Commonwealth would own and operate the facility; and that the Commonwealth should offer to form a joint venture with the NSW Government to operate all the NSW loaders.[52] Cabinet agreed to approach the NSW Government offering to fund the new loader and to establish a joint authority for coal ports as recommended by Jones; it also agreed that if the funding offer was rejected by NSW, it would seek legal advice on its constitutional power to assume direct responsibility for ports. The Acting Prime Minister then wrote to the NSW Premier in January 1975 offering to fund and operate the loader; that offer was rejected by the Premier as the NSW Government had already approved of Gollin's involvement. Following the rejection, Jones sought legal advice from the Attorney General's Department which indicated that the Commonwealth could acquire land for a coal export facility but also warned that there could be lack of cooperation from the State Government and potentially a challenge to the High Court. Charlie Jones' bid to get the Commonwealth involved in coal export ports ground to a halt.

While the Whitlam Government had announced its more flexible foreign investment policy in November 1974, and the PMA had been killed off by the High Court in June 1975, the Government's policies had created serious concerns amongst the Australian industry and its overseas customers and investors. According to John Menadue: "There were serious impediments to resource development..... Importers and investors in Europe, Japan and North America were confused about Australia's policy and intentions. Projects were dying on the vine.... To get around the logjam, Whitlam agreed to my proposal in August 1975 that there should be a Resources Committee of Cabinet. It consisted of

Whitlam, Connor, Hayden, as well as public servants....The Connor-Hewitt axis was being outflanked. As a result, the Government did start to get important resource developments under way. The new crude oil pricing policy provided encouragement for explorers and developers. The foreign investment guidelines were liberalised to get dormant mining projects started. New coal projects in Queensland received the green light. It all flagged to the business community that, on terms which advanced Australia's interests, resource projects would go ahead."[53]

The coal export duty – an early super profits tax

The mining industry had suffered from the shocks of the Labor Government's export controls introduced in early 1973 and the controls on foreign investment, but a direct hit on the coal industry then came in the August 1975 Budget. The Federal Government, eyeing the handsome profits being made in the coal industry, particularly by Utah, announced that a tax would be levied on the industry's exports. High quality coking coal, which commanded the highest prices in the market, would be subject to a duty of $6 per tonne; lower quality coking coal and steaming coal were hit with a $2 per tonne duty. The $6 duty was aimed squarely at Utah's operations, but it also affected the higher cost NSW coking coal sector. All coal with a carbon content of 85% whether produced from open cut mines or from underground mines would now have to pay the higher rate of duty.

The export duty was an initiative of the Treasury and not Connor or his Department. Connor's preferred way to tax the coal and minerals industries was through a profits related tax, a tax which he claimed to have advocated throughout his period as Minister for Minerals and Energy, and which he said would apply to those best able to pay. But Connor also understood the "dilemma of the Treasurer of the day" (Bill Hayden) who, he accepted, "wanted to do something and to do it in the Budget."[54] This duty was a super profits tax by another name, albeit a fairly crude one, and was designed not only to generate some needed revenue for the Government, but also to ensure that a greater proportion of the value of the coal produced accrued to Australia, rather than to Japanese and other buyers who were seen as under-paying for the country's valuable coal resources. The coal industry reacted with shock to the new tax, as did Queensland Premier Bjelke-Petersen who said it was a "two faced policy", with coal companies on the one hand criticised and condemned for their new projects, and then hit with company tax and now the new export duty.[55]

Charles Copeman, chairman of Bellambi Coal Company, a NSW south coast coking coal producer, said the tax would be a major deterrent to the establishment of underground mines, and also have a major impact on the viability of his company's proposed extension to the Bellambi mine. Bellambi's General Manager, Ted Ryan, said that the levy would be equal to around 16% of the selling price, and that if the industry could not recover the tax from the Japanese, it would cost the industry around $150 million a year. "I can't believe it," Ryan said "perhaps it's only a 60 cent levy, not $6?"[56] Keith Gale, CEO of Gollin Holdings, said that the export levy was a "remarkably selective, unfair and almost vindictive tax on one of the few industries which has significant capital investment programs under way, and is enjoying, after many lean years, a brief period of comparative prosperity. In all my 17 years of experience in the mining industry I have never known anything like it."

Two days after the Budget, Utah announced that it had secured agreement with the Japanese for a retrospective price increase for all the coal sold under long term contract from its Blackwater, Goonyella, Peak Downs and Saraji mines. From 1 April the price had been lifted by US$15 per tonne, and from 1 July by US$16.50 per tonne.[57] Utah was affected by the export duty, but clearly not too seriously, and one coal industry source commented that it was ironic that "the shot gun blast at Utah by the Government should so completely miss its target while wounding all the innocent bystanders."[58] Utah was expected to pay around $75 million per year in export duties, but within a few weeks of the Budget it announced its quarterly profit results for the three months to the end of July. Net profit had risen to just over $42 million, compared with around $17 million in the same period in 1974. For the nine months to July, Utah had earned almost $85 million compared with only $25 million in 1973-74. The company was now set to take over as Australia's most profitable company, pushing BHP into second place. BHP had just set its own record, with a profit of around $110 million in the year to May 1975, the first time it had broken through the $100 million barrier.[59] The Japanese market for coking coal however was soon to be cut back and the export duty would begin to look more damaging to the industry's prospects.

With the record Utah profit result, and the prospect of continuing high profits, Connor said that it was time once again for the company to look at increasing its local equity from the modest 10.8% held through the listed Utah Mining Australia Limited. Connor took much of the credit for Utah's healthy earnings, telling Parliament that Utah could thank the support of the Australian Government for its present situation because the contracts it entered into with seven-year pricing terms with the JSM did not provide for annual price reviews. Utah however was dismissive of Connor's push to increase local equity, blaming his export duty for the company postponing any consideration of such an increase.[60]

The Federal Opposition was also critical of the duty, with National Party Leader Doug Anthony saying that it would be very damaging and would wipe out nearly half the profit being earned on high quality coking coal, and all the profit on steaming coal. It was incredible, he said, that the Government has attacked the one industry which held out real prospects for large-scale expansion. "Not only will planned new coal developments now be abandoned, but some existing mines producing steaming coal are likely to go out of business."[61] Anthony said that it was clear that Connor was aiming the new tax "at the major Queensland producers (ie Utah and TPM), and that in his desire to get at these companies he has attacked the whole of the export coal industry," adding that the implications for the new tax for the Bowen Basin were especially serious. Anthony's comments were well directed, but even though he would take ministerial responsibility for the industry later in the year, the coal export duty would be a lingering impost for years to come.

Connor met senior Japanese industry representatives including representatives of EPDC and the managing director of Nippon Steel, Mr Tanabe, in Canberra on 9 September. Connor reassured the delegation of Australia's commitment to be a reliable supplier to Japan. It was also revealed that day that Connor had pressured Australian coal exporters not to seek to pass on the coal export duty to their Japanese customers. How exporters were meant to contain the duty so that it was not reflected in prices was not clear, but with Connor soon gone as Minister and the Government close to losing office, that dictate effectively went nowhere. The Government however did listen to the steaming coal exporters' concerns about the duty. The tax on steaming coal which commanded a much lower price than coking coal was particularly severe, even if only $2 per tonne. In response to industry concerns, the rules were amended in October 1975 to exempt steaming coal with an ash content of over 14%.

Bowen Basin development stalls

The massive Goonyella mine and Hay Point port and export coal loading facility officially opened in November 1971. The official opening was attended by the Governor General, Sir Paul Hasluck, the Queensland Premier, Joh Bjelke-Petersen, and the chief of Utah Development International, Ed Littlefield. The Goonyella project involved the development of two major open cut mines, Goonyella and Peak Downs, the development of a new township at Moranbah, the construction of a railway to the port, and the development of the port and loader. The Goonyella and Peak Downs mines were already in production, Goonyella with an annual capacity of 4 million tonnes, and Peak Downs 5 million tonnes. Littlefield said that plans were proceeding for the development of the Saraji and Norwich Park mines, but warned that markets had to be assured for these new mines and appealed for industrial peace in the industry. He also said that Australia was not the only country with coking coal reserves and that Australia was still on trial as a stable and continuous supplier of coal. Littlefield added that large-scale mining operations, using the best equipment, were needed to help carry the infrastructure costs not applicable to some competitors.[62] The development of Saraji, with a capacity of 4.5 million tonnes per year began in 1972-73 and the first shipment of coal was planned for early 1975.

While Goonyella and Peak Downs were now operating as planned and Saraji was well under way, new developments in the Bowen Basin hit a roadblock in Canberra in 1973, with Utah's plans for Norwich Park stalled by Connor and Lennox Hewitt. Utah needed Connor's approval for an export permit for Norwich Park, but Connor was refusing to issue one. Utah attempted to negotiate with Connor and Hewitt but made no headway for reasons which became clear to the public in March 1975 when Connor made a detailed statement to Parliament.[63] Connor stated that Utah was already operating four mines quite efficiently and was seeking approval for a fifth. As far as Connor was concerned, it had approval to mine and export up to 300 million tonnes of coal and his Government told Utah that it would honour their rights to extract that tonnage and no more. Utah had told Connor that it could expand production from its current mines, but that it would be cheaper to develop the new mine. Connor told Parliament that in Queensland only 10% of coal was capable of being mined by open cut methods, and Utah's approval to mine 300 million tonnes meant that it would be responsible for extracting "half of the easily won coal." Connor went even further, with a not too subtle threat to kick Utah out of the industry once it had mined the approved tonnage: "This Government will honour to the letter commercial commitments entered into, but if and when the company has extracted its full quota of 300 million tons it will then be at liberty, under a Labor administration, to sell its plant to an Australian consortium. There are plenty of them who are willing to take over the operation from there. Buying back the coal is just one of the methods by which we can buy back the farm and restore ownership to the Australian people."

Another complication for Utah in gaining development approvals was the policy of the Queensland Government on foreign investment in new projects. In 1974 Mines Minister Ron Camm announced that the State's policy was to have at least 50% Australian equity wherever possible in future projects. While this was consistent with the Commonwealth's policy, the problem for Utah was whether the 50% would apply to all the CQCA coal assets (the mines and the Hay Point terminal) or just to Norwich Park. The Queensland Government wished to see the local equity share averaged across all the mines, a policy which would have pushed local equity in CQCA from around 10% to around 20%. As outlined below, the Coalition Government essentially maintained the Whitlam Government's 50% local equity requirement; Doug Anthony advised in January 1976 that the requirement would apply to Norwich Park for Utah to receive an export permit. Norwich Park was finally approved under the Fraser Coalition Government, but not until March 1976 when the Treasurer Philip Lynch and Anthony announced that Norwich Park could proceed on the basis of 55% local equity by the introduction of a new Australian joint venture partner. Utah would also have the option of that new partner taking an interest in all Utah's coal projects through CQCA.[64]

With Norwich Park and other Queensland projects on hold in 1975, a test of the Whitlam Government's approach to foreign investment came in October of that year, only a month prior to its dismissal by the Governor General. Two days before the resignation of Rex Connor as Minister for Resource and Energy, and after a review of the coal industry by Cabinet's resources committee, Whitlam announced that there would now be a prima facie presumption that export approval would be given to projects which had obtained sales contracts which were acceptable to the Government, including in terms of prices. Whitlam said that new projects would need to be consistent with the Government's foreign investment guidelines announced the previous month, and said he believed that the new projects expected to commence in the near future would be sufficient to meet the likely demand for coal.[65] The Prime Minister's announcement was received cautiously by the industry, with the Australian Mining Industry Council's executive director, G Paul Phillips, saying that the new policies were a step forward. Whitlam said that the Hail Creek project had already been approved subject to foreign investment conditions being satisfied, and that subject to the same conditions, the Government was prepared to approve the Nebo project.

NSW port capacity holds back the industry

In late 1973 and early 1974 several new export contracts were signed for coal to be shipped through the port of Newcastle. This was good news, but there was a major problem – the port's capacity was now insufficient to be able to load all the coal planned for export. By January 1974, a queue of ships had formed off the port waiting for the all clear to enter and load; with more ships having notified the port that they were on their way, the situation was serious. That month the JCB decided to act and issued Order number 30 which required all exporters to obtain its consent before they could load coal for export. The JCB also said that it was not confident that export tonnages would be able to be increased beyond current levels before 1977. With the Board also stating that it was apparent that control was needed at the three NSW coal export ports, in March it amended Order 30 to extend it to Balmain and Port Kembla; its realistic assessment of loading capacity at these facilities was 9 Mtpa for Newcastle, 4 Mtpa for Port Kembla and 2.5 Mtpa for Balmain.[66]

With pressure from the exporters, the JCB and the Japanese customers, the NSW Minister for Public Works, Leon Punch announced approval in principle for a new coal loader to be built in Newcastle in June 1974; the loader would have a capacity of 10 Mtpa and would at least double the port's capacity. Gollin & Company won the tender to construct the loader, but as we will see in the next chapter, the company's financial problems saw it wound up and a new industry consortium, Port Waratah Coal Services (PWCS), was established to take over construction and operation. The new PWCS Carrington loader was completed in 1977, and the capacity of the port increased to 20 million tonnes. However as events would prove, that would be inadequate for the expansion in exports that was to come.

In 1974 consideration was being given to a new loader in Port Kembla to be built outside the existing Inner Harbour loader, and to another new loader in Botany Bay. A privately funded loader in Botany Bay was proposed by the NSW Colliery Proprietors' Association in 1974 on behalf of Coalex, Austen & Butta and Clutha.[67] Coalex and Austen & Butta had mines in the western district around Lithgow and were eyeing opportunities to expand as overseas demand for steaming coal began to grow. Clutha, with its Burragorang Valley mines, was also strongly behind the Botany Bay proposal. The Colliery Proprietors' Association was also supporting a new loader in Port Kembla and did not see the Botany Bay proposal as the only major new facility needed. On the other hand, the coking coal producers in the southern coalfield, including KCC and Bellambi, did not want Botany Bay taking business away from Port Kembla, fearing it would lead to higher loading charges.

The NSW Liberal Government supported a new Botany Bay loader, and in November 1975, the NSW State Pollution Control Commission gave formal approved for the new loader which would handle ships of 130,000 DWT and have an initial capacity of 7-8 million tonnes per year. A second stage of the loader would be able to accommodate ships of 200,000 DWT, with an annual capacity of 13-14 million tonnes. A change of Government in 1976 then derailed the process, with the Wran Labor Government finally rejecting Botany Bay in June 1977 and favouring instead a new loader to be built in Port Kembla, together with an upgrading of the Balmain coal loader. The new Port Kembla facility would take another five years to be completed and began operating in late 1982 with a capacity of 7.2 million tonnes.

The community and political opposition to a loader in Botany Bay had been extremely strong and the loader's rejection by the Wran Government is understandable, particularly when viewed from the perspective of more recent community opposition to major projects. However, for the Western coal exporters, Port Kembla would mean a longer and more expensive freight haul, and for Clutha it would mean that the Government would have to build a new rail line to link the Burragorang Valley mines to the main southern line. The investment in new rail capacity did not eventuate, and the Wollongong area saw a major increase in the transport of coal by road to the port, an outcome which created a backlash from the community. The Government also had to compensate the Western district coal producers by subsidising the freight cost they now had to pay for railing their coal to the port. That subsidy would prove to be a significant cost to the Government during the 1980s and 1990s.

The Fraser and Anthony years

In its last year in government, the Whitlam government was clearly in serious trouble with the electorate, with Rex Connor one of the major problems. Connor was forced to resign as Minister on 14 October 1975 following revelations about his dealings with a dubious financier from the Middle East, dealings which had continued even after his approval to seek loan funds overseas had been revoked. With the Opposition refusing to pass legislation to fund the Government's budget, Whitlam was sacked by the Governor-General Sir John Kerr in November 1975. A new Government was elected and took power in December 1975, with Malcolm Fraser the new Prime Minister and Doug Anthony the Deputy Prime Minister with responsibility for the important Trade and Resources portfolio. The minerals industry welcomed the change of government after a turbulent three years. Development of the industry had been held back by Connor, export approval had been extended across the sector, its reputation had been damaged by the Fitzgerald report, taxation concessions had been removed and taxes increased. With the new Government, the spectre of massive Federal control of investment eased, but while there would be significant change under the new government, in terms of foreign investment policy, export controls and the coal export duty, change would be slow. The new Fraser government moved to correct what it saw as some of the major problems created by the previous government. But the early period of the Fraser Government was marked by what could be described by some as timidity or by others as a careful and measured approach to policy measures regarding the minerals and energy industries.

Doug Anthony had given the mining industry real hope in September 1975 that the foreign investment policy constraints which applied under the Labor Government would be eased if the Coalition won the next election. As leader of the National Country Party and Opposition spokesman on Trade and Resources, Anthony acknowledged that there would not be a return to the open door policy which applied earlier, but said that the Liberals and Nationals would welcome foreign investment and that under a new Coalition Government each project would be assessed on its merits. Anthony admitted that it was not desirable or possible to have watertight rules for foreign investment and that the aim of a new Government would be to encourage the maximum level of Australian ownership while recognising the need for foreign capital to supplement the reluctance of Australians to invest in major development projects. Where the Coalition would differ from Labor, he said, was that it would not be prepared to bring new development to a halt in order to achieve its objectives. "As long as foreign investors obey the laws of the land…there is nothing to fear from foreign capital, and a great deal to gain."[68]

Only a few months into the term of the new government, Treasurer Phillip Lynch announced a new foreign investment policy designed to "dispel the atmosphere of confusion and uncertainty in which foreign investors have had to operate over the last few years". Lynch said the Coalition Government's objectives were "to encourage foreign investment in Australia because of the considerable contribution it can make to Australian development and prosperity but, at the same time, to see that such investment is on a basis of fair sharing of net benefits as between the foreign investor and the needs and aspirations of the Australian community."[69]

These were fine words and were certainly more encouraging for the industry than some of the statements uttered by Whitlam or Connor when Labor held power. However, apart from announcing that the Government would legislate to establish a foreign investment review board to advise it on the merits of foreign investment proposals, little had changed. Bill Hayden, then Opposition Shadow Treasurer, congratulated Lynch, saying that his statement was, in his opinion, "a very good one because it is almost exactly the same as the one which we made in September last year." Hayden noted that the only significant difference related to uranium which would now have new projects requiring a minimum 75% local equity, down from the 100% that Labor had required.[70]

Coal prices in the 1970s

The first oil shock of the 1970s began in October 1973 and by March 1974 oil prices had jumped to $US12 per barrel compared with around $US3 only months before. With Japan so dependent on imported oil for its energy requirements, this was a major blow to the country and saw its economy go into reverse in 1974 after years of very strong growth (although GDP had managed an increase of only 0.4% in 1970 before growth picked up in 1971). Japan's GDP recovered in 1975, up 3.1%, and continued rising but at a much more moderate pace than before. With growth in the economy stalling or only moderate, Japanese crude steel production also proved to be volatile, falling in 1971, before recovering again in 1972 when output rose 9.4%. This was followed by a surge of 23% in 1973 when production hit 119.3 Mt. However 1973 would prove to be the peak for the Japanese industry, with production down by almost 2% in 1974, and down a further 12.6% in 1975 to only 102.3 Mt. Steel production did recover somewhat, reaching almost 112 Mt in 1979 when the second oil shock hit the world's major economies. With the steel industry growing rapidly in the 1960s, Japan's imports of coking coal had been rising dramatically, up from around 15 Mt in 1965, to 45 Mt in 1970. Coking coal imports rose to 55.6 Mt in 1973, before levelling off over the next three years, rising again in 1977 and falling again to only around 53 to 54 Mt in 1978 and 1979.

This was the background to the coal price negotiations and outcomes in the 1970s. Following the surge in steel production in 1973, coal prices leapt in 1974. Negotiations for JFY 1974 were not finalised until July and were backdated to 1 April, the start of the Japanese financial year; the new prices were 50-70% higher than for 1973. This was great news for the exporters, but with steel production about to fall, the JSM were soon far from happy. For the Japanese steel mills, the increase in prices in 1974 represented a major cost increase; in terms of the landed cost of coking coal in August 1974, for example, the average cost per tonne in yen was 95% higher than the average for the period from January to June that year.[71]

The JFY 1974 negotiations were led by Ted Ryan from Bellambi Coal in his capacity as Australian Coal Association chairman. Negotiating under the guidelines of Connor's Department, Ryan was representing his own company, as well as KCC and Clutha, and in March it was reported that the JSM had agreed to major increases in prices, although the finalisation of the arrangements for JFY was not announced until July as already mentioned.[72] Separate negotiations by Utah and by TPM followed, with both companies successful in obtaining even higher increases. And after the Department of Minerals and Energy instructed Thiess to renegotiate prices for South Blackwater, it too won a big increase in price.[73]

In 1975, Connor was again prominent in the negotiations when he visited Japan in June following lack of progress in reaching a deal by the exporters and the JSM. The agenda for Connor's visit was broader than coal, but coal dominated the media coverage, with Connor able to claim credit for a deal under which prices for hard coking coal would now reach $US50 a tonne, with soft coking coal rising to $US38 per tonne. As these prices were not what Connor had been pushing for, the chairman of Nippon Steel, Yoshihiro Inayama, undertook that Japan would up its purchases of Australian coking coal from 27-29 Mt to 44-49 million tonnes by 1980.[74] Importantly, the commitment also contained the proviso that if through economic circumstances there was to be a reduction in Japanese steel production, there would not be a reduction in coal purchased from Australia under contracts with the mills.[75]

Rex Connor's role in assisting companies to secure higher prices was recognised by the media and by the JCB which said in 1974 that the oversight of contracts by the Australian Government had been "an important factor in Australian producers obtaining contracts at improved prices." Almost all contracts over one year were now providing for annual price negotiations, escalation clauses for cost increases and exchange rate protection (allowing exporters to see a constant return in Australian dollar terms.)[76] However Inayama's commitment to a major increase in purchases of Australian coking coal was to prove unrealistic; and if individual exporters believed that their contract tonnages were safe, they were soon to be confronted with the reality. Australian coking coal exports did not grow as predicted and by 1980 were virtually static at around 27Mt.

Nippon Steel's Saburo Tanabe had sounded a clear warning for coal exporters in May 1975, saying that following the slump in the export market for steel in November 1974, the JSM were operating at only 75% of capacity, and were losing money. Tanabe said that production would have to be cut further, and that if the Australian Government was determined to secure higher prices for coal exports to Japan, this could only "severely damage Japan's steel industry."[77] Tanabe also warned some months later that with current steel production down by 30-40%, there was "naturally a possibility that we will have to ask for a 15 per cent to 20 per cent reduction in shipments." At this rate, he said "we expect our stock yards will be full to capacity by March or April (1975)."[78]

With the downturn in Japanese steel production in 1975, the JSM overrode the contracts they had written with Australian exporters and reduced imports of coking coal well below the contracted tonnages. In JFY 1976 for example, Australian exporters shipped about 26 Mt to the JSM, compared with contract tonnages totalling around 32 Mt. And these cutbacks in tonnages were not uniform across the industry, with some Australian exporters feeling the full effect of reduced purchases from Japan, while at the same time the mills were signing export contracts with new projects in Queensland. The major cuts to contracted tonnages were hitting NSW exporters hard, with deliveries only 73% of what had been contracted for JFY 1977. The following years were slightly worse; the percentage fell to 69% in JFY 1978, rising back to 72% in JFY 1979.[79]

By August 1975, all the major suppliers had achieved major increases in contract prices (f.o.b.) for their coking coal sold to the JSM. Australian prices ranged from $US37.60 (for soft coking coal) to $US50.50 for hard coking coal. Canadian hard coking coal prices ranged from $US49.50 to $US50.50 and the US hard coking coal prices ranged from $US57.58 to $US59.97.[80] As the JCB noted that year, all the major exporting countries selling to the JSM had been seeking large increases for their coal exports in 1974 and 1975, with negotiations protracted.[81] Australian hard coking coal prices had risen from around $US15 to $US20 in 1973 to around $US50 in 1975, but while there was parity with Canadian coal (at least on an f.o.b. basis), all the work by Connor, his Department and the Australian producers still had not achieved parity with the prices for US coal.

With Rex Connor's resignation as Minister in October 1975 and the election of the Coalition Government in December 1975, the following year saw a more hands-off approach from Doug Anthony as the new Minister responsible for mineral exports. The change of government in Australia did not, however, mean that the JSM would be any less demanding in the price negotiations. Warwick Parer, then a senior executive with Utah, and chairman of the coal industry's export committee (the formation of which we will come to shortly), said that in the most recent round of negotiations the JSM succeeded in playing Australian exporters off against each other, and as a result, Australia had secured a lower price than may otherwise have been obtained.[82] "In recent negotiations with the Japanese steel mills," Parer said, "the Government went to great lengths to avoid interfering. It attempted in good faith to repair the damage caused by Government interference between 1973 and 1975 and it expected that both sides would act honourably. Regrettably it did not entirely work out as planned, as extreme pressure was placed on certain exporters in the Port Kembla area by buyers restricting the number of scheduled vessels. It was apparent that the continuation of this tactic would result in capitulation by the individual companies. If these sorts of tactics are repeated, the Government may be compelled to involve itself to ensure that a fair price is paid."

The Australian Coal Association's export committee met Anthony and senior officers from his Department in early 1977 and gave a presentation on the export market. At that time Japan was importing around 27 million tonnes of coking and thermal coal out of Australia's total exports of 34 million tonnes. The committee noted its sympathy for the situation in Japan which had forced the steel mills to cut back their coal imports, with cutbacks to more expensive US coals of around 18 million tonnes the absolute maximum the Japanese believed was appropriate, and reductions in the intake of other coal by around 20%.

The Japanese steel industry in 1978 was still suffering; production was down, and the steel mills were carrying high stocks of iron ore and coal and were losing money. In April, at the inaugural Australia Japan Coal Conference on the Gold Coast (a meeting of the Japanese buyers and Australian exporters), Nippon Steel's senior executive in charge of raw materials, Saburo Tanabe, delivered the major speech from the Japanese side.[83] Tanabe said that with Japanese steel production well down from its 1973 peak, and with the mills incurring a loss of over $US300 million in the first half of JFY 1977, it was time for Australia to bear its share of the risk in the market. Tanabe said that Australia was becoming a major coal supplier in Asia, and claimed that with its ability to be flexible with production levels, Australia could not remain immune from the fluctuations in the balance between supply and demand. He went on to point to the high profits which he said producers had earned in recent years, reasoning that prices now needed to reflect the pain that the steel mills were feeling. The future was still bright, he said, predicting that when the current hardships in the market had been overcome, Australia would be able to enjoy sustained prosperity. Doug Anthony opened the 1978 coal conference, encouraging the exporters to diversify away from Japan. Anthony said that Australia had the resources and was competitive enough to meet Japan's needs while also increasing its sales to Europe and other markets such as South Korea. In perhaps a subtle warning to Japan not to push Australia too hard on price and contract terms, Anthony pointed to the opportunities existing for Australian suppliers to secure long term contracts with other countries.[84]

The next day, Paul Keating as the Opposition Resources spokesman, attacked the Government and Doug Anthony over their attitude to the Japanese coal and iron ore trade.[85] Keating quoted from Nippon Steel's letter to Connor in 1975 assuring Australia of being able to continue to supply contracted tonnages in the event of economic downturn, and accused the Government of failing to "manoeuvre the Japanese steel mills into a continuing acceptance of the philosophy of the Inayama letter, namely that the producers who were first into the market on a long term basis were the last to be penalised." Keating said that the latent over capacity in the world iron ore and coking coal industry was the direct consequence of the steel mills' own policy, and that the mills must take responsibility for the excess mining capacity that had only recently gained access to the Japanese market. "The mills wanted the over-capacity as a weapon to depress prices in good times; they should now take responsibility for their folly in the bad times," Keating said. Anthony responded to Keating's attack, noting that when he was in Japan in 1976 Inayama had explained that Nippon Steel had been forecasting steel production of 140 to 150 million tonnes per year by 1980. Once in power Anthony said that the Australian Government had removed roadblocks to the development of coal and iron ore to meet the expected demand from Japan, but that currently the Japanese steel mills were operating at only 70% capacity and were "knee deep in raw materials."[86] Anthony was playing the statesman role, and was not intent on damaging the relationship with Japan, but later that year would take a tough line when announcing new export control guidelines for Australian resources.

Later in 1978, Gordon Jackson, the CEO of CSR Limited, addressed the annual Japan Australia Business Cooperation Committee meeting in Osaka. These meetings brought together senior executives from both countries and were a forum for each side to place major issues on the agenda. Jackson, while diplomatic, made a hard hitting speech, warning the Japanese that their failure to observe long term contractual obligations was making it difficult for Australian mineral resource companies to secure loan funds for development of projects based on the security of sales contracts with the Japanese.[87] Jackson detailed a number of specific concerns including the tendency for Japanese buyers to be over optimistic in the tonnages specified in contracts leading to over supply; the practice of the Japanese unilaterally varying quantities specified in contracts, and even holding back ships which were destined to collect cargoes; and Japanese resistance to changes in annual pricing arrangements in long term contracts arising from changes in exchange rates between the Japanese yen, the Australian dollar and the US dollar. Jackson also accused the Japanese of resisting proposals that would allow Australian suppliers to divert supplies on a long term basis to other markets when Japan did not take its contracted tonnages.

In 1978 the JSM had been attempting to impose price reductions on the Australian coking coal exporters, and in fact forced Thiess Peabody Mitsui in February to accept a cut of $US1.30 per tonne for Moura coal and the removal of cost escalation provisions in the contract. Utah had been attempting to negotiate with the JSM on the contract for Blackwater coal, but broke off negotiations in October. The JSM wished to impose a similar price and the removal of escalation provisions for two years on Utah. With a number of contracts up for renegotiation for JFY 1979, the Australian Government had approved the Moura deal for only a twelve month period to prevent the JSM from forcing all the exporters to accept their terms. The JSM had also pressured the Canadian exporters to agree that escalation provisions would be removed from contracts, although the exporters did manage to secure a modest increase in prices of $C1.40 per tonne. The JSM were also "telegraphing through trade journals that Australian coal (was) more expensive in yen terms than Canadian coal."[88] In NSW there were a number of export contracts due for completion by March 1979 totalling 7.8Mt and in the absence of an improvement in the market, the exporters were expected to see the JSM and other customers demanding reductions in tonnages and changes in conditions in new contracts that would suit the buyers.

Given the pressures on Australian exporters, Doug Anthony made a major announcement on 24 October 1978 on export controls. The new policy now involved a tighter control regime in relation to coal, iron ore, bauxite and alumina. For coal exports it was a return to something more like the Connor regime. Anthony emphasised that Australia was faced with "a fundamentally changed market situation" which involved "depressed demand and of serious overcapacity among the major suppliers of bulk minerals - particularly iron ore and coal - to the world's consumers", commodities which had no transparent pricing such as on the London Metal Exchange.[89] He went on to clearly put the Japanese on notice that the prices in the market were not reasonable: "The Australian Government's concern is that individual Australian companies face buyers who are co-ordinated or who have a high degree of consultation and who, as a result, can and do successfully play one seller off against another. The result is prices, terms, and conditions which do not reflect a fair and reasonable return. This is particularly so in the current state of world markets with Australian companies seemingly fighting each other for available tonnage often at the expense of fair and reasonable returns. Recent developments have indicated that buyers are taking full advantage of this situation."

Anthony went on the say: "This is a most unsatisfactory situation and I have already expressed my concern at the price settlements that were agreed to this year for iron ore sales to the Japanese market. A situation can arise where contract conditions may be accepted by one company, albeit reluctantly, but can have grave implications if imposed on other exporters. This can lead to an overall result which is not in the national interest. Based on evidence in recent coal negotiations there is every reason to believe that unsatisfactory results could flow from the forthcoming coal negotiations which encompass approximately 80 per cent of our coal exports to Japan." Anthony also pointed out that the new procedures were designed to ensure a continuing flow of information between the Government and the exporters and to reduce the potential for the Government continuing to be faced with having to accept or reject a settlement.

About 6 weeks after Anthony's export control statement, Utah signed a new Blackwater contract for 6 Mt. Anthony welcomed the contract, saying that the outcome had shown that commercial negotiations could be carried out successfully within the context of the revised export control procedures, and that it was fair and reasonable for all parties. Anthony also said that he was pleased to see that Utah had been able to retain the escalation clause and that the contract did not contain onerous features such as a "recession clause" (which would have provided for cuts to tonnages in the event of a downturn in the Japanese steel industry).[90]   In JFY 1979 NSW coking coal exports to Japan were almost unchanged on JFY 1978 (11.5 Mt vs 11.4 Mt), while Queensland's exports were reduced to 12.9 Mt (down from 14.4 Mt in JFY 1978). The following year saw both states recording increased exports to the JSM – NSW to 12.2 Mt and Queensland to 15.1 Mt. It appeared that there had not been any major adverse impact from Anthony's new export controls, although Australia's share of the Japanese imported coking coal market was cut from 48% in JFY 1978 to 45.4% in JFY 1979. The USA was rewarded with an increase in its market share from 20.9% to 23.5%.

While Australia had emerged in the 1970s with the largest share of the Japanese coking coal market, with the traditional supplier (the USA) well back in second place, and the price gains won in 1974 and 1975 had been essentially maintained, by the end of the decade there was still a significant differential between the prices the JSM were prepared to agree with Australian exporters and the prices for US exporters, and to a lesser extent for Canadian exporters. Rex Connor had helped to achieve a boost for Australian prices in 1974 and 1975, but the reality by 1980 was that Australian producers were still discriminated against. In June 1980, Australian metallurgical coal export prices ranged from $US 42.55 to $US 54.00 per tonne f.o.b. (the lower end of the price scale applying to soft coking coal). The prices for Canadian coal ranged from $US54.05 to $US 58.78 per tonne f.o.b. and for US coal from $US52.16 to $US77.26 per tonne f.o.b.[91] Price discrimination would continue into the 1980s, a decade which would also see the JSM invest in two large Canadian coking coal mines and reward them with high prices. The support for the new Canadian projects was evidence that the export control policies and foreign investment policies of Australia's governments in the 1970s had not been forgotten by the JSM.

Australian Coal Association's Export Division – a short life

Following a suggestion from Doug Anthony, the Australian Coal Association (ACA) established a committee to represent the coal exporters in early 1976. Up until that time, the ACA's role had been restricted to education, environment, finance and taxation, legislation, and national energy policy. Industrial relations matters were handled by the two state associations which comprised the ACA (the NSW Combined Colliery Proprietors Association and the Queensland Coal Owners' Association).[92]   Following its formation, the new export committee, the Australian Coal Export Division, or ACED, wrote to the Trade Practices Commission and explained that it had been formed following a suggestion from Minister Anthony and that its chairman was Warwick Parer, commercial manager for Utah Development, with John Doherty, Kembla Coal & Coke's marketing manager, as deputy chairman. The ACED had two broad aims – to have meaningful discussions with the Federal Government on the attitude of coal producers to the terms and conditions of coal exports, and to formulate principles to be adopted for coal exports. The new committee was not intended to act as a cartel, membership was not compulsory and decisions were not binding on members.   The ACED also told the Commission that Anthony had requested that "the (coal) industry take the responsibility of settling its negotiations on an industry basis in a spirit of cooperation amongst themselves, and to assist the Government in its administration of export controls." The ACED also noted that the actual responsibility for the export control system remained with the Government.[93] At its second meeting in February 1976, the ACED was starting to clarify what it saw as its role (rather than the role that Anthony had proposed); the committee agreed that its role was to communicate the industry's position to the Minster for National Resources and his Department in relation to the forthcoming negotiations with the Japanese (ie for coal contracts for JFY 1976). The ACED also agreed that as a basic philosophy, the Japanese request for a return to "cost plus reasonable return" as the basis for coal prices was not acceptable.[94]

In 1973, Rex Connor announced that the coal exporters had agreed to adopt a united approach to overseas buyers. The Canberra Times heralded this as the coal industry planning "to end rivalries and work out common principles for negotiating sales to Japan."[95] The announcement from Connor came after meetings between the two state coal associations (the Queensland Coal Association and the NSW Combined Colliery Proprietors' Association), and after the recent iron ore negotiations with Japan had seen the Australian iron ore exporters cooperate to secure price relief following the impact of the currency changes which had hit the earnings of Australian exporters. Connor said that the associations had agreed to work to agree on basic principles for the orderly export marketing of Australian coals, and that the concept of a national coal industry would be promoted by the Australian Coal Association.[96]

The industry did hold discussions with the Japanese steel mills which simply reaffirmed that the previous trade principles would remain unchanged. The ACA and the Japanese steel, gas and coke industries issued a statement signed by the ACA chairman, Edward Ryan, head of Bellambi Coal, and Saburo Tanabe, the senior managing director of Nippon Steel. The statement reaffirmed that "negotiations for contracts should be based on the principle of individual negotiations by both parties" and that the coal trade between the two countries had always been conducted on a private basis... Negotiations have been conducted on an individual company-to-company basis, with both parties respecting and honouring the contracts… When problems have arisen, they have been settled by the parties through sincere and fair negotiations. Both parties hereby confirm that this basic trade principle will remain unchanged."[97]

However as detailed by David Lee, the negotiations for JFY 1974 contracts were carried out within the policies set down by Connor and his Department, and did lead to significant increases in prices. While Ryan negotiated on behalf of his own company, KCC and Clutha, the major Queensland coking coal exporters (Utah, TPM and Thiess) undertook separate negotiations, although still under the umbrella of the Department's guidelines.[98] The Australian Financial Review was glowing in its praise for Connor in 1974 following the coal price increase saying that he had "presided over a radical change in public acceptance of government involvement in energy policy" and warned against any return to the "laissez faire days of a few years ago".[99] What the AFR and others glossed over, however, was that the price increases were won in the context of a continuing strong market for coal, despite the Japanese steel industry's production being cut back in JFY 1974. The Japanese actually increased total coking coal imports in the period between JFY 1973 and JFY 1975 from 55.6 million tonnes to almost 57 million tonnes; it was domestic Japanese coal production that suffered in that period. Coal producers in other major coking coal exporting countries also won significant increases.

Connor's idea of the industry negotiating as a united group with the Japanese was always going to be problematic given the competition between companies for export contracts, the widely varying cost structure of the producers and the potential legal threat from US anti-trust legislation. And it was doubtful that the Japanese would have accepted an ongoing sales group negotiating on behalf of the whole Australian industry. The Japanese steel mills were the customers and were not prepared to see their negotiating power weakened.

In May 1978 the ACA Export Division was disbanded following detailed legal advice on the problems which could arise for Australian coal exporters from US anti-trust legislation. However in August, Anthony as Minister for Trade and Resources advised the ACA that he considered that it was essential that the industry should establish a representative group in which industry consultation could occur, and which could meet with him or his Department with a view to keeping him informed. He requested the ACA to consult with its members and advise him of arrangements it had decided on.[100]

Anthony had developed a close working relationship with the industry, particularly through the meetings of the ACA Export Division under the chairmanship of Utah's Warwick Parer. With the difficulties in the coal market and the potential for the Japanese to inflict pain on Australian coal exporters, he was clearly unhappy to have no such ongoing forum for meeting with the industry and was now pressuring the industry to have an export group set up to replace the disbanded Export Division. The ACA met on 6 September and agreed to set up an export committee to operate along the same lines as other ACA committees, as an advisory body to recommend policies to the ACA. The ACA chairman, Dick Austen, wrote to Anthony on the same day as the meeting advising him of the new committee, which he said would have its first meeting on 11 October.

Following Anthony's statement on export controls, Austen also wrote to Tanabe of Nippon Steel advising him of the new committee that Anthony had requested be established. He said that the committee would provide the ACA with "recommendations to enable replies to be given to requests from Government and external parties seeking the Association's advice on matters relating to coal exports." Austen said that the committee's operations would be "low key" and entirely advisory, with no statements to emanate from the committee and that an article in the Financial Review on 15 September was "fallacious in its entirety" with the exception of the names of the office bearers of the committee. He concluded the letter by saying that the committee would have no role to play in the commercial arrangements between buyer and seller and would in no way constrain individual negotiations.[101]

The Export Committee met on three occasions in September and October 1978 and was involved in two meetings with Anthony. Anthony was told that the ACA was not planning to participate in any "market regulatory function" (ie export controls). He said he welcomed the formation of the export Committee, but said that in his view there was little coherence in coal marketing activities of companies. He said he believed that individuals, groups or the ACA itself should be prepared to provide the government with policy advice and that he did not wish to decide on policies without the industry's assistance. Some coal producers continued to express their concerns about the potential impact of US anti-trust legislation as a result of their participation in the Export Committee. Anthony had suggested that his invitation to the ACA to form a committee be interpreted as a type of government directive.

However, this did not appear to have eased those concerns.[102] The ACA's Export Committee would continue to operate into the 1980s and 1990s as a forum to discuss export issues, although it was careful to steer clear of issues which could be construed as collusion on pricing or related matters.

BHP snaps up Peabody's stake in TPM

Thiess would be the subject of a major takeover battle in 1979 and 1980, but in 1976, a decision in the USA provided an opportunity for a major company to buy a majority stake in Thiess Peabody Mitsui. The US Federal Trade Commission in 1975 ordered Kennecott Copper to divest its interests in the Peabody Coal Company which it had acquired in 1968; Kennecott had been charged with restricting competition in the coal business and was given until April 1976 to sell out of Peabody. TPM, in which Peabody was the major shareholder with a 58% equity, had large leases in the Bowen Basin on which it was planning to develop the huge Nebo project. However Thiess' 22% share of the company was insufficient to gain approval for the project under the Federal Government's foreign investment policy. TPM had signed sales agreements with the JSM which would provide the basis on which to proceed to develop the project which involved a number of mines, including Riverside (with sales of around 40 million tonnes over 14 years), and Poitrel (33 million tonnes over 16 years). Rex Connor however told TPM "point blank" that until its local equity was increased there was no point in talking about the future of the project. [103] Connor was reported as seeing the sale of Kennecott's Peabody subsidiary as an opportunity for new owners to become involved in TPM, with BHP, Western Mining, CSR and Thiess mentioned in the media as potential buyers.[104] Connor was correct, and BHP joined a consortium of companies led by Newmont Mining to bid for Peabody Coal. In June 1977, the US Federal Trade Commission approved the sale of Peabody Coal, and this in turn gave BHP approval for its purchase of Peabody's Australian coal assets. Thiess launched court action in Australia in March to head off the sale to BHP, claiming first rights to buy Peabody's assets under the 1962 TPM agreement. However, it withdrew the claim a few days after hearings commenced in May.

For a modest sum of $A81 million, BHP, through its subsidiary Dampier Mining, became the dominant shareholder in what became Thiess Dampier Mitsui (TDM), as well as in the Moura/ Kianga mines which were exporting around 2.5 million tonnes per year. BHP had been proceeding with the development or planning of some of its own mining developments in Queensland in the 1970s, including the Cook and Leichhardt underground collieries and the Gregory open cut, but with the leases held by TDM it was now in a position to become the second force in coal mining in Queensland behind Utah. In August 1976, BHP had secured a major stake in the North West Shelf oil and gas project when it bought Burmah Oil's 41% share in Woodside Burmah NL. The purchase of Peabody assets in 1976 was heralded as the second major "buying back of the farm" in just two months.[105]

Due to difficult market conditions, the development of the first mine in the Nebo project at Riverside was delayed until 1981, with the first shipment to Japan under the company's long terms contract with the JSM made in October 1983. Events in the early 1980s would see BHP take over Utah and move from number two in Queensland to become the dominant producer. The Utah purchase would dwarf the purchases made in 1976-77 and set BHP up as the king of the international coking coal market.

Export duty continues under the Coalition

The coal industry had high hopes that the election of the Coalition Government in December 1975 would see an end to the coal export duty, but it was soon to be disappointed. Shortly after the last Labor Budget in 1975, Anthony had said that the duty was an iniquitous tax which would halt new development, and committed to reviewing it when in Government. Anthony also correctly pointed out that the tax would be particularly damaging for steaming coal exports, and would wipe out the profits of some exporters.[106] The new Government, responding to pressure from the industry, did act to lower the rates of duty in its first Budget in August 1976, reducing the tax on high quality coking coal to $4.50 per tonne and the tax on lower quality coking to $1.50 per tonne, and exempting steaming coal from the tax. Treasurer Lynch also promised to phase out the duty entirely in three years, a promise that would again fail to be fulfilled.

In the 1977 Budget, the tax rates were reduced, high quality coking coal down to $3.50 and other coking coal to $1.00. However there was no change to the rates in the 1978 Budget, with Treasurer John Howard deferring the phase-out due to financial pressures, and not committing to the future abolition of the tax. Utah's John Wruck slammed the tax as "infamous and ill-conceived".[107] In 1979, when a full phase-out would have been expected, the rates were maintained. To rub salt into the wound, Utah was astounded to learn that the tax on high quality coking coal would be reduced from $3.50 to $1.00 per tonne, but only for mines commencing operations after June 1980. Utah's Norwich Park mine was expected to commence production in December 1979, and so would be subject to the $3.50 rate. Utah slammed the Government, saying that the new arrangements were discriminatory. It later claimed that the Government's apparently arbitrary date of July 1 1980 was a deliberate decision to ensure that Norwich Park would be subject to the higher rate.[108] The export duty continued to be applied in the 1980s, and Utah would again have reason to complain about the discriminatory tax, with changes made to target only its own mines.

The saga of the export duty in the 1970s also reveals some of the inner conflicts in the Coalition Government, with Doug Anthony fighting in Cabinet to do away with the tax. In a Cabinet submission prior to the Budget in 1976, Anthony recommended that the duty be lowered to $4 per tonne for high quality coking coal, and to zero for lower quality coking coal and steaming coal. His submission was contained as an attachment to a submission from the Australian Coal Exporters Division of the Australian Coal Association, listing all the major coal export companies, including Utah, TPM, Thiess, Coal & Allied and Clutha.[109]

Anthony told Cabinet that numerous submissions had been made by companies culminating in the industry submission and these stressed "the deleterious nature of the duty on existing operations and the threat it posed for new projects currently estimated to cost some $2 billion and estimated to employ 10,000 people during the construction stage and some 4500 during production." He said that "The main thrust of the industry submission is that because of the duty new projects are at risk and the majority of coal exporting companies are receiving unacceptable returns on investments. Stress is laid on the unique and discriminatory nature of the duty and its complete disregard for the financial, economic, geographic and marketing differences between companies. Many companies believe that they are being penalised because of the profits of one company – Utah. The industry claims that Australia's natural advantages …have largely been eroded by the imposition of the duty and, with regard to new projects, makes the point that Australia does not have a monopoly in supplying new demand for coal."

The exporters' submission argued for the duty to be completely abolished and also contained data compiled by consultants Arthur Andersen on the profitability of twelve export producers for the six months to 30 June 1976, expressed as the ratio of net profit to gross assets. The data showed a wide variation in profitability, with one company making a loss, two making less than 10%, four making between 10 and 20%, two between 20 and 30%, and the three most profitable making 38%, 41% and almost 48% returns on assets. The submission pointed to the median return of 13.9%, but no doubt the hungry eyes in Canberra focussed on those most profitable companies in the list. The returns were estimated actuals for the half year, and of course were significantly lower once the duty was factored in. However, even with the duty, three companies were shown as earning over 30%, and one over 20%. The median return on assets with the duty fell to 12.3%, and two companies were loss making. Export prices had risen sharply in 1974 and 1975 and clearly some exporters were able to earn handsome profits. The most profitable producers, most notably Utah, were not representative of the industry on the NSW South Coast where costs were high, or producers in the Newcastle and Hunter Valley regions exporting lower cost steaming coal.

Anthony in 1977 recommended a faster phase-out of the duty than was agreed by Cabinet, proposing that the duty of high quality coking coal fall to $2.25 and on other coking coal to $0.75. He clearly lost out again to the Treasurer and the Treasury department, with the rates only reduced to $3.50 and $1.00.[110] Treasurer Lynch reaffirmed that it was the intention to remove the duty in 1978, however John Howard as the new Treasurer in 1978 deferred the phase-out. Anthony had been overruled again and had proposed reducing the duty to $1.75 and $0.50 and eliminating it from July 1979. The coal export duty would continue to be levied on the industry until it was finally abolished in July 1992, by which time it applied only to coal being mined by BHP in the Bowen Basin, but was still a significant cost at $3.50 per tonne.

The West Moreton's future in the balance[111]

With much of the focus in the 1970s on the Bowen Basin coal developments, it is easy to overlook the West Moreton region, the birthplace of the Queensland coal industry. In the early 1970s, the region was confronted by its own problems, with widespread concern about the state of the coal industry and its future. The industry was still dominated by relatively small under-capitalised producers who relied on the supply of coal to local power stations and industrial users. The development of the Swanbank power station 10 kilometres from Ipswich had given the district's coal mines a boost, with Swanbank A (6 by 66MW) commissioned in 1967 and Swanbank B (4 by 120MW) commissioned in 1971. But the old power stations in Brisbane (Tennyson and Bulimba) had only a limited life. A major new power station at Gladstone was also planned and this was expected to have implications for the West Moreton coal mines as, once it was commissioned and linked to the grid, the demand for power from Swanbank would be affected.

On 31 July 1972 an explosion at the Box Flat mine near the Swanbank power station led to the loss of 17 lives. The mine had opened only in

1969 to supply the power station, and now was largely crippled. The Box Flat disaster was a huge blow not only for the local community, but also for the industry. The Box Flat group produced 658,000 tonnes in 1971-72, but with the explosion knocking out the two major tunnels from which the mine produced most of its coal, production fell to only 282,000 tonnes in 1972-73. Production from other West Moreton mines made up most of the shortfall, but with demand increasing, there was a serious run-down in coal stockpiles in the district.

In January 1974 the Brisbane and West Moreton region was hit by massive floods which devastated the area, causing significant loss of life. The floods were another blow for the coal industry, with four underground mines in the West Moreton (Aberdare 8, Moreton Extended, Haighmoor and Westfalen), accounting for almost a quarter of the district's production, flooded. All the West Moreton open cuts were also affected by the floods for varying periods. Of the four underground mines, only Westfalen was back in production in 1974-75. To its credit the industry in the district was able to boost production in 1974-75 by around 12%, but only because of a boost in production from the open cuts. Following the floods, the Queensland Government approved the shipment of coal from Central Queensland to Swanbank. While this was not to be a long term solution to the problem, it was another danger sign for the future of the coal mines in the West Moreton.

The West Moreton produced 2.5 million tonnes of coal in 1974-75, 1.5 Mt from underground mines and 1 Mt from open cut mines. The underground sector's production was the lowest since 1956, and around 30% below the peak level of 1966. The open cut sector began in the region in 1967, but was characterised by small operations with limited lives, with mines ceasing to produce being replaced by other small operations. The combination of the Box Flat disaster, the 1974 floods and the general state of the industry was a cause for concern: could the West Moreton coalfield continue to be a viable and significant part of the industry, particularly at a time when its major consumer, the Swanbank power station, would soon face competition from major new sources of electricity further north?

In October 1974 Queensland's Cabinet decided that there would be a major review of the long term future of the West Moreton coalfield to be carried out by the QCB and the State Electricity Commission (SEC). The review was to look at the problems of both the producers and the coal consumers, the demand for coal over the next 15 years, the efficiency of the coal producers, the scope for coal contracts and what the electricity industry could do to assist to achieve stability on the coalfield. The Southern Electricity Authority, responsible for generation in the south east of the State, was concerned about the cost of coal and was looking to a rationalisation of the industry to achieve a more efficient coal industry at a time when doubts were widespread about the ongoing viability of the coal industry in the district. The coal producers were looking for long term contracts, and said that their failure to be able to supply short term peaks in demand was due to a lack of guaranteed long term markets. The Colliery Employees' Union also wanted to see long term contracts for the producers, and was concerned about future demand, and, of course, the potential impact on the jobs of members. Clearly the terms of reference for the review had been structured to achieve an outcome which was likely to give the industry what it needed to continue to operate and invest. If the coal producers and their financiers were to have the confidence to continue to invest, this could only come from the electricity generators being prepared to offer long term contracts for coal supply.

Following a submission by the QCB and the SEC, Cabinet then approved the employment of consulting firm Mineplan to investigate a range of technical issues, including the adequacy of geological data and the viability of underground mines, and the scope for increasing efficiency and reducing costs. The terms of reference for the Mineplan review were also no doubt structured with certain objectives in mind – forcing the producers to lift their game and become more efficient, and forcing the union to accept the need for rationalisation. Mineplan's report, completed in May 1975, recognised the reality of the situation, that the continued long term operation of the West Moreton coalfield was not essential. Queensland had massive coal resources in other areas of the State which could be mined to supply all the needs of the electricity industry.[112]

The QCB and SEC reported to the Government in July 1975. Their report, which drew on the work by Mineplan as well as consultations with the industry and other stakeholders, concluded that there would be a steep reduction in demand for the district's coal when the Gladstone power station was completed and that the coal producers would have to adjust to this change while still being prepared to maintain a significant level of production through continued investment in their operations. The report also endorsed the negotiation of long term contracts between the coal producers and the electricity consumers, noting that the Southern Electricity Authority could provide the stability needed by the coal industry, but also that it may need financial assistance to ensure its customers were not adversely affected. In fact one of the recommendations of the review was that contracts for the supply of coal to Swanbank should be re-introduced. The clear message to the Government here was that if it wanted to extend the life of the West Moreton mines, it would have to accept that the price of coal would be higher than if coal was bought from Central Queensland, and some form of subsidy to the electricity generators may be needed.

The review had obviously forecast excess capacity in the district's coal mines. The QCB and SEC report concluded that this excess capacity could be managed by closing open cuts, regulating (ie restricting) the intake of new workers, and stockpiling coal in times of over-production. Although construction of the Gladstone power station was behind schedule, commissioning was not far away, and in January 1975 the coal producers were officially advised by the QCB that they needed to take into account the "certainty" that demand for their coal would fall once Gladstone was up and running. Producers were warned that any decisions on capital investment would need to bear this in mind. This may have just been the QCB protecting itself from any criticism, however it seems difficult to believe that the coal companies were not well aware of the potential impact of Gladstone, but presumably were confident that production from the district would be underwritten before too long by contracts with the local electricity generator.

West Moreton disputes see power rationing

As if there were not problems enough confronting the West Moreton coal industry, 1975 also saw major industrial disputes, and reductions in coal production which led to electricity rationing in southern Queensland. With the coal industry still not fully recovered from the 1974 floods, industrial disputes in the June quarter of 1975 caused another reduction in coal production in the region, cutting supplies to the Swanbank power station. Coal supplies from Blackwater were arranged to supplement the local supplies and commenced in May 1975. But then in June, the mining union imposed bans on overtime, and by the end of the month, Swanbank had less than 140,000 tons of coal stockpiled, equivalent to just over 3 weeks' consumption. With a mining holiday period coming up in August, arrangements were made for coal supplies from Blackwater to increase to 20,000 tons per week. However there was further industrial action in July which cut coal production, with overtime bans extended, and a major strike in the Queensland and NSW coal industry over a log of claims on the employers. The log had been served on employers in January and the unions took their members out on strike from 28 July.

To ensure supplies to Swanbank, and in a move which had the industry recalling the 1949 intervention in NSW open cut coal mines by the Army, Queensland Mines Minister Ron Camm threatened to ask the Federal Government to send in troops if the strike was extended for more than 5 days. Mine workers in Queensland voted overwhelmingly to extend the strike until the end of the week. State-wide, the vote in favour of extending the strike was 2019 to 421 against, with the vote in the West Moreton 416 in favour to 65 against. A newspaper report of the West Moreton meeting stated that a union official asked the meeting: "Is there one man here who is prepared to come off his five day strike to ensure supplies of coal to Swanbank?" The report then stated that not one hand was raised. Instead, the meeting recommended a ban on the transport of coal from Blackwater to Swanbank.[113]

By the end of June, power station stocks in the south were down to around 2 weeks' consumption. The Queensland Government moved to impose electricity rationing, with the first round commencing at midnight on 25 July, and additional restrictions coming into force on 30 July. Supplies from Blackwater recommenced on 5 August and electricity rationing was eased, but when the Blackwater mine workers refused to send further coal south, supplies from Blackwater ceased again on 25 August. Severe electricity rationing was then imposed from 29 August, and industries in the region were forced to operate on a 3 days per week basis, with households also required to cut their power usage. Coal supply from Blackwater resumed on September 1 and power rationing was lifted from 4 September.[114]

While coal stocks at the southern power stations were built up to a more reasonable level by the end of 1975, 1976 saw supplies still being supplemented from Blackwater. And in August 1976 the Southern Electricity Authority was still railing coal from Blackwater to Swanbank, with the power station's coal stocks down to a modest 6 weeks' supply. In August the Minister for Mines and Energy stated that: "The Ipswich collieries have not been able to guarantee the total requirements of the Swanbank Power Station over the next twelve months, although on recent performances it is expected that they will go close to supplying the total needs of the power station. The limited railings of Blackwater coal are only being continued to help to build up stockpiles after the miners' August vacation and until such time as it is clear that the local coalfield will meet all the needs of the power station."[115] This was hardly a ringing endorsement of the ability of the West Moreton coal producers to satisfy their major customer.

It was also in August 1976 that the Gladstone power station commenced generating power and the QCB was then advised what coal requirements would be for the Swanbank and Brisbane power stations for the period January to July 1977. The QCB in turn advised the coal companies that supply to Swanbank would be down by just over 13,000 tonnes per week (a cut of around 24%) which it said would have meant retrenchments of miners and the possible closure of some mines. But the QCB negotiated a revised temporary arrangement with the Southern Electricity Authority involving an increased tonnage to commence from

January 1977, and advised the coal producers that no additional workers were to be employed. Further increases were then advised to the coal producers taking the weekly tonnage figure to be supplied to a little more than the level that was in place at the start of the year.[116] The West Moreton mines knuckled down in 1976-77 and were able to boost production to 2.8 million tonnes, up almost 14% on the previous year, the highest annual increase in production in the district since the QCB commenced compiling data. The QCB's annual report for that year acknowledged "the cooperation of proprietors, miners, and contractors for supplying the additional coal requirements of the electricity industry."[117]

It took some time, but the 1974-1975 investigations into the West Moreton and the subsequent recommendations by the QCB and SEC were finally bearing fruit. In 1977-78 the Southern Electricity Authority called tenders for the supply of 16 million tonnes of coal over 15 years from suppliers in the Bundamba area. After negotiations involving the coal producers, the SEC, the Queensland Electricity Generating Board[118] and the QCB, contracts were signed with New Hope Collieries, Southern Cross Collieries, W M McQueen, Rhondda Collieries and Westfalen Colliery. A short term contract was also signed with Aberdare Collieries. Supply of coal under these contracts was to begin in January 1978. The QCB said that these contracts would give Swanbank an appropriate supply of coal and give the producers the confidence to effectively plan for the future. These contracts were a lifeline for the coal industry in the district and would give the industry a firm base on which to operate, although at not much more than 1 million tonnes per year, would not eliminate all the problems in the industry.

In fact production from the district was on the way down. The 2.8 million tonnes achieved in 1976-77 would prove to be the peak, and by 1979-80 production had dipped below 2 million tonnes. There were now only 11 underground mines operating, and another 11 open cuts. Employment was down to 1043 in June 1980, around half of the level twenty years earlier. In 1977-78 the power stations at Swanbank and in Brisbane consumed around 2.2 million tonnes of coal, and so there would be significant scope for coal purchases outside the long term contracts to be reduced. As the capacity of the Gladstone power station increased, the demand for coal from the Swanbank and Brisbane stations did fall, and the allocation of contracts to the coal producers involved lower tonnages. The 1980s would see continuing problems for the industry in the West Moreton, but also the development of significant exports through Brisbane which breathed some new life into the district.

New towns spring up in the Bowen Basin

With the major new export coal projects developed during the 1960s and particularly the 1970s, central Queensland also saw new urban development, with two completely new towns – Moranbah and Dysart – developed largely by Utah, and other towns such as Blackwater receiving a major boost in population. Moranbah is around 150 kilometres south west of Mackay, and Dysart is around 80 kilometres south east of Moranbah. Moranbah was developed under an agreement with the local Shire council as the dormitory town for Utah's Goonyella and Peak Downs mines. Utah began to build the town in 1970, and by 1971 the population was 1050, rising to just over 4000 by 1976, and almost 6900 by 1986. Utah built around two thirds of the homes in Moranbah for its workers who paid a very nominal rental to the company of around $5 per week. By the late 1970s there were around 900 homes in Moranbah, of which Utah had built around 600 and the Queensland Housing Commission the remaining 300. The initial development of Moranbah was planned on the basis of building 435 homes for married couples, with single workers housed at the minesites. As an indication of the efficient planning and development of the town, the local school opened its doors to students on the first day of the school year in 1971.[119]

Moranbah is located south of the Goonyella Mine and north of Peak Downs mine; Dysart is located between the Saraji and Norwich Park mines. Dysart was also developed in conjunction with the local shire council under a similar agreement that applied to Moranbah, with the companies responsible for the town's infrastructure not including water supply, which was the responsibility of the council. Utah employed the same consultants to design the town. By 1976, Dysart had a population of 1585. "Utah's grand strategy for developing its CQCA leases was to build two sets of paired mines, each serviced by a separate town. Utah's proposed mine sites were stretched along a line running north-north-west for about 150 kilometres and parallel to the coast which (was) about 160 kilometres to the east. Utah worked systematically from north to south, building the Goonyella mine, the town of Moranbah and the Peak Downs mine; then the Saraji mine, the town of Dysart and finally the Norwich Park mine. This strategy allowed sequential development and the steady expansion of rail and port infrastructure, and ensured that no mine workforce had to travel more than 20 to 30 kilometres to work."[120]

The existing town of Emerald was used as the main town for the Gregory mine from around 1979. The Emerald Shire saw its population grow from 5639 in 1976 to 8435 in 1981, with much of the expanding population settling in the town of Emerald. Other towns to see a significant expansion in the 1970s included Blackwater and Moura. Blackwater, the centre servicing the Blackwater and South Blackwater mines, saw its population grow from 1984 in 1971 to 4638 in 1976 and 5434 in 1981. Moura, the town servicing the Moura and Kianga mines, saw its population grow from 1093 in 1966, to 1902 in 1971 and 2694 in 1976. Over time, of course, as other mines were developed, these towns also served as the centre for the expanding workforce, or at least those who lived in the region.

The very nominal rental paid by Utah's employees living in company-built housing led to a major confrontation with the Federal Government in 1980, when the Taxation Commissioner indicated that he intended to levy tax on the benefit workers in remote areas received from their subsidised company housing. The Tax Commissioner's stance saw a major dispute erupt in Queensland, with the mine workers striking in protest in late June, and the Federal Government initially refusing to accommodate their demands. Doug Anthony, Deputy Prime Minister, accused the miners of wanting to pay no tax at all on the subsidised housing and to be treated differently from all other Australians.[121]

With the strike already four weeks old, John Howard, then Treasurer, and Anthony, together with Queensland Treasurer Llew Edwards, went to Blackwater on 1 August to meet with the union leaders. When he left the local hotel where the meeting was held, Howard was reported to have been "grabbed in a headlock and almost dragged to the ground" before police intervened. "As police rushed to Mr Howard's assistance, more demonstrators poured toward him. Police managed to get Mr Howard to a waiting police car and out of the way of the angry mob."[122] In response to the union protests, Howard proposed a change to the arrangements to reduce the base for tax on subsidised housing in remote areas. However, Howard's proposed concession still left workers exposed to additional tax and did little to ease the concerns of the miners. The new arrangements were also to be phased in over the subsequent two years, and Howard said that the Government was not prepared to consider further changes. "An important taxation principle is at stake", he said. "We have been reasonable. It is now only fair and reasonable that the miners accept the Government's offer".

The strike in Queensland lasted for ten weeks before further concessions by the Federal Government saw the union recommending to the workers that the strike end. Bill Stone, on behalf of the combined mining unions, said that the settlement would see no tax paid in the 1980-81 financial year, and he also said that he believed that it would be some time before any tax was paid at all.[123] At five mass meetings of mine workers in Queensland there was a strong vote to end the strike (1524 in favour, 503 against), although the meeting at Blackwater saw an overwhelming vote to continue the strike (468 to continue, 170 to end the strike).[124] With the Blackwater workers out-voted work resumed in the Bowen Basin.

Oil shocks open the way for a boom in demand for steaming coal

With the rapid expansion of the Japanese economy the major driving force behind the growth of the Australian coking coal industry in the

1960s and beyond, it was the complex web of events in the Middle East in the 1960s and 1970s which was to be the impetus for the expansion in Australia's steaming coal industry in the late 1970s, and in the 1980s and subsequent years. These events also had a major impact on the coking coal market in the 1970s and 1980s, but whereas the impact on the steaming coal market was positive, the impact on the coking coal market was the opposite.

The Middle East, the source of a major proportion of the world's oil, has been an unstable region for much of its history, and in particular since the end of World War One and the breakup of the Ottoman Empire. The post-World War Two era has proved to be no exception. One of the major events in the lead up to the oil shocks of the 1970s was the nationalisation of the Suez Canal by Egypt in July 1956, which led to Israel invading the Sinai region of Egypt, and the subsequent landing of troops along the Canal by the UK and France. The Canal, the waterway along which much of the region's oil was shipped, was closed to shipping from October 1956 to March 1957. Pressure from the USA, the United Nations and the Soviet Union forced a humiliating backdown by the UK and France and saw Egyptian leader Nasser emerge as one of the strongest leaders in the Arab world.

OPEC – the Organisation of Petroleum Exporting Countries – was formed in 1960. Its founding members were Iran, Iraq, Kuwait, Saudi Arabia and Venezuela; membership expanded in the 1960s and early 1970s to include Qatar in 1961, Indonesia and Libya in 1962, the United Arab Emirates in 1967, Algeria in 1969, Nigeria in 1971 and Ecuador in 1973. In 1967 the Six Day War between Israel and Egypt, Syria and Jordan resulted in a humiliating defeat for the Arab countries, and the loss of territory to Israel including the Gaza Strip, the West Bank and the Golan Heights. In 1969 Colonel Muammar al-Qaddafi seized power in Libya and warned the oil companies that they had to increase the "posted price" of oil, the price against which the producing countries calculated their taxes and royalties. In 1971 a number of countries nationalised oil assets – Algeria took 51% ownership of French oil interests; Venezuela legislated for oil concessions to revert to the government on their expiry in the early 1980s; Libya nationalised BP's interests and later also took 50% of the operations of Italy's ENI, 51% of Occidental Petroleum and 51% of other companies' operations; Iraq nationalised the Iraq Petroleum Company's interest in the large Kirkuk oil field; and Iran nationalised the remaining private operations which had survived the nationalisation in 1951. The Shah of Iran, on getting full control of the nation's oil industry, said: "Finally I have won out... Seventy two years of foreign control of the operations of our industry has ended."[125] In Saudi Arabia, Aramco (the consortium of American producers – Exxon, Mobil, Texaco and Standard of California) - decided that, to avoid outright nationalisation, it would agree to a partial de facto nationalisation with the participation of the Government in its operations.

The 1973 Yom Kippur War between Egypt and Syria and Israel ran from 6 October to 20 October. The war initially led to pressure by the Saudi government on the USA to repudiate its pro-Israel policies, but the USA proceeded to provide aid to Israel. OPEC, now in an extremely strong position, with its members in control of a high proportion of world oil supply, raised the posted price of oil by 70% to US$5.11 per barrel (although this was also recognising the market reality of rising spot prices). The Arab oil ministers also agreed to an oil embargo on the USA and other countries deemed friendly with Israel. The ministers also agreed to reduce oil production by 5%, with further cuts to production until their objectives were met. However countries deemed to be friendly to the Arab countries saw their oil supplies maintained. On 20 October Saudi Arabia announced that it was cutting off all oil exports to the USA.

The combination of war, cutbacks in supply, embargoes and uncertainty forced up the price of oil even higher. In December 1973 the Shah of Iran proposed that the posted price be raised to US$11.65 and this was agreed by OPEC. Crude oil prices which had averaged only US$1.80 per barrel in 1970, had risen to an average of US$3.29 in 1973 (this figure only partially reflecting the OPEC oil price hikes). In 1974 the average was US$11.58, a figure which more accurately reflected the 1973 price hikes.[126]

The 1973 price hikes and embargoes were a major shock to Japan. Just before the turmoil of October 1973, the Japanese Agency for Resources and Energy (part of MITI - the Ministry for International Trade and Industry) released a white paper on energy that attempted to address the insecure oil market and the importance of policies which would help the country cope with an emergency. The oil embargo hit Japan particularly hard, dependent as it was on imports for all of its oil. "In Japan, the embargo came as an even more devastating shock. The confidence that had been built up with strong economic growth was suddenly shattered; all the old fears about vulnerability rushed back. Did this mean, the Japanese asked themselves, that, despite all their exertions, they would be poor again?"[127] Japan was included by the Arab oil ministers in the friendly to Israel camp. With almost half of Japan's oil coming from the Arab Gulf states, and with 77% of its total energy needs supplied by oil, Japan was extremely vulnerable.

The USA was also hard hit, with President Nixon making a major announcement in November in which he proposed that "in the spirit of Apollo, with the determination of the Manhattan project …by the end of this decade we will have developed the potential to meet our own energy needs without our depending on any foreign energy source."[128] Nixon did not survive the Watergate scandal and resigned in August 1974. The USA was never likely to get anywhere close to becoming energy independent by 1980, but the years ahead would see major new policies such as fuel efficiency standards for automobiles, and a steady increase in energy efficiency in the USA.

The oil embargo on Japan was relaxed once Japan removed itself from the pro-Israel camp in November 1973, and the embargo was lifted on all other countries in 1974. Japan, however, had been badly shaken by the events of 1973 and was now having to bear the cost of the much higher oil prices which were feeding into its cost structure, damaging its international competitiveness. Japan's economy went into reverse in 1974, GDP falling in real terms by around 1%, the first year of negative growth since the war. While the recession in Japan was short-lived, with the economy resuming growth in 1975, the Japanese Government and major industries were now looking to new policies to help insulate the country from the vagaries of the international oil market, and in particular the uncertainties resulting from its dependence on volatile Middle East suppliers.

Japan's responses over the coming years would come to include a switch from oil as the major energy source for electricity generation to coal, uranium and natural gas (and increasing reliance on imports of these energy sources), and diversifying its sources of supply of oil to increase the share purchased from non-Middle East suppliers. While Japan relied on imported coking coal to feed its steel industry, to protect its domestic coal industry for many years it had an official embargo on the import of steaming coal. Japan was self-sufficient in steaming coal, although its embargo on imports was denying it access to much cheaper quality coal. But the most significant problem for Japan was that its power generation industry was largely dependent on oil, with around two thirds of electricity generated from oil fired power stations. The oil crisis forced a change of policy and in January 1974 Japan announced that it was lifting the embargo. The lifting was at first only partial, with Government initially saying that imports of up to 850,000 tonnes of steaming coal would be permitted in JFY 1974, although this was later reduced to 500,000 tonnes. In JFY 1975, the limit was a very modest 220,000 tonnes. However the door was now being prized open for imports and Australia was well positioned to capture a major share of the Japanese steaming coal market dominated by the power utilities and the cement manufacturers.

Japanese show interest in Australian steaming coal

One of the first manifestations of the new policies in Japan was the arrival in Australia of a Japanese-government sponsored delegation in June 1975 to investigate the potential for the supply of steaming coal from Australia and opportunities for joint ventures to develop mining projects. The Japanese were also planning to send missions to other potential suppliers, with a second mission to visit the USA in July, a third to visit South Africa in September, and a fourth to visit India in October.[129] A few months later Nippon Steel's senior managing director, Saburo Tanaba, and representatives of the Electric Power Development Company met with Rex Connor in Canberra. Connor in one of his final acts as Minister assured the Japanese that Australia would always meet Japan's coal needs.[130]

In February 1976, Deputy Prime Minister Doug Anthony visited Japan for discussions with senior Japanese Ministers on the supply of resources from Australia. Following these discussions he said that Australia was in position to play a significant role in Japan's $300 billion energy expansion program. While the major component of this program ($160 billion) would be allocated to a massive expansion of Japan's nuclear power generation, Anthony said that the outlook for steaming coal exports to Japan was bright. He was told by Japan's Minister for International Trade and Industry that it was definitely part of Japan's plans to expand the role of coal fired power generation. Japan's imports of steaming coal had only been around 40,000 tonnes in 1974, but were expected to reach 5 million tonnes by 1980 and 15 million tonnes by 1985. Anthony said that by 1980 Japan was prepared to take around 55% of its steaming coal imports from Australia.[131]

Coal and Allied Industries secured the first major steaming coal supply deal with Japan in 1974 when it won a contract with cement manufacturer Ube Industries for the supply of 400,000 tonnes per year for 7 years. This contract was extended by another 20 years in 1977.[132] In June 1977, Australia's first major steaming coal contracts with the Japanese electricity generation industry were confirmed. RW Miller secured a contract with Japan's EPDC for the supply of 2.5 million tonnes over 5 years commencing in 1980 to supply a new power station in Nagasaki prefecture; Oakbridge secured a similar contract with EPDC for 1 million tonnes per year.[133]

British company Costain was part of a consortium involving HC Sleigh, Mitsubishi and the Australian Resources Development Bank, which had won the rights to develop a major coal deposit at Warkworth in the Hunter Valley in 1976. This deposit contained an estimated 500 million tonnes of coal, around 350 million tonnes of which was soft coking coal, and 150 million tonnes steaming coal. Costain was already operating the Ravensworth mine in the Hunter Valley under contract to the NSW Electricity Commission, and was the manager of the new Warkworth mine. The Warkworth tender, the first major coal deposit which the NSW Government had ever put up for what was effectively an auction, attracted 8 other bidders. NSW Mines Minister Wal Fife said that the Costain consortium had submitted a considerably higher offer than the other 8 which included an up-front cash payment and an additional royalty of $1.05 per tonne of coal mined, with that rate to escalate over time. Fife said that the NSW Government could earn up to $170 million over the initial 21-year lease period from the Warkworth consortium, a highly attractive revenue stream at that time.[134]

The NSW Government, very happy with the terms of the winning Warkworth bid, then adopted the super royalty rate and the up-front payment as elements it now expected in new coal leases. However the Government seemed to have done a poor job of communicating the new policy to coal companies, with journalist Alan Mitchell in March 1977 reporting that although the NSW Mines Department believed that the industry was well aware of the "Costain formula", a number of coal company executives he contacted "virtually had a seizure when they heard it was to become a general policy."[135] Mitchell also wrote that Coal and Allied was also told that it would have to pay the super royalty and the "right to mine" payment once it had been agreed what lease it would have. C&A went on to be awarded the lease for Hunter Valley 1 mine.

The year 1977 also saw other developments as coal producers positioned themselves to win part of the new steaming coal market in the years ahead. British Petroleum, keen to gain a stake in the expanding industry, signed a deal with Oakbridge in November 1977 for a 50% share of the Clarence mine near Lithgow. Clarence was already being developed, but the entry of BP and the added investment it made resulted in an acceleration of development and an increase in planned production from 800,000 tonnes per year to 2 million tonnes.

The involvement of BP in the Clarence mine depended on Federal Government approval, and this was forthcoming in December with the FIRB announcing its approval. However, with the NSW Government now also in the foreign investment approval business, its approval was also necessary, but that took a little longer. The NSW Government needed to approve changes to the leases for Clarence to recognise the proposed new owners and so was able to impose its own foreign investment rules. It was not until February 1979 that NSW approval was forthcoming, but it was on the basis of BP having a 49% share rather than the 50% share originally announced by the two companies. The delay was due to the NSW Government taking some time to decide what its own policy should be.[136] Another new development involved RW Miller which was granted a major lease in the Hunter Valley in July 1977 at Mt Thorley where it planned to develop an open cut mine producing 2 million tonnes a year of both steaming and soft coking coal, expanding to 4 million tonnes per year over time.

In Queensland, while the major focus was on coking coal projects, steaming coal was also emerging as an area of interest. In 1977 MIM acquired the Newlands steaming coal deposit when it purchased Griffin Exploration, although Newlands would not be developed until the early 1980s as part of MIM's coordinated development of Newlands and the Abbot Point export terminal and expansion of the Collinsville mine. In 1978 MIM and Thiess announced plans to develop the Wandoan deposit in the Upper Dawson Valley, around 350 kilometres north west of Brisbane. This project was not aimed at the export market, but the local electricity generation market, and the new mine's development would depend on whether MIM and Thiess were successful in securing the contract to supply the next major new power station in Queensland. In 1980 CRA's Subsidiary Pacific Coal won the 66 million tonne 18 year contract to supply the new Tarong station. Wandoan remains undeveloped to this day.

Second oil shock in 1979

There were a number of major developments which helped to ease the oil market in the 1970s including supply from the North Sea and the completion of the Alaskan pipeline linking the oilfields to the shipping port at Valdez on Alaska's southern coast. New discoveries in Mexico were also significant. However, in the late 1970s trouble was brewing in Iran, which would lead to what became known as the second oil shock. The Shah left Iran in January 1979, never to return, and the revolution saw Ayatollah Khomeini return to become the new de facto leader of the country. Oil production in Iran was already in a mess and oil exports had ceased at the end of 1978. While oil production from other countries made up much of the gap caused by the situation in Iran, demand had been growing since the mid-1970s. This, combined with uncertainty over supplies and confusion in the market, caused panic buying and saw prices skyrocket. In 1979, with prices for crude oil and petroleum products surging, OPEC allowed its members to charge what they thought was justifiable; in other words, the oil market became a free for all. Like most countries, the USA was worried and the words of President Carter's Energy Secretary, James Schlesinger, no doubt reflected the general feeling of fear for the future: "Today we face a world crisis ... there is little, if any, relief in prospect. Any major interruption – stemming from political decision, political instability, terrorist acts or major technical problems – would entail severe disruptions ...The energy future is bleak and likely to grow bleaker in the decade ahead."[137]

As if the situation in 1979 was bad enough, the following year saw the outbreak of war between Iran and Iraq which "shook the Persian Gulf yet again and threw the oil supply system into jeopardy, threatening a third oil shock."[138] By 1980, the second oil shock led to oil prices rising to around US$36 a barrel, 20 times the level in 1970. Countries were now forced to consider how to replace oil as a fuel source for power generation, and to reduce the role oil played in their economies. And coal was now about to play a much more important role; the 1980s were expected to see demand for coal surge and prices were expected to remain strong.

The International Energy Agency (IEA) was established in 1974 as an autonomous body under the framework of the OECD. Its establishment was in response to the first round of oil shocks in 1973, and it provided a vehicle for the major countries in Western Europe, together with the USA, Canada, Japan, Australia and New Zealand to cooperate in relation to energy security and other energy issues. The membership has broadened over the years to include some countries in Eastern Europe, together with Turkey and the Republic of Korea. As the IEA's own historical document says, "From the time of the founding of the IEA, coal was already identified as the principal energy sector to be developed as an alternative to oil. While domestic oil production, natural gas, nuclear energy, hydroelectric power, and renewable energies were also strong elements, coal was the leading alternative energy source. Coal benefited from its ample supply availability in many industrialised countries and from its favourable transportation and trade prospects. Environmental considerations were taken into account, but early IEA policies were developed before climate change became a public policy concern."[139]

In 1977 the IEA adopted its Principles for Energy Policy which included policies supporting the progressive replacement of oil in electricity generation, a strategy for utilising steam coal and promoting international trade in steaming coal, concentrating the use of natural gas on "premium requirements" (which did not include its use in electricity generation) and the steady expansion of nuclear generating capacity.[140] The following year the IEA published a major study on the outlook for coal up to the year 2000, Steam Coal Prospects to 2000, which concluded that even with a modest rate of economic growth the world would need to see a huge switch from oil to coal. This study was an influential one, and had a significant impact on the IEA's own policy development and on thinking in a number of countries. Then 1979 became "the year of coal" when the IEA adopted major policies on coal which set the framework for the 1980s and 1990s.[141]

The IEA was now looking to the consumption of thermal coal in its member countries almost doubling between 1976 and 1990. The principles for coal adopted by the IEA members in 1979 required members to "ensure that an economic, fiscal and investment climate prevails which is conducive to development of coal production, trade and utilization". The principles also referred to establishing energy pricing principles to allow coal to "develop its full competitive power", and in relation to electricity generation, one principle committed members to preclude "new or replacement base load oil-fired capacity", to confine oil use progressively to "middle and peak loads", and to make "maximum use of fuels other than oil in dual-fired capacity".[142]

In May 1980 at the meeting of energy ministers of member countries the IEA Secretariat tabled its annual review which conclude that even stronger actions were needed to expand coal production from Australia, Canada and the USA, and to expand coal use in Germany, Italy, Spain, Japan and the UK. The Review also concluded that in relation to trade, greater attention needed to be focussed on long term contractual arrangements in order to provide the stability and confidence for new mines to be developed and for investment in transport facilities.[143] The coal principles also covered a range of other areas including transportation systems, port facilities and other infrastructure necessary to handle higher coal tonnages, coal mining methods, R&D programs and commercialisation of coal technologies. The 1980 year also saw the IEA establish the Coal Industry Advisory Board to advise it on coal related issues, with the board comprised of representatives of coal producers, coal users, traders and other energy related organisations. For the member countries of the IEA, coal was now to be a fundamental component of the energy and policy mix necessary to meet the challenges of a world in which energy demand was growing, and a world in which oil had become too expensive and oil supply too unreliable.

The decade in review

The 1970s was one of the most turbulent periods of the post war era. Major negatives such as the oil shocks, currency realignments and inflationary pressures were countered by many positive factors, at least for a commodity exporter like Australia. The Japanese economy, despite some difficult periods, continued to grow, although its steel industry reached peak production in 1973. Australia became the dominant supplier to the Japanese steel industry with around 45% of its coking coal imports. Coking coal was also being exported in increasing quantities to other markets, including various countries in Europe and Asian countries including Korea and Taiwan.

Japan's concerns about the high cost of oil and its vulnerability to interruptions in supply led it to lift its embargo on imported steaming coal and start planning for a switch to coal, nuclear and gas for its power industry. Other countries too were now looking to coal to become a major fuel for power generation and Australia was well placed to benefit. Australia's steaming coal exports grew rapidly, but were still modest by 1980 at around 7 million tonnes, with the bulk of tonnage taken by European customers. The major growth period for our steaming coal exports to Japan and other Asian markets was still to come in the 1980s and 1990s.

By 1979-80, coal had become a significant component of Australia's export income, with the value of coal exports at around $1.7billion, compared with total merchandise exports of $18.9 billion. The 1980s was shaping up as a boom period for Australia, with demand for coal and many other minerals expected to grow rapidly. Coal producers were investing heavily in new mines and expanded capacity, and the State Governments in NSW and Queensland were investing in rail and port infrastructure to cater for the boom in exports. The States were also about to embark on major expansions of their electricity generation through a number of new power stations. What could go wrong with such a positive outlook?

Dragline at Blackwater mine, Queensland 1970. NAA A1200 L89693

Shuttle car loading in Awaba mine, NSW 1968. NAA A1200 L69683

2

From Boom to Gloom

The 1980s began with a general feeling of optimism and even euphoria that the decade ahead would see the Australian economy booming on the back of surging demand for our mineral commodities and investment in industries including power generation and aluminium smelting. Setting the scene for Australians, in September 1980 in his election speech Prime Minister Fraser held out the prospect of massive development ahead based on our resources: "In my Policy Speech in 1977, I said Australia could look forward to $6,000 million of development. Some amazement was expressed at this — even disbelief. Because the Labor Party had stopped development dead in its tracks. Yet in the two years after that, more than $6,000 million was invested in mining and manufacturing. And now, prospective investment is $29,000 million. This development promises to be as important to Australia and individual Australians as anything in the last 35 years…Already, new aluminium smelters and mines are being established in Australia along with the associated new towns, railways, roads and port facilities. The benefits of this will be felt nation-wide. We are not just talking about development for development's sake – we are talking about development because of what it means to people. And it means jobs, prosperity and security for Australian working men and women."[144]

Fraser cautioned the electorate about the "world being a harsher place than anyone expected five years ago", but then went on to give another example about development ahead: "The increase in electricity generation through the 1980s will be almost equal to that which occurred over the last 30 years. We are going to do in ten years what previous generations took 30 years to accomplish. Modern new industries are coming on-stream with modern plant and equipment involving the most advanced technology. These new developments will give Australia in the 1980s a much stronger international base. New markets are opening up for Australian industries." Fraser also referred in his speech to the huge Rundle Shale oil project in central Queensland as likely to be "the largest ever undertaken in Australia, and one of the largest in the world". He was, of course, not the only one predicting rapid growth in our resources industries, and consequent benefits for the average Australian, but the sentiments expressed in his speech did reflect a view by many that the 1980s would be prosperous years – even boom years – which would be of great benefit to resource-rich Australia.

Dark clouds form

The over-hyped resources boom started to look shaky early on in the decade and signs from the USA and other countries should have alerted Australia that the outlook was going to be difficult. The events of the 1970s had formed the basis for many of the economic problems of the 1980s. The combination of the breakdown of the international monetary system, the way in which the US Government funded the Vietnam War, the oil shocks of the 1970s and the inflation which ensued meant that the time of reckoning would be not long in coming. The US Federal Reserve had started to raise its funds rate in mid-1978, and it continued to push the rate higher in 1979 and 1980. By March 1981, the rate averaged 17.2%, and by June 1980, 17.6%. The Fed then eased back, but hiked the rate again in December 1980 to an average of almost 19%. The Fed was intent on reining in the inflation beast, but its success in this regard also helped to push the US economy into a shallow recession in 1980, with a deeper recession to follow in 1982.

A major problem for the US economy from around 1981 was also its rapidly appreciating exchange rate. The surge in interest rates in the US also pushed up the exchange rate, causing problems for export industries, including the American coal industry. Between 1980 and February 1985, when the US dollar value peaked, it had appreciated by almost 60% in trade weighted terms. In September 1985 the Plaza

Agreement saw the major Western countries (USA, Japan, Germany, UK and France) agreeing that Germany and Japan would intervene to revalue their currencies against the US dollar. The US dollar/ Japanese yen exchange rate halved between a peak of 260 in early 1985 to around 125-130 in early 1988, causing major problems for the Japanese economy; Japanese steel production fell by over 6% in 1986. The Australian dollar was worth $US1.15 in mid-1981, before falling to $US1.02 in June 1982, and to $US0.67 by June 1985. This devaluation of the Australian dollar of course assisted our coal industry, generating higher $A returns for contracts written in US dollars, although the annual price negotiations also took account of this factor.

In the 1980s, the US economy was by far the dominant economy, accounting for around 24% of global GDP. US economic policies, together with the impact of the OPEC oil shock of 1979, saw world GDP growth fall to around 2% pa in 1980 and 1981 and to only 0.4% in 1982. The expected boom in demand for resources in the 1980s was now looking uncertain, to say the least. The Japanese economy, highly dependent as it was on imported energy commodities, had been hard hit by the oil shocks of the 1970s. Its economy went into reverse in 1974 after the first oil shock, although it recovered to see GDP growing by over 5% pa in 1978 and 1979, before the second oil shock pushed its growth down to around 3% in 1980.

The Japanese steel industry, sensitive to domestic economic trends, and particularly to trends in export markets for steel and for steel intensive exports such as motor vehicles, saw significant variations in production in the 1980s. The 9% fall in production in 1981 was not a happy start as far as the Australian coal industry was concerned, and was followed by further falls of 2.2% in 1982 and 2.3% in 1983. So in three years, Japanese steel production was down by over 12%, from 111.4 million tonnes to 97.2 million tonnes. Production jumped back in 1984 and 1985 to over 105 million tonnes, before another retreat in 1986 and 1987 to below 97 million tonnes, and another surge in 1988 to almost 106 million tonnes.

These variations in production, together with severe changes in exchange rates, and international competition in both the steel market and the coal market, had a profound effect on the Australian coal industry in the 1980s, and with issues such as industrial disputation in Australia, were the basis on which many of the price negotiations played out between the Australian hard and soft coking coal exporters and the Japanese steel mills. The development of new coking coal mines in Canada in the early 1980s with the financial and price support of the JSM and other Japanese companies was also a significant factor in the way the market operated in the 1980s. The steaming coal market internationally was also impacted by some of the same economic events, but with the oil shocks of the 1970s having fundamentally changed many countries' policies in terms of power generation, steaming coal would see some good growth during the 1980s and strong prospects for further growth in the 1990s and beyond. But competition would be fierce and prices would remain low.

Japanese contracts bring new Canadian mine developments

In January 1981 an announcement was made which would trigger ongoing bitterness and concerns by Australian coal producers throughout the 1980s and 1990s. The Japanese steel industry had agreed to new contracts for two Canadian companies to supply it with 94 million short tons of metallurgical coal and 16 million tons of steaming coal over a 15 year period commencing in October 1983. Two major new open cut mines were to be developed – Quintette by Denison Mines, and Bullmoose by Teck Corporation. The development of the mines and the associated infrastructure was described as the largest industrial development and construction project up until that time in the province of British Columbia, and was actively supported by the Canadian national and provincial governments hoping to see that region opened up for a range of new developments.[145] The cost to develop the mines and infrastructure was estimated at around $US1.9 billion.

The new mines were developed in the Tumbler Ridge region in the north east of the province, with the British Columbia and Canadian governments taking responsibility for developing a new town, the electrified rail line to the coast and a new coal export terminal at Ridley Island near Prince Rupert. The development of Quintette was financed by Denison Coal Limited, Charbonnages de France, Mitsui Mining, Tokyo Boeki, Sumitomo Trading and the nine major Japanese steel mills. Japanese equity in Quintette was 38%. Bullmoose was developed by Teck Resources, with 10% equity from Nissho Iwai. The mines came into production in 1983 with a capacity of around 6.5 million tonnes.

Not only was the development of these new mines adding significant capacity to the global coking coal trade, but it was underwritten by contracts that gave the companies developing the mines prices well above the prices being received by other exporters. The cost of Quintette and Bullmoose mines and associated transport infrastructure required significantly higher prices than the prevailing prices being received by Canadian producers in the south of the province and by Australian producers. The contract prices for Quintette and Bullmoose were subject to escalation clauses, and by JFY 1983, were around $C94 to $C95 per tonne, or around $US85. That same year the JSM contract price for the coking coal from Canadian producers Luscar and Balmer was $C69.83, or around $US63. The contract price for Utah's Goonyella coal was $US60.69 and for US producer Pittston, $US59.05.[146]

The Japanese soon realised that these prices were not sustainable at a time when its steel mills were struggling against new competitors and when the international steel market was weak. In September 1983 there were reports about possible debt re-scheduling by the companies which financed the new mines, with the JSM unwilling to continue to pay the contract prices.[147] However, under the escalation provisions of the contracts, the prices received by Quintette and Bullmoose actually rose during the 1980s, and by 1987 the JSM were seeking dramatic cuts of the order of $C40 per tonne. Luscar, Balmer and Goonyella were receiving only $US46.90 in JFY 1987, compared with $C95.28 (around $US71) for Quintette and $C93.24 (around $US70) for Bullmoose. Quintette's contract price hit a high in JFY 1989 of $C99.69; the peak for Bullmoose was in JFY 1990 at $C97.91 per tonne.

In 1987, with the pressure on Qunitette Coal to agree to a price cut to $C57.85 per tonne, the company's president Paul Kostuik warned that lower prices would force the closure of the mine. Kostuik called the requests by the Japanese "unconscionable" and in relation to the closure said that this was "not a scare tactic...The Japanese knew from the start that northeast British Columbia would never have been developed if the price of coal were $57.85 per tonne. But they encouraged the participants to proceed with the project and promised to pay the price necessary to get a secured source of high-quality coal."[148] In November 1987, after 10 months of discussions between Quintette Coal and its Japanese customers, the parties at the request of the Japanese agreed to take the issue to be settled by an arbitration panel in British Columbia. The arbitration was followed by court challenges, and resulted in slow reductions in the contract prices the mines received in the 1990s. However by JFY 1998, Quintette's price was still high at $83.89 and Bullmoose's price was $C91.77. In 1991-92, with Quintette's debt being restructured, Teck bought a 50% interest in the mine and took over as the operator. With the 15-year contract period ending in 1998, the JSM were then able to slash the prices for coal from Quintette and Bullmoose. For JFY 1999 the new contract price, this time written in US dollars, was down to $US41.12 for both mines, broadly in line with Goonyella $US41.90, Luscar $US41.10 and Balmer, now called Elkview, $US40.38.

Quintette closed in 2000, and Bullmoose closed in 2003. The Japanese steel industry and the companies financing the development of these mines had paid a heavy price in terms of the development of these mines and in terms of the annual prices paid for the coal. However, it can also be argued that the overall benefit to Japan was actually positive if, by helping to develop these mines, the JSM were able to pay lower prices for a number of years to their other major suppliers in Australia, Canada and the USA.

Steaming coal prospects looked bright

In 1980, Japan was looking to Australia to play a key role in its plans to diversify its electricity generation away from oil. While nuclear energy was to be the major component of Japan's power generation mix, coal and gas were planned to also be major sources of fuel. Following a visit to Australia in January 1980 by Prime Minister Ohira, a senior Japanese steaming coal study mission came to Australia in April. The mission was led by Mr Toshiwo Doko, the chairman of the Keidanren, Japan's powerful federation of economic organisations and its peak business body. Australian Ambassador John Menadue said at the time that there were few more important recent missions that had left its shores. The Tokyo correspondent for the Sydney Morning Herald, Hamish McDonald, wrote that "The mission set out to probe Australian intentions...the Japanese are nervous, particularly of what they see as a bloody-minded trait in us." He referred to industrial relations problems in Australia, but also to the mission trying to better understand the decision-making process in Australia, which, for example, had knocked back the Japanese equity in the Blair Athol project only days earlier.[149] Clearly, the senior Japanese industry leaders were keen to better understand Australia, and assess how reliable we would be as a supplier of critical resources over the long term. For Japan to achieve its goals in electricity generation, the Australian coal industry would need to be seen as an industry Japan could rely on and work with.

Signs that the April mission returned back to Japan more confident in Australia were to follow. In June, the chairman of TEPCO, the Tokyo Electric Power Company, Mr Hiraiwa, said that the Keidanren had agreed to set up a coal council within its operations. Hiraiwa referred to this council as the Australian Coal Deliberative Council which would prepare forecasts of long term demand for coal and assess infrastructure requirements.[150] At the same seminar, Mitsui's chairman, Yoshizo Ikeda, forecast that Japan was likely to be importing 22 million tonnes of steaming coal by 1985, rising to 55 million by 1990 and 80 million by 1995. Australia, he said, was likely to account for 50% of these tonnages, the USA 15%, Canada 15% and South Africa 10%.

However an issue which would plague Australia Japan relations on and off for years – industrial disputation – soon reared its ugly head. Phillip Lynch, then Minister for Industry and Commerce, said that when he visited Japan in August there were many questions and "a sense of bewilderment as to their perception of what is happening in Australia".[151] There had been a coal strike in Queensland over tax arrangements for remote area housing which had disrupted the industry for 10 weeks and also other disputes including on the NSW South Coast as BHP looked to make retrenchments at its coking coal mines. The Queensland strikers' argument was with the Federal Government, but was still a black mark on the industry as far as the Japanese were concerned.

Of course, the industrial situation in Australia was not unique, with a seven-week strike in the US coal industry ending in September 1980 following the expiry of the old employment contract in March that year. However, the Japanese were not backward in using our poor industrial relations as a stick to beat us with. John Menadue, who had just finished his term as Ambassador to Japan, said that while Japan was prone to exaggerate the problems linked with strikes, they "were very uptight and very worried" because of the unpredictability it caused for coal producers. By comparison, Menadue said that the Japanese domestic economy was "well-ordered and predictable....And when they place themselves in our hands with such large percentage of things such as coal and iron ore, they are going to be very conscious of stability of supply." Menadue also said that he believed it was very likely that this had led to the Japanese writing smaller contracts with Australian suppliers than would otherwise have been the case, adding that with new contracts for steaming coal, diversification of supply "will be uppermost in their minds."[152]

One of the major problems for producers, however, was how much reliance they should place on the various forecasts for coal demand. It was generally accepted that local and international demand for Australian coal would grow during the 1980s and 1990s, but there were significant variations in the forecasts, with some forecasts covering a wide range of tonnage. For example, in 1980 the Joint Coal Board published forecasts for demand for Australian coal covering the domestic market and exports. The JCB's forecasts, with high and low level projections, covered a wide range of possible demand; from around 40 million tonnes in 1979, the JCB's high level export forecast for 1990 was 180 million tonnes; its low level forecast was 115 million tonnes. The gap between the forecasts was indicative of the uncertainty of the times. The actual Australian export figure for 1989-90 turned out to be around 105 million tonnes.[153] The Australian Department of Trade in late 1981 also got into the forecasting business, its forecasts putting steaming coal exports by 1990 at between 50 and 70 million tonnes and coking coal exports between 50 and 60 million tonnes.[154]   Total Australian exports according to Trade would reach between 100 and 130 million tonnes. Trade was less optimistic than the JCB on its high level forecast, largely due to the timing of the forecasts which were developed as the reality of the likely demand was beginning to be better understood by governments, coal producers and coal consumers.

While Japan was expected to continue to be the dominant coal importer for steaming coal in the 1980s and 1990s, other markets were also looking extremely promising. There was the potential for countries to export to Europe, although the longer distances would put Australia at a disadvantage compared with South Africa and the USA as a major supplier to that part of the world. However, by the early 1980s prospects in Asia were improving, with South Korea, Taiwan, Hong Kong, the Philippines and Malaysia expected to be growing markets. A study issued in 1983 by the Institute of Energy Economics in Japan (IEEJ) had demand for imported steaming coal by those five countries growing from under 5 million tonnes in 1981 to 36 million tonnes by 1990, with South Korea and Taiwan accounting for 25 million tonnes of the total.[155] Those forecasts for South Korea and Taiwan would prove to be fairly accurate, but competition for these growing markets would prove to be much more intense than many producers had expected.

Japanese early forecasts prove unreliable

Forecasts by organisations such as the JCB and the Department of Trade relied in part on the forecasts the Japanese were publishing on coal demand in the 1980s and beyond, for both coking coal and steaming coal. Of course, it was one thing to make a forecast of demand for coal in a particular country such as Japan, but for Australia and other suppliers the other critical part of the equation was that country's market share. And other factors which had to be considered for steaming coal included an assessment of how realistic the timetable for the construction of new power stations might be, with the power utilities having the ability to delay construction or final commissioning of stations depending on the market outlook, and with those utilities also subject to delays in approvals due to community opposition and drawn out environmental assessment processes.

Following the June 1980 Mitsui forecast for steaming coal demand mentioned above, at the annual meeting between the Australian producers and the Japanese steel mills, power utilities and cement companies in October, the Japanese tabled more optimistic forecasts for steaming coal demand, with demand for Australian coal rising to 18-21 million tonnes by 1985 and 32.5 million tonnes by 1990. The Australian executives attending the meeting expressed some suspicion about the forecasts, and were concerned that the Japanese had a strategy which would create an excess supply of coal and therefore a buyer's market.[156] Until that time the official forecast which the industry had been relying on was the 1979 forecast by Japan's Ministry of International Trade and Industry (MITI – now the Ministry of Economy, Trade and Industry) - which had total steaming coal imports into Japan at 22 million tonnes in 1985 and 53.5 million tonnes in 1990; the Australian share was assumed to be in the range of 40 to 50%. Nobuyoshi Teranishi of Nippon Steel said that the estimates of power demand by the utilities and cement companies were the most realistic available, and Japan Coal Development executives said that the estimates had been based on individual company data, with one JCD executive denying that there had been any "cheating".[157]

By 1981 the tenor of the Japanese statements and forecasts changed significantly as the economic outlook became more uncertain. In October at the annual meeting with Australian suppliers, Teranishi delivered a bombshell, telling the producers not to expect any new coking coal contracts to be signed over the next few years. The contracts which had been signed in 1980-1981 involving Thiess Dampier Mitsui's Riverside mine and Capricorn Coal's German Creek mine would be the last until 1985 or later. Coking coal imports from Australia were expected to be around 27 million tonnes in JFY 1981, rising to around 33 million tonnes in 1985 and 36 million tonnes in 1990.[158] The November 1982 annual meetings did nothing to lift the air of gloom, with the Japanese telling the Australian producers that established producers would have to suffer to enable the most recent contracts from new suppliers to be fulfilled, with those contracts including the large new mines in Canada.[159] The Japanese steel industry was struggling to grow, but it had to make good on its recent supply contracts; whether it was intentional or not, there was a major surplus looming in the international coking coal market. Australian coking coal exporters were now looking at a market share in Japan of only 40% rather than the expected 45%, and if this occurred, their total exports by 1990 would actually fall to around 24 million tonnes compared with around 26 million tonnes in 1982.

MITI, as the department of the Japanese Government which oversaw the energy industry and energy policy, produced a new set of forecasts in 1983 for energy demand, including coal demand and coal imports. The forecasts were presented as "tentative estimates" and gave a range of possible outcomes, although it was reported that MITI had also produced "private estimates" which were more precise.[160] MITI now had Japanese demand for all the major energy sources, including coal, gas and nuclear, lower than previously forecast; total coal demand was now expected to be in the range of around 107 to 113 million tonnes by 1990, with imported steaming coal between around 45 and 48 million tonnes compared with around 53 million tonnes previously. At the public release of the forecast, a senior MITI official admitted that previous MITI estimates of trends in energy demand had been incorrect, while also pointing out that MITI was not alone in this respect, with organisations such as the OECD also producing forecasts which proved unreliable.

By the 1984 annual meeting, the message from the Japanese side was for a "severe" outlook for the energy market in Japan, with the Australian producers told not to expect any growth in the market for coking coal; in addition the Japanese were not optimistic about growth in the market for steaming coal.[161] The Australian producers were reported to have expressed concern about the forecasts for steaming coal, with John Yeowart, then chairman of the Australian Coal Association, saying that Australian producers were still looking to growth in the Japanese market and hoped that more appropriate longer term contract arrangements would be implemented. On behalf of the producers Yeowart also stressed that current prices were not adequate to provide the basis for producers to confidently fund new steaming coal projects. A joint statement issued at the end of the meeting contained some soothing words for the producers, with the Japanese undertaking to maintain Australia as the major supplier, provided, of course, that the producers remained competitive and reliable suppliers.

New power stations create excess capacity

Growth in the export market was clearly expected to be the major opportunity for Australian steaming coal producers in the 1980s, but the domestic market was also expected to grow strongly. A significant part of the expected growth in demand in the local market was from electricity generation, with the JCB's forecast for sales to this sector expected to range between 48 and 61.5 million tonnes by 1990, compared with just over 22 million tonnes in 1979.[162] The actual figure for the power generation sector in 1989-90, however, proved to be significantly below the forecast range, with demand only 39 million tonnes. There was significant growth in power generation in Australia, but the early forecast proved too optimistic; by the end of the decade there would be significant excess capacity in power generation which would take some years to reduce to a more reasonable level.

In November 1979 the Fraser Government, keen to ensure that there was sufficient generation capacity for the decade ahead, invited the States to submit proposals for the funding of new power station projects based on coal. Expansion in the aluminium industry was expected to be the major factor generating growth in the local market for coal, with smelting capacity expected to increase from less than 300,000 tonnes to 1.2 million tonnes by 1985, and making Australia one of the largest aluminium producers in the world. The Government was pushing the States to submit proposals which would need the approval of the Loan Council.

At the start of the decade, Queensland was already expanding its generation at Gladstone and Tarong. The Gladstone power station had four 275 MW units generating by 1980, consuming almost 2 million tonnes of coal per year, and its fifth unit was commissioned late that year. Unit six at Gladstone was commissioned in early 1982, bringing the total capacity of the station to 1650 MW. Construction works for the new Tarong power station had been speeded up and the first of its four 350MW units was planned to be operational by mid-1984, with the other units operational by 1986. In November 1980 the Queensland Government then made a major commitment, approving a multi-billion dollar program to expand its power generation capacity by 2100 MW during the 1980s. The new stations were approved to meet the surge in demand that the State Electricity Commission (SEC) was predicting, with the SEC forecasts showing demand rising from 2230 MW in 1980 to just over 6000 MW by 1990, and with the 1990 level potentially as high as 6690 MW. The growth in power demand would be due to growth in sectors such as aluminium smelting, alumina refining and mining, and from growing household and commercial demand. There were two new power stations to be developed at Callide and in the region near the Curragh coal deposits. The new 700 MW Callide B station would be supplied from the nearby Callide open cut mine owned by the Shell/ Thiess/ AMP joint venture, with the supply contract for around 40 million tonnes over 22 years. Callide B was expected to begin to supply power by June 1986.The second new station, which in due course was named Stanwell B and which would be located west of Rockhampton, would be supplied by the Curragh mine to be developed by the joint venture involving ARCO, R W Miller and Mitsui. Stanwell B was expected to commence to supply power by mid-1987. The Curragh mine would also supply coal to the Gladstone power station under a 65 million tonne contract. The Government also announced that the State Electricity Commission would investigate the feasibility of the private sector building and operating a coal-fired power station as part of plans to have additional capacity available by the late 1980s.[163]

By 1984-85 it was apparent that the SEC's forecasts were too optimistic and the construction timetable for the Callide B and Stanwell stations was slowed. Callide B was now expected to commence operating in early 1988 and Stanwell was in the early stages of its development, with its flexible construction programme allowing for power generation to commence by the early 1990s.[164] The second unit at Callide B was not commissioned until March 1989 and the first Stanwell unit was not commissioned until 1994, with the other units planned for commissioning in each of the following three years. In 1981 the SEC, with a new power station of four by 350 MW units in mind, invited applications from the private sector from companies to develop such a station. However, with the expected lower in growth in power demand, consideration of a new station was shelved.

In NSW, major expansion of the State's coal-fired power generation was also underway, centred on the Eraring station on Lake Macquarie, south of Newcastle, the Bayswater station near Muswellbrook and Mt Piper near Lithgow. By 1984, when it was clear that a major excess capacity in power generation was looming, the NSW Electricity Commission was committed to an expansion program planned when there was widespread optimism about the likely growth in power demand. Eraring's fourth 660MW unit was to be operational by the end of 1984 and Bayswater's four 660MW units were expected to be operational by 1986, lifting the State's capacity to over 10,000 MW. In addition, the Mt Piper station's two 660 MW units were planned to be available by 1990.

The growth in the aluminium smelting industry was one of the major factors driving the demand for electricty in the 1980s, but with the cancellation of several large proposed smelters, the industry was also the cause of much of the excess power supply which existed by the late 1980s. In NSW the Tomago smelter north of Newcastle was commissioned in 1983 and was expanded in the 1990s, but in 1982 BHP withdrew from a consortium with Alumax which was planning to build a 236,00 tonne smelter at Lochinvar near Maitland; Alumax later decided not to proceed with the project. In Queensland Comalco's 210,000 tonne Boyne Island smelter near Gladstone was developed between 1979 and 1982, but Alcan's proposed 296,000 tonne Goodwood smelter project was cancelled; Boyne Island was expanded in the 1990s to 540,000 tonnes and is today the largest smelter in Australia. In WA a 240,000 tonne CSR/ Reynolds smelter also did not proceed, although that project would not have been expected to draw on coal from NSW or Queensland but rather local coal from the Collie region.

A major review by consulting engineer Gavan McDonell in 1986-87 recommended major changes to the way in which the NSW Electricity Commission operated, and also recommended that the Commission's plans for new power stations be scrapped, saving the Government billions in capital expenditure. McDonell said that he believed that better management practices within the Commission and better utilisation and expansion of existing power stations were the appropriate way forward.[165] Following that report, the Commission was able to delay the construction of Mt Piper, that station becoming operational in 1992-93. Mt Piper, with two 700MW units, was the last large coal-fired power station constructed in NSW (not including the small 120MW Redbank station constructed near Singleton in the 1990s), although Mt Piper was designed for four units, and so in theory could be expanded at some stage. McDonell also recommended that responsibility for the Commission's coal mines should be removed from the Commission; reform of the mines, however, would not come until the 1990s, following the corporatisation of the Commission and its division into several new entities.

By 1990, Queensland power stations were consuming 11.2 million tonnes of coal a year compared with only 4.9 Mt in 1980. Coal consumption by the power stations in NSW reached 20.4 Mt by 1990, up from 14.4 Mt in 1980. For Australia as a whole, black coal consumption by power stations grew from 24.5 million tonnes in 1980 to 38.8 million tonnes in 1990.[166] Power generation continued to grow in the 1990s and saw national black coal consumption increase to 50 million tonnes by 1999-2000, just above the lower end of the range forecast by the JCB in 1980.

For both Queensland and NSW, the growth in the power station demand had been a huge boost to the coal industry throughout the 1980s, even if the forecasts at the start of the decade had been too optimistic. Supply to local power stations also meant a number of coal producers were less exposed to the severe competition in the international steaming coal market, with local contracts not subject to foreign exchange rate uncertainty and contracts generally providing for longer term supply. It is doubtful if any major coal producer with such contracts lost money in the 1980s through supply to the State-owned power generators.

Australian companies stand firm in 1981 JSM negotiations

A few months after the Japanese announced the new Quintette and Bullmoose developments to take place in Canada, the NSW coking coal producers from the Hunter Valley had a major win. The change in the market for steaming coal as a result of the second oil shock in 1979 saw Japanese cement companies and power utilities look increasingly to steaming coal to replace oil. In JFY 1980 the contract price for Hunter Valley steaming coal to Japan was $A39.00; the JFY 1981 price rose to $A45.00. The soft coking coal producers had been in a relatively weak market position, but with steaming coal demand now stronger, the producers were able to negotiate from a position of strength. A group of six producers from the northern district exporting through Newcastle – Coal & Allied, CSR's Buchanan Borehole Collieries, Gollin Wallsend,

Clutha, R W Miller and Bloomfield Collieries – formed a united front in the negotiations with the JSM in 1980 and 1981 for a new soft coking coal contract. The soft coking coal produced in the Hunter Valley was from mines which were also able to sell the same coal to power utilities and cement companies as steaming coal. To which market the coal was channelled was up to each company, with the soft coking coal market requiring additional washing to meet the JSM requirements.

The story of how these producers had won the day in Tokyo, battling Nippon Steel's powerful vice-president for raw materials, Nobuyoshi Teranishi, and Katsushige ("Kats") Tanaka, general manager for fuel and ferrous materials, was told in an article in the Sydney Morning Herald by its Tokyo correspondent, Hamish McDonald.[167] The battle plan had been developed in the first half of 1980 by several executives believed to have included John Doherty of Wambo, John Smith of Peko and Charles Copeman, then with Consolidated Goldfields.

The negotiations began in December 1980, with Tanaka one of those on the Japanese side, and Ian Coddington from CSR, Sid Grover from Coal & Allied and Alex Morrison from Clutha on the Australian side. Coal specialists from Mitsubishi, Mitsui and C Itoh (now known as Itochu) were also present, acting as translators. Interestingly, as part of their negotiating tactics the Australians did not nominate a leader for the negotiations and did not advise who would speak first at negotiating sessions. The tactic was clearly designed to minimise the chance of one of the companies being singled out for reprisals by the Japanese. With no price increase for two years, the Australians argued that a major price increase was justified on the basis of rising costs over that period and the fact that new markets for steaming coal had now emerged. They proposed $A53 per tonne, an increase of around 40% on the old price. The Japanese were only offering a $A5 per tonne increase, arguing that the boom in steaming coal demand was only temporary, with the country heading into an economic downturn and demand for coking coal falling.

This was just the first round of the negotiations and as McDonald noted, "it was clear that this was no ordinary negotiation." Nippon Steel, through the major steel mills, had enforced controls on the various Japanese buyers of coal and according to one source had "put an embargo on all steaming and coking coal negotiations with Australia" and had ordered the power utilities and cement manufacturers not to negotiate until after the coking coal negotiations had been finalised. And any price above $A40 per tonne was not to be agreed. A sale of steaming coal from Queensland to Idemitsu Kosan in March 1981 had "incurred the wrath of Nippon Steel" and "caused panic at Idemitsu headquarters" and a denial soon appeared in *The Tex Report*, the Japanese English language newsletter which reported on coal, iron ore and related matters.

The negotiations dragged on into 1981, with the Australians now offering $A55 per tonne for the first year, with reductions in the price in subsequent periods. The Japanese countered with an offer of $A45 per tonne. By late March the difference in offers was narrowing, with the Australians offering $A51.50 and the Japanese $A47.30. The new Japanese financial year was about to kick off and there were concerns about the possibility that Australia could suspend shipments of coal if negotiations dragged on for too much longer. However in early April the Australians agreed that they would continue to ship coal on a provisional basis and in mid-April a provisional price of $A47.30 was agreed. Final negotiations in May took place in the context of an international coal market in turmoil, with China and the Soviet Union having difficulties supplying coal, strikes in the USA and problems in Poland. The Japanese steel mills were in retreat. Agreement followed, with the Australians obtaining a price of $A49.50, an increase of around 32% on the old price of $A37 to $A37.60. The victory was sweet and the producers had had a major win, but it would prove to be the last for some time.

WWF and TWU fight leads to Wran's Navy

One of the major obstacles to the development of the NSW coal industry in the early 1980s was the delay to the desperately needed third coal loader in Newcastle. For once this was not a problem caused by the coal industry itself, but nevertheless did the industry's reputation significant unnecessary harm and it helped to reinforce the Japanese concerns about industrial relations in Australia.

The involvement of the private sector in the port of Newcastle's coal export trade began in 1958 with the construction of the Dyke Loader at Carrington, site of the current PWCS Carrington loader. In 1968 Canwan Coal developed stockpile operations at Carrington which were later linked by conveyor to the Maritime Services Board's Basin Loader. In 1974 a new privately owned loader was approved, with Gollin and Company given the nod to develop and operate the project. However, with Gollin's subsequent financial implosion, the partly completed loader was taken over and completed by Port Waratah Coal Services (PWCS), a consortium whose major shareholders were the Hunter Valley coal export producers and Japanese interests. The Basin loader was owned and operated by the Maritime Services Board (MSB) and its employees were largely members of the Waterside Workers' Federation (WWF). The MSB was also the owner and operator of the coal loaders in Port Kembla and Balmain. The PWCS Carrington loader's employees were members of the Transport Workers Union (TWU), although the final loading operations at Carrington were the responsibility of the MSB, with the employees involved members of the WWF. PWCS had an industrial agreement with its employees which was described as a "union organiser's dream", with a 35 hour week, generous overtime rates, a 50% annual leave loading, long service leave of 8 weeks after 13 years' service (the coal industry standard), and other benefits. It was reported that there had never been a strike at the PWCS site, and that there was a waiting list of 2,000 people keen to secure a job at the loader.[168] Clearly the TWU and its members at PWCS were happy with the arrangements.

The PWCS Carrington loader was completed in 1977, lifting the coal export capacity of the port to 20 million tonnes per year. That figure looked comfortable at the time, but as the 1980s unfolded, with optimistic forecasts of coal demand, 20 million tonnes capacity would soon be totally inadequate. The NSW Government and the NSW Colliery Proprietors' Association had discussed proposals to build a third coal terminal in November 1979. However, the unions only became aware of those discussions in March 1980 when the NSW Minister for Public Works, Jack Ferguson, asked PWCS and the MSB to cooperate in the planning of the new loader. The unions covering the MSB employees and the Newcastle Trades Hall Council wanted to see a new loader owned and operated by the Government. The TWU's position in 1980 was that if the Government decided to allocate the new loader to the MSB it would support that decision provided it received guarantees for the throughput of the PWCS loader, such guarantees then providing job security for its members at PWCS.

Some of the unions which had members employed by the MSB were concerned that the Government would allow the private sector to own and operate the new loader, and this led to the Newcastle Trades Hall Council authorising one day stoppages in July and August 1980, with a view to pressuring the Government to give the MSB the responsibility for the new loader. Compounding the effects of the stoppages that year were other disputes, including an 11-day strike by mining unions, stoppages by the crews of tug boats which were essential for the operation of the port, and a strike by railway shunters; there was also a closure of the port for 5 days for dredging of the channel.

With a firm Government decision still not announced, the TWU wrote a savage letter to Ferguson in January 1981, slamming the Government for the delay and opposing the allocation of the loader to the MSB, which it said had "one of the worst industrial records of any State Government department." The TWU's letter also distanced the union from the Trades Hall Council, and contained a not-too-subtle threat, saying that "Threats by others unions to take industrial action in Newcastle and other ports if the loader is not built, owned and operated by the State, are easily countered by similar action being contemplated by this union. No coal need be loaded out of Newcastle. In fact, no coal need be transported to the Port Kembla coal loader or to any rail siding in this State if transport workers are of a similar attitude."

The TWU letter seemed to galvanise the Government into action.

Two days later the Premier, Neville Wran, announced that a company would be set up to finance and build the new loader, with the NSW Government holding a 20% equity, and the other 80% to be offered to companies shipping coal through the port.[169] The Opposition said that the Government should not be involved in such projects which should be left to the private sector. The Government said that "traditional maritime unions" would operate and maintain both the stockpile area and the shiploading facilities, with the loader designed to handle a throughput of 15 million tonnes per year initially, but capable of being expanded over the long term to 50 million tonnes. In view of the urgency, the Government directed the MSB to continue with design work and environmental studies.

The saga continued and in March the Government issued a "preliminary memorandum" for potential shareholders in the loader which had the MSB as the manager of the loader, with employees to be "members of the traditional maritime unions at the port of Newcastle", the words the Premier had used in January. Dredging work in the channel next to Kooragang Island began at the end of March, but the TWU put a ban on the coal loader site, declaring it out of bounds. Work on Kooragang could not now proceed as the TWU wanted coverage of the loader, arguing that it was to be 80% owned by the private sector and that it should operate in terms of union coverage in the same way as the PWCS loader.

By April 1981, 35 coal ships were in a queue off the port waiting to load, with the coal exporters picking up the cost of demurrage for an average delay of around 20 days for each ship. What came to be known as "Wran's Navy" would still be a problem in Newcastle for some time. That month was also important as it was the time when the Premier had been forced to clarify what he meant by the "traditional maritime unions" which would be involved in the new loader, telling a union delegation that these were the unions involved in the MSB's operations. The TWU would therefore be largely frozen out, with only a handful of its members to be involved. The MSB was to be given responsibility for management of the loader, but Wran was not to be swayed on the ownership structure, saying that the 80% private equity would remain.[170] In May 1981, with the stand-off between the TWU and the WWF and other unions continuing, and with rail transport disputes hitting deliveries to the port, Wran accused the unions of putting the future development of the Hunter region in jeopardy through their irresponsible industrial action. Wran also suggested that disputes were forcing Japanese coal customers to look elsewhere for their coal.[171] The Premier said that the loader was no longer a matter for conjecture and foreshadowed an imminent announcement which would put it beyond doubt. However it took over a year from the time of the initial announcement that a new company would be formed to build and operate the new loader for a final decision to be made, and in April 1982 the Government announced the details of the shareholding arrangements for the new company. BHP would be the largest shareholder with 30%, the NSW Government would have 20% as originally planned, Howard Smith 12.5%, a consortium of coal companies 27.5% and Japanese companies 10%. As the major shareholder BHP would now have the management rights to the loader.

By the end of 1981, concerns had been expressed that, even with the new loader, the port may not have the capacity to meet long term export commitments. The Kooragang loader, when completed in 1984, was expected to lift the port's capacity to 25 million tonnes. However the general manager of CSR's energy division at that time said that there appeared likely to be a shortfall of 4 to 6 million tonnes in capacity in 1984 and 1985. This statement followed criticism by the Japanese in October about Australia's infrastructure planning and industrial disputes. There were 25 ships waiting to berth in the port, incurring around $375,000 a day in demurrage.[172]

The demand for coal was still strong, but because of industrial disputes the port of Newcastle had seen only around 13 million tonnes of coal exported in 1980-81, well under its then capacity of around 16 million tonnes. In 1982 PWCS upgraded its capacity to 28 million tonnes. The Kooragang Coal Loader (KCL) was completed ahead of schedule and took delivery of its first trainload of coal from BHP's Saxonvale mine just before Christmas 1983, and loaded its first ship in May 1984.

With the upgraded Carrington loader, the port's capacity was now around 43 million tonnes, well in excess of the current throughput. The concerns over capacity in the mid-1980s would prove to be unfounded, with the higher capacity in Newcastle, an easing in demand and the capture of some contracts by new mines in Queensland. Nevertheless, the port of Newcastle's coal exports increased over the decade, reaching 30 million tonnes by 1989-90. In 1990 KCL became part of Port Waratah Coal Services Pty Limited, with the combined capacity of the Carrington and Kooragang loaders reaching 46 million tonnes. That year also saw the MSB Hunter Ports Authority withdrawing from coal operations in the port, with all loader maintenance and operations becoming the responsibility of PWCS. The Basin loader was closed and decommissioned in December 1988.

Queensland port development and rail electrification proceed smoothly

In contrast to the problems experienced in the early 1980s in Newcastle, port development in Queensland was planned and implemented smoothly. The coal export terminals at Gladstone – Auckland Point, Barney Point and Clinton – were capable of handling around 23 million tonnes per year. Utah's two Hay Point berths had a capacity of 25 million tonnes, Brisbane's terminal 1 million tonnes, and Bowen's terminal 0.5 million tonnes. Abbot Point's terminals, north of Bowen, handled coal from a number of mines. The Auckland Point Terminal, operated by the Gladstone Harbour Board, handled coal from Utah's Blackwater mine, and from the Cook and Leichhardt mines. Barney Point was operated by Thiess Dampier Mitsui to handle its Kianga and Moura coal. The Clinton Coal Facility (later to be called the RG Tanna Terminal), also owned and operated by the Gladstone Harbour Board, was opened in May 1980 and had a throughput of 4.3 million tonnes in its first full year of operation, initially handling mainly Utah coal.

The development of a new facility at Abbot Point by the Harbours Corporation of Queensland and MIM to handle the coal from MIM's

Collinsville and Newlands mines was planned to be completed in early 1984 with an initial capacity of around 6.5 million tonnes per year; construction began in mid-1981. At Gladstone dredging was underway to deepen and widen the access channel to cater for ships of up to 140,000 DWT. Utah developed its Hay Point Coal Terminal south of Mackay in the 1970s to cater for its CQCA mines, including Goonyella, Peak Downs, Saraji and Norwich Park. Hay Point shipped its first coal in October 1971, and by the end of 1975 the terminal's capacity had been expanded to 20 million tonnes per year. A new terminal close to the Utah facilities called Dalrymple Bay Coal Terminal was under construction in 1981. DBCT was developed by the Queensland Government to handle exports from the northern section of the Bowen Basin, including from German Creek, Oaky Creek, Riverside and Blair Athol. The first stage of this facility was planned for 15 million tonnes capacity and for completion in mid-1983. In the south, the Port of Brisbane Authority commenced negotiations with Queensland Bulk Handling Pty Ltd in 1981 for the new terminal at Fisherman Island which was being developed for coal producers located in the West Moreton district and also for possible future mining developments in the Darling Downs. The new facility was planned for an initial capacity of 3 million tonnes and for completion by the end of 1982. The second stage expansion to 5 million tonnes capacity was planned to be in operation by 1984-85.

The new DBCT facility loaded its first ship in November 1983, just ahead of Abbot Point in February 1984. Construction began at Fisherman Island in March 1982 and was completed by the end of the year, and the first ship was loaded in February 1983. By 1984-85, around 33 million tonnes of coal was exported through the Queensland ports. With other developments completed since the early 1980s, the capacity of the ports now exceeded 70 million tonnes. Queensland's ports had significant excess capacity, were positioned to cater for the expected growth in exports for the years ahead, and all the ports had plans in place to cater for further expansion.

Port development in Queensland proceeded in the 1980s alongside a major expansion of the capacity of the rail system and the electrification of the system. In 1983 the Queensland Government approved a major program to electrify the coal rail system, covering around 1,000 route kilometres, and almost 1500 kilometres of track. The first two stages of the work were to cost $600 million, with the first stage covering the system between Blackwater and Gladstone; this section of the system was completed in 1986. The second stage was the Goonyella system which included the rail line from Hay Point to the Goonyella mines, and then the line south west to Blair Athol and the line west to Gregory where it linked up with the stage one work. Stage two was completed in 1987. The full electrification program was completed by 1990 with stages three and four involving the electrification line from Blackwater to Emerald and the line between Brisbane and Rockhampton, and cost around $A1 billion. Stages three and four were largely to cater for sectors other than coal. The electrification program also involved the purchase of 146 electric locomotives for the coal industry and the system serviced 15 coal mines by 1990.[173]

Industry restructuring in the early 1980s

Employment in the NSW coal industry increased by over 3000 between 1980 and 1982 to reach almost 21,000. However, over the rest of the decade major retrenchments and mine closures resulted in employment falling to just over 17,300. In May 1982, Clutha, then a BP subsidiary and suffering significant losses, closed two of its Burragorang Valley mines, reduced production from a third and retrenched 216 of its 1700 workforce. In September BHP cut over 400 jobs from its south coast mines. In October 1982, mineworkers and families travelled to Canberra to demonstrate against recent and planned job cuts, particularly in the Newcastle, the Illawarra and Burragorang Valley areas. One group descended on the old Parliament House, smashing through the glass doors and occupying Kings Hall for a short time.

BHP stunned the industry in March 1983 with a further round of retrenchments from its south coast mines, its collieries division head office and from its John Darling Colliery and Stockton Borehole mines near Newcastle. On the south coast, Wongawilli, Metropolitan, Bulli, Cordeaux and Corrimal collieries all saw cuts. That same month CSR cut 94 jobs from its Buchanan Borehole mine near Newcastle, and R W Miller also announced cuts totalling 181 from its Delta No.1 and Preston mines. In announcing the March round of job cuts, BHP cited the major reduction in demand for coal from its steelworks, with Newcastle demand down 20% and Port Kembla down 40%.[174] The job cuts announced by BHP from September 1982 to March 1983 totalled 1077, with 851 from its south coast mines. The major trigger for the BHP retrenchments was the downturn in the local steel industry, which was hitting BHP's Port Kembla and Newcastle steelworks very hard. Steel imports into Australia had increased significantly and local demand was weak. In the ten months to March 1983, BHP's steel production was down by 29% (4.2 million tonnes compared with 5.9 million tonnes the previous year); and in the 9 months to February 1983, its steel division lost $361 million before tax.

The problems in the steel industry in 1982 saw the Fraser Government direct the Industries Assistance Commission to review the steel industry; the IAC's initial report was released to the public in March 1983, and its final report in May. The IAC recommended against any long term assistance to BHP by way of higher protection against steel imports or tax assistance. BHP, however, had said that whatever the outcome of the IAC inquiry, it would be shedding around 10,000 jobs over the following ten years. John Button, the Hawke Labor Government's Industry Minister, announced the Government's new steel industry plan in August 1983. The Government rejected the IAC recommendations and committed $100 million for economic development projects and industry training in the three steel regions, and introduced a scheme involving bounties and quotas to provide greater protection against import competition. BHP had sought a guaranteed 85% of the local market, but this was not part of the plan. However, under the plan BHP committed to the continued operation of its steelworks at Newcastle, Port Kembla and Whyalla and to job security for its current steel industry workers, and to investing between $500 and $800 million over 5 years to upgrade its facilities. The plan helped to stabilise the steel industry during the 1980s, with local steel production holding relatively steady after 1983 between 6 and 6.7 million tonnes per year, compared with the peak of 8.1 million tonnes in 1979 and the low of 5.7 million tonnes in 1983.

There was further restructuring of BHP's southern collieries during the 1980s, with the closure of three of its mines announced in 1985 – Bulli, Wongawilli and Corrimal. BHP said that the closures would take place over the following three years, with no retrenchments planned. There were still reserves of coal to be mined at these collieries, but the company planned to access the coal by extending tunnels from nearby collieries.[175] At a meeting convened by the JCB (which had to approve any closures) BHP gave a written undertaking to the unions that there would be no retrenchments provided that the unions cooperated with its rationalisation plan.[176] This rationalisation of the southern BHP mines, however, was not the last for this group as pressure on the steel industry to improve its productivity remained strong throughout the 1980s and another round of retrenchments and restructuring was undertaken in BHP's collieries in the early 1990s.

In Queensland the 1980s did not see the sort of dramatic restructuring that was experienced in NSW, although the West Moreton saw over 200 job losses between 1982 and 1985, not a large number, but nevertheless a reduction of around 17%. For the Bowen Basin, the major problems were the state of the international market, low prices and the cost of government-provided rail freight services. There were a number of major mines developed in the period between 1980 and 1984, with the decision to proceed taken when the outlook was positive and companies were keen to cash in on the expected strong demand for steaming and coking coal. Among the mines which came on stream in this period were Newlands (first coal produced in 1984), Blair Athol (first railing of coal of the newly developed mine was in 1984), German Creek (coal produced from 1982), Gregory (development completed in 1982), Oaky Creek (open cut mining commenced in 1983), Riverside (first coal produced in 1984), and Yarrabee (production commenced in 1982). Further south, the Tarong mine near Kingaroy, developed to supply the new power station, made its first deliveries in 1984. As already outlined, development or expansion of the export terminals at the ports of Brisbane, Abbot Point, Gladstone and Dalrymple Bay and Hay Point also occurred during this period. However by 1990, development of other major projects which had been under evaluation in the early 1980s had not gone ahead. Major projects still listed by the QCB as potential projects included Acland, Baralaba, Clermont, Ensham, Gordonstone, North Goonyella, Rolleston, Theodore and Wandoan.

In the West Moreton, the long term contracts for supply to the Swanbank power station had been a godsend for the producers in the late 1970s, but their future was still far from secure. However, with the growth in the export market for steaming coal from the late 1970s the prospects for the district were starting to look a little brighter. West Moreton coal had never had major export markets, and certainly nothing of significance in the major or emerging Asian economies. And there were still major infrastructure issues to confront, notably the lack of rail access to the Port of Brisbane and the lack of a modern export terminal capable of handling greater tonnages. In 1977-78, six West Moreton coal companies formed a consortium, the West Moreton Coal Exporters, to look for export markets for their steaming coal. In 1978-79 West Moreton exports were a very modest 16,000 tonnes, mainly shipments by Rylance Colliery to Fiji for use in cement production. In 1979-80 the tonnage was a little higher, and included around 18,000 tonnes to Japan by Rylance and New Whitwood. Those shipments to Japan were trial shipments to test the suitability of the coal and laid the foundation for the growth that was to follow. By 1980-81 the companies' efforts were beginning to show significant results. That year West Moreton exports through the Port of Brisbane reached almost 400,000 tonnes, and grew to just over half a million tonnes the following year. But with a new export terminal, West Moreton exports would grow to 2.5 million tonnes by 1989-90, with Japan the major customer.

A major development occurred in 1981 when the Queensland Government announced that the Port of Brisbane Authority was to commence negotiations for the development of a common user export terminal at Fisherman Island, at the mouth of the Brisbane

River. Queensland Bulk Handling (QBH), a joint venture between New Hope Collieries and TNT, won the right to develop the terminal. New Hope had purchased an old sugar loader in 1979 and stored it in one of its underground mines in preparation for the construction of the new facility.[177] QBH loaded its first shipment of coal destined to Japan in February 1983. By 1985 the new export terminal was operating as a common user facility with a capacity of 3 million tonnes per year and could accommodate vessels of 80,000 tonnes. Now the West Moreton producers had a modern facility which could accommodate reasonably large vessels of Panamax size, and the industry was set to take advantage and expand exports over the next few years. As Bruce Denney notes in his history of the New Hope Group, while the new facility was not large compared with the major coal ports in central Queensland, "the establishment of QBH as a critical part of the coal chain allowing exports was absolutely key to the survival of the Ipswich coalfield."[178] Denney also notes that New Hope wisely continued to barge some of its coal down the Brisbane River to the new port, allowing it to bargain with QR for lower rail freight rates.[179]

With the West Moreton producers struggling in the early 1980s, in 1984 the Queensland Government also moved to ease the pain for these producers, announcing cuts to the rail freight rates. The current rate of $5.67 a tonne was reduced by up to 74c a tonne, or around 13%, with the new rates backdated to January 1984. The Deputy Premier, Bill Gunn, said that the reductions were an attempt by the Government to provide assistance to the West Moreton producers who were facing hard times.[180] The major producer in the West Moreton was New Hope Collieries, a company which commenced operations in 1952. Washington H Soul Pattinson Ltd purchased 50% of New Hope in 1970 and remains its largest shareholder. New Hope expanded through acquisitions of Tivoli Collieries in 1978 and Southern Cross and Haighmoor Collieries in 1979. By 1980 New Hope was producing over 250,000 tonnes from its underground operations. New Hope bought Rylance Collieries in 1981, and then commenced development of the new Jeebropilly open cut mine near Amberley air base. By 1982-83 the group was producing almost 1.2

million tonnes. New Hope went on to expand its interests into Indonesia in 1985, in 1995 purchased Rhondda Collieries, and in 1999 bought Oakleigh Collieries and the Acland deposit north- west of Toowoomba.

While the underground sector was contracting, with Westfalen and Box Flat ceasing operations in 1986-87, the open cut sector was expanding. Another new operation was the Ebenezer open cut which commenced production in 1986-87. In 1988-89 the big Japanese trading company Idemitsu Kosan already held a 22.5 stake in the Ensham project and added to its Australian investments with the purchase of the Ebenezer mine from Allied Queensland Coalfields. Ebenezer's production grew to 1 million tonnes of saleable coal by 1990-91.

At the end of the 1980s, the West Moreton industry was producing 3 million tonnes of coal per year. But there were now only 5 underground mines producing a very modest 440,000 tonnes per year. The open cut sector had 7 mines, with 4 of those – New Hope's Jeebropilly, Ensham's Ebenezer, Allied Queensland Coalfield's New Whitwood, and the privately owned Oakleigh - accounting for 90% of open cut production. Further changes were ahead for the industry in the 1990s, with the closure of the last of the district's underground mines, New Hill and Oakleigh No.3, in 1997.

Government charges, royalties and freight rates

When the mining industry is earning healthy profits, or is believed to have strong prospects for healthy profits, governments often can't help themselves and move quickly to raise royalties and charges for government run services including rail freight transport and port services. The period between the late 1970s and the early 1980s was no exception, with the boom predicted for the mining industry as it entered the new decade expected to see mining companies enjoying high profits. The companies were ripe for plucking, or so governments thought.

In 1975, to cream off what it saw as windfall profits, the Whitlam Government hit coal exporters with the coal export levy - $6 per tonne on high quality coking coal, and $2 per tonne on lower quality coking coal and steaming coal. By 1980 the levy on coking coal had been reduced to $3.50 for high quality coking coal from certain open cut mines (those mining below 60 metres), and to $1 per tonne for other high quality coking coal. The levy on lower quality coking coal and steaming coal had been reduced to zero. However, in the 1981 Federal Budget steaming coal was hit again with a $1 per tonne levy, but the 1982 Budget saw steaming coal exempt once again. The only mines paying the levy from that time were the Queensland open cut mines operated by Utah, CQCA, Thiess, TDM and BHP. The levy was now just a Queensland coal tax. It would not be until 1991 that the export levy, still at $3.50 per tonne, would be abolished.

The coal export levy was a Federal Government resource tax designed to cream off some of the industry's profits, but State governments were also not backward in wanting their share of the coal industry's profits, meagre or non-existent though these profits were at times. The Queensland rail freight regime dates back to the 1968 CQCA agreement when the Government required the company to pay for all the costs of the railway line construction and rolling stock, but retained ownership of the assets. The royalty rates for the early TPM and CQCA agreements were, prima facie, ridiculously low at 5 cents per tonne. But the Queensland Government was intent on getting its return from the coal industry through rail freight rates rather than royalties.

As Roger Stuart wrote in an analysis of the Queensland rail freight policy experience: "From the government's point of view, the major advantage of this choice was that the tax take remained hidden….. this meant that prospective coal developers would be attracted into the exploration/ development stage by the low royalties before facing negotiations over freight rates. Moreover, the fact that the size of the profit component was not known, even to the companies involved in the negotiations, made freight profits much more difficult than increased royalties for the companies to criticise publicly. In all agreements after the Moura negotiations, the profit component was indexed to an inflation factor. By the end of the decade, the profit was about $1 per tonne."[181] Stuart went on to explain that "In addition to the profit component, there were two other elements in the freight rate – the operating cost and the capital cost (or 'security deposit') elements. The operating cost component was calculated to cover the actual costs of the rail authority of operating the service. It also contained an escalation clause… The capital cost component played several roles: the primary function was to provide the funds from which the government repaid the company over a fixed term for the initial financing of the railway infrastructure. The loan was repaid at the same time and at the same rate as the companies paid the capital cost component of the freight rate. In effect, the companies were paying to the government the instalments which the government would then use to repay the loan…" There was also a financial advantage for the coal companies in this arrangement as they were able to deduct all of the freight rate payments as expenditure for tax purposes.

The CQCA freight rates were set under a complex formula which made it difficult to determine how it was structured in terms of its components – the operating costs, capital amortisation and profit. However, the Premier stated that while the profit margin accruing to the state was "many times the royalty rate", he was determined to keep the profit element secret.[182] Bjelke-Petersen also said that the profit margin would increase substantially after 12.5 years when the capital cost was paid for. The CQCA agreement provided for freight rates to reduce by 20% after 12.5 years, but the Premier said that the profit component was greater than 20%.[183]

In fact the 20% reduction was not implemented, with the Government implementing a new royalty regime for coal and for minerals in 1974. The 5 cents per tonne royalty was now applied only to coal produced for the domestic market. Export coal was now subject to an ad valorem royalty of 5% free on rail for open cut coal and 4% for underground coal. Treasurer Gordon Chalk in announcing the new rates said that at current prices the royalty on export coal would average around $0.80 per tonne. When combined with the profit element in rail freight rates, Chalk said that the total return to the Government would be around $2 per tonne, and with export production of around 20 million tonnes, the total revenue was expected to be $40 million in 1974-75.[184] As coal prices firmed and production increased, Chalk's estimate of $40 million was soon out of date.

By 1980, Queensland Rail's revenue from coal haulage had increased to $183 million, with $85 million from the Utah operations alone. Those figures soared in the next few years to $294 million and $122 million respectively in 1982, and to $642 million and $256 million respectively in 1985-86. The estimated profit component of the coal revenue rose from $117 million in 1980, to $200 million in 1982 and to $423 million in 1985-86.[185] Utah's estimates for the rail profit component for its own operations, including CQCA, put the 1980 figure at $54 million, 1982 at $83 million and 1985-86 at $170 million.[186] The profit components of Queensland Rail's coal revenue were estimates, but indicate the very significant benefit the Government was extracting from the industry. Queensland Rail and the State Treasury were becoming increasingly dependent on coal revenue, and by the early 1980s despite the profit from the coal industry, Queensland Rail was still recording overall deficits. The first overall surplus would only be recorded in 1984-85.[187]

Queensland freight rate battle

In the early 1980s the Queensland coal industry was feeling the harsh effects of the headwinds in the Japanese economy and the international economy generally, and the coal industry began to negotiate with the Government over rail freight rates. Up to that time, the coal companies had tended to accept that freight rates contained an element of tax, but when the market turned down, they started to look for some relief from what they saw had become an unfair tax component.

According to Queensland Rail "Prior to 1992 coal rail freight rates were determined and negotiated directly between Queensland Treasury and mining companies, with Queensland Rail basically having the non-commercial role of providing the rolling stock and infrastructure required by the mine and implementing the terms of Treasury's rail contracts. The level of the freight rate set by Treasury was usually based on the mine's ability to pay. In many instances, this resulted in freight rates being significantly in excess of the costs of haulage."[188] With the Treasury assessing a mine's ability to pay, it is unlikely that the freight rate negotiations were what could be termed normal commercial negotiations. The Queensland Government did respond to the industry's concerns and announced rail freight rate concessions in its 1983 budget which it said would save the companies around $280 million over four years. The companies disputed this figure, claiming the saving would be only $40 million. The concessions on offer certainly did not settle the matter.

The issue came to a head in May 1984, when Joh Bjelke-Petersen addressed the Queensland Chamber of Mines' annual general meeting. Bjelke-Petersen had taken over the Treasury role in 1983 from Dr Llew Edwards who had argued that the state could not afford cuts in its rail revenues. The Premier had reportedly been sympathetic initially, but said that he was not able to overrule his Treasurer. Now holding both portfolios, Joh was under pressure. The coal producers, seeing cuts in export prices and tonnages, had been looking to the Government to share some of the pain. At the Chamber's AGM, its president, Gavin McDonald, a senior executive of Utah, said that government had to share some of the burden of the difficult times.[189] No doubt aware that some coal producers had been delaying payments to Queensland Rail as a form of protest, Bjelke-Petersen did not take the Chamber's advice meekly, reacting as if the government was being unfairly attacked while the industry was "going easy" on the Federal Government and the unions. "What about the unions? What about asking them for concessions?" Joh is reported to have said. The producers' position on this sort of argument was that wage costs amounted to a modest proportion of total costs (around $4 to $6 per tonne), while rail freight rates averaged around $14 per tonne, with around half of this figure being "profit" accruing to the State Government.[190]

Bjelke-Petersen told the meeting that the coal companies had until May 31 to agree with Treasury to pay the new freight rates and to settle outstanding payments (some companies had been withholding payment in protest against the Government's refusal to negotiate). Failure to sign up and pay up, said Joh, would mean companies would lose the concessions offered in 1983.[191] Representatives of the industry met the Premier a few days after the Chamber AGM, but to no avail. Joh demanded that they accept the rates on offer by the end of the month or the offer would lapse. The companies then went public and through the Queensland Coal Association called a media conference to put their case. They warned that new coal projects in the Bowen Basin were struggling to be viable financially; they planned to meet again the following week to plan their next move.[192]

The next part of this saga came in late May when the Premier wrote an eight page letter to coal producers, which played down the industry's problems, arguing that Queensland coal was still competitive on world markets and that export volumes were still expanding. Joh's letter also said that there was no conclusive evidence that the producers were not capable of withstanding the current downturn in the industry. The letter also referred to how favourably rail freight rates in Queensland compared with rates in other countries, quoting rates in in terms of dollars per tonne. He said that the average Queensland rate was $13.30, while it was $9.74 in South Africa and $17.90 in Canada. And missing from Joh's comparison was any mention of the payment by coal producers for rail capital costs of around $6 per tonne. On the basis of cents per tonne kilometre, a generally accepted way to compare rates, Queensland rates averaged around 7.8 cents, South Africa 2.1 and Canada 1.6.[193] The Premier's letter and his refusal to negotiate won the battle. By the end of May it was reported that a number of companies had advised the Government that they would accept the rates on offer, just beating the end of month deadline he had set. The Queensland Coal Association, which had coordinated the industry's negotiations, could do no more; it had advised the Government that individual companies would be responding to the Government.[194]   Later that month the West Moreton coal producers won freight concessions from the Government worth up to $1.44 per tonne and which they had to accept by the end of June. In announcing the concessions, Treasurer Bill Gunn said that they were offered to offset the "hard times in the coal industry" and that the Government had "virtually cut freight rates to the bone" with savings on offer of up to 25% depending on tonnages railed.[195]

The Queensland Government continued to extract a very healthy resource tax from the Bowen Basin coal industry through its rail freight charges, and it was not until the early 1990s that arrangements changed. QR was commercialised in 1991 and then adopted a policy of setting freight rates for new contracts solely on the basis of providing a rail haulage service. But the Queensland Government in 1993 announced a new rail freight rate regime to be phased in from January 1994 with a higher royalty for new mines and for existing mines on the expiry of their contracts. With rail freight rates down, but royalties up, the Government was still on a winner.

NSW producers push back on rail, port and electricity charges

NSW coal producers in the late 1970s were becoming increasingly frustrated and concerned as government charges escalated along with labour costs, but at a time when international markets were extremely tough. In 1979 the NSW Combined Colliery Proprietors' Association (soon to adopt the name NSW Coal Association) wrote to the Premier to urge the Government to constrain all its agencies to ensure that coal industry charges did not increase more than was justified by actual increases in labour and materials costs. The Association also requested that there be no increase in the standard royalty rate which had been steady since 1975, although the effective royalty rate for new mining projects had been raised to over $2 per tonne.[196]

The Association's letter advised the Premier of the results of a survey of 13 coal producers by accountants Arthur Andersen which found that two in three were operating with profit margins of less than $2.50 per tonne, with half of these companies operating in 1977-78 with margins of less than $1 per tonne. Over the previous two years, while revenue per tonne had risen by around 7% per annum, the survey found that operating costs had risen by around 9.5% per annum; charges by the

Public Transport Commission (the body which ran the rail system), the Electricity Commission and the Maritime Service Board rose by over 23% per annum. The Association also pointed out that the survey results did not reflect a number of recent factors which had impacted profits, including price reductions for hard and soft coking coal, further increases in government charges and a log of claims by mining unions currently before the CIT which had the potential to have a significant impact.

The Association's arguments had no lasting impact on the Government's policies, and in 1981 there was an increase in royalties from $1 per tonne to $1.70 per tonne. That year also saw the NSW Government lift rail freight rates by 25%, electricity charges by a similar amount, and port handling charges levied by the Maritime Services Board were hiked by 47%. Coal Mining unions were also demanding an increase in wage rates of 20%.

In the late 1970s the NSW Government began to realise it was missing out on another potentially lucrative source of revenue due to the fact that around 30% of the state's coal production was from mines on land where the coal rights were held by the landowner. In some areas such as the Upper Hunter, private ownership of coal accounted for close to 100% of coal mined. This situation dated back to the 1800s when, in 1850 following the end of the AACo monopoly, the Governor issued a proclamation which transferred the coal rights to land owners. From 1884 the NSW Parliament legislated to exclude coal rights from all future grants except those issued for mining purposes.[197] In 1982 the Wran Government passed the Coal Acquisition Act to change the ownership of coal underlying privately owned land from the land owner to the Government. Some coal companies were affected, although the Government's move also affected many individual land owners who had been fortunate enough to own land over good quality coal seams. Some coal companies had paid private landowners to gain access to the coal for new mines. The 1982 Act did not endear the Wran Government to the coal industry, although compensation was paid during the 1980s to the landowners who lost their coal rights.

The delays in the completion of the Kooragang coal loader due to the dispute between the TWU and WWF, industrial disputes in the coal industry, and the costs of demurrage caused by the lack of loading capacity in Newcastle were having a negative impact on the industry in NSW. Demurrage costs were estimated at around $40 million in 1981, and NSW Coal Association CEO Barry Ritchie said that the serious industrial relations problems were clearly having an effect on the industry's prospects, "driving customers away overseas, both for spot cargos to make up for our contract performance, and also to diversify sources of supply when new contracts are being awarded."[198] Independent analysts were also querying the state of the industry in NSW, including the unflattering comparison between NSW and Queensland, with the climate for development in Queensland seen as much friendlier. Macquarie University mineral economist Professor Don Barnett was one of the more prominent commentators, in June 1982 saying that "The great tragedy for NSW is that we had the opportunity over the past two years to really commit countries such as Korea, Taiwan and Hong Kong to conversion, and to lock them in to us as solid, long-term buyers of steaming coal. Instead we have caused them to have second thoughts and to think about other suppliers."[199]

In 1982 the deterioration in the competitive position of the NSW industry was highlighted by some estimates compiled by the AMP Society. The estimates related to the costs of a typical new open cut producer of steaming and coking coal in January 1980 and January 1982, the two-year period having seen the average market price for the two coal types increase in Australian dollar terms from $35 to $49 per tonne. Total production costs including an allowance for depreciation and interest charges were estimated to have risen from $29.60 to $47.20 per tonne; the profit margin therefore had dropped from a modest $5.40 per tonne in 1980 to only $1.80 per tonne in 1982.[200]

In early 1982, with mine closures and retrenchments in NSW putting pressure on the State and Federal Governments to act, Deputy Prime Minister and Minister for Trade and Resources Doug Anthony began to put pressure on the NSW and Queensland Governments to reduce their taxes and charges to assist the coal industry. Anthony was critical of these taxes and charges, pointing out that with many mines becoming marginal, there would be major social problems with the resulting unemployment, and calling on the states to face up to the problems in the industry.[201] He queried whether new coal mines in Queensland would be deferred because of the state of the industry, although ruled out any Federal action to block developments in Queensland to protect NSW underground mining operations.

Anthony met NSW Premier Neville Wran in May and then travelled to Brisbane and met with Queensland Premier Joh Bjelke-Petersen, hoping to find some common ground which would take pressure off the NSW industry. Bjelke-Petersen, however, was in no mood to make any concessions, telling a separate media conference that Queensland had nothing to do with the NSW coal industry. Joh said that Anthony wanted to curtail coal production in Queensland to help stabilise coal markets, but stressed that he did not believe in that and was "going flat out to sell Queensland coal."[202] Three days earlier, following Anthony's criticism of the states, the Premier had also signalled Queensland's refusal to consider any reductions in government charges, rejecting claims by Anthony that state taxes and charges had made many mines marginal. Bjelke-Petersen's argument was that the Federal Government had actually made conditions worse in the industry by the re-imposition of the export levy on steaming coal in the last Federal Budget. Joh said Queensland was "not marking time on coal" and that he and his deputy, Llew Edwards, would be shortly travelling to Korea and Japan and would be "making every effort to seek out new market opportunities for Queensland and giving every encouragement to private enterprise wishing to expand coal operations or undertake new ventures", adding that they would be "be distancing Queensland in every way from the difficulties being faced in NSW."[203] Clearly, Queensland was not going to be drawn into the problems of the NSW industry; any actions to relieve cost pressures in NSW would have to be taken by the NSW Government, the Federal Government or the producers.

Anthony delivered a win for the industry at the end of July 1982, announcing that the $1 per tonne export duty on steaming coal (which had only been reinstated in 1981) would be lifted, saving the industry around $42 million per year. On the same day, the NSW Government announced that it was freezing the royalty on coal for the 1982-83 financial year at the rate of $1.70 per tonne which had applied since February 1981; under the State's policy, the rate would have been increased to $1.96 in line with inflation. Further concessions by the NSW Government would be made in 1983 and 1984, and although welcomed by the producers, they would be seen as inadequate to meet the industry's problems.

In March 1983 a meeting between representatives of the NSW coal producers and the NSW Government saw relations hit a new low point. The NSW Coal Association was represented by its chairman, Dick Austen, CEO Barry Ritchie, and other company executives; the NSW Government by the Premier Wran (who was late for the meeting), Deputy Premier Jack Ferguson, Transport Minister Peter Cox, and Mineral Resources Minister Kevin Stewart. The Association made a presentation to the Ministers on the state of the industry and the need to reduce government charges, but Wran then spent some of his time at the meeting accusing the Association of using mineral economist Professor Don Barnett of Macquarie University to blacken his name.[204] Barnett had been interviewed on ABC radio the day before regarding work he had done on the impact of state charges on the viability of the industry. Wran also told the Association representatives that "You have never forgiven me for the Botany Bay coal loader", a reference to his Government's decision in 1976 to block a decision of the previous government to construct a coal loader in Botany Bay to service coal mines in the Western district and in the Burragorang Valley.

In Parliament the same day the Opposition Leader Nick Greiner asked whether the Government was prepared to freeze all state charges at the levels prevailing in 1982, a question which supported the Association's line that government taxes and charges were damaging the industry. Wran accused the Association of providing information to Greiner, but the Association's representatives denied doing so. Ironically, it would be Greiner as Premier in 1989 who would hike coal royalty rates at a time when there was hope that the coal market had improved, and would react savagely when criticised by the coal industry. In answer to Greiner's question in Parliament, Mineral Resources Minister Kevin Stewart pointed out the downturn in the coal industry in both NSW and Queensland and in other countries, but did acknowledge that the Government was undertaking a review of charges and that he and the Premier had met with company representatives and discussed the need to remain internationally competitive. Stewart also referred to the meeting later that day with the Association at which he said that the discussion of the industry's problems would be "in a most positive and objective way".[205]

The rates paid by coal companies to the State Rail Authority (SRA) in the 1980s became a major bone of contention between the NSW coal producers and the Government. The producers believed that they were being hit with unfair freight rates by an inefficient organisation. An article in the Sydney Morning Herald in 1982 put the level of government charges (including royalties, freight rates, port charges and Federal government taxes) into perspective. Compiled by industry analyst Ian Story, it estimated that in NSW just over 32% of every dollar of revenue earned by coal producers was made up of government charges, with the figure even higher in Queensland, a whopping 44%.[206]

A major problem for the Government, however, was the fact that the Railways were losing money, and lots of it. In the 1981-82 financial year the SRA's deficit had climbed to $387 million, up almost 30% on the previous year. The 1981-82 result was announced two days before the Association met the Premier and other Ministers, and Wran would also have been aware of the major commitment being made by the SRA to upgrade its capacity to haul coal. The SRA was planning to increase its capacity to haul coal to around 60 million tonnes a year, or around three times its 1981-82 capacity.[207] That plan involved a major financial commitment, and the Treasury and the SRA would have been warning the Premier not to be too generous to the coal producers. Despite Wran's anger towards the Association and the coal producers, his Government did move to restrain charges, although it took until September 1984 when it announced a 12-month freeze on rail freight rates, port charges and royalties.

The industry continued to argue for lower charges, and in particular for lower rail freight rates, and focussed on the high rate per tonne kilometre in NSW. The SRA argued that on the basis of the rate per tonne of coal, NSW compared favourably with other countries. The weakness in this argument, of course, was that rail transport of coal in NSW was generally over relatively short distances, particularly in the Hunter Valley. On the basis of the rate per tonne kilometre (tkm), NSW rail freight rates were much higher than in countries including South Africa, Canada and the USA. For the typical Hunter Valley haul from mine to port of around 100 kilometres, the cost was 8.2 cents per tkm. For the 343-kilometre haul from Blackwater to Gladstone in Queensland the rate was 2.9 cents per tkm; in South Africa a 500-kilometre haul to Richards Bay cost 1.5 cents per tkm; in Canada a haul of almost 1200 kilometres to the port cost 1.6 cents per tkm, and in the USA a 640-kilometre haul cost 3 cents per tkm.[208]

Behind the raw rail freight rate comparisons, of course, were a host of factors, including topography, which made comparisons difficult. The short distances in the NSW Hunter Valley, the inefficient loading equipment at some mines and the complex nature of the Hunter Valley system supported the SRA's arguments to a certain extent, but it was becoming increasingly clear that rates in the Hunter Valley were well above what was justifiable for an efficient system earning a reasonable but not excessive rate of return. The Hunter Valley was a system handling a significant tonnage each week, with no steep gradients to traverse, unlike the line from Lithgow; the system was generating an excess rate of return (also known as monopoly profits) for the SRA. The debate would continue into the 1990s and would see more pressure on the SRA and the NSW Government as studies revealed the inefficiencies in the organisation and the need to overhaul its charges.

The SRA in the early 1980s was investing large sums into the coal rail system, with new locomotives and wagons and upgraded track. The coal industry was the only profitable component of the SRA's business, providing a significant profit to at least partially offset the losses in its passenger and general freight businesses. In 1982-83 and 1983-84 the SRA spent close to $250 million on upgrading the coal rail system and in the decade to 1990 it invested around $1 billion in the system. Coal freight rates in NSW were held steady between 1986 and 1988, with the SRA reducing rates in 1988; the SRA stated that rates in real terms had fallen by 30% between 1986 and 1989 and in nominal terms by over 10%.[209] The producers saw the reining in of freight rates as modest and would continue to pressure the SRA and the Government to force SRA to become more efficient, and to eliminate the monopoly profit element in rates.

Coal industry profitability poor

Fom 1982 the Australian Coal Association began to publish the results of surveys of the profitability of the industry by independent accounting firms. The results of the survey for the year to June 1982 were published in December 1982 and showed that the coal industry accounted for around 90% of the profits earned by the whole mining industry. The results for the coal industry were based on the accounts of 40 companies which accounted for 79% of the industry's production.[210] The $180 million dollar profit for the coal industry may not have seemed unreasonable, but as it included what was believed to be the most profitable major mining company in Australia (Utah), the survey actually did a disservice to the many coal producers who were struggling to just break even, particularly in NSW. Utah Development earned a net profit after tax in the year ended December 1980 of $122 million and $133 million in the 1981 calendar year; for the rest of the coal industry, there were meagre pickings. The survey for the 1983-84 financial year showed a net profit for the 39 coal companies reporting their results of $193 million; 17 of these companies made losses for the year and the average return on shareholders' funds was a modest 5.8%, down from 7% the previous year. With the takeover of Utah by BHP, Utah's profits were now not reported separately, but it can be safely assumed that that company still accounted for the major portion of the industry's profits.

In 1985 results for the NSW coal industry were published for the first time and showed the dismal state of the industry. The results for the combined NSW and Queensland coal industry that year had seen profit rising to $354 million, an increase of over 80% on the previous year; Utah's profit was believed to have accounted for around 70% of that total and 19 companies reported losses for the year.[211] In 1984-85 the NSW coal industry incurred an aggregate loss of $51 million, (or $73 million including extraordinary losses such as write-downs in the value of assets) with losses also having been incurred in the previous three years.[212] By 1985-86 the picture had improved, with that year showing a net profit for NSW before extraordinary items of $92 million; however there were also extraordinary losses of $161 million, producing a bottom line loss of $69 million. Investments in coal mines in NSW were being written down in value to reflect the dire state of the industry. The survey results also showed that there were significant borrowings by NSW companies from parent companies which were non-interest bearing; if those loans had been financed by banks of other lenders, the interest payments would have made the profit result even worse.

The problems in the coal industry in 1987 saw major retrenchments, mine closures and continuing and strike action by the unions. As we detail later in this chapter, the unions' priority was to secure Government action to set up a national coal authority and so remove the responsibility for coal marketing from the hands of the producers. In a meeting with unions convened by the JCB, the NSW Coal Association rejected the need for such an authority, with its chairman Ian Dunlop telling the meeting that the coal industry had been "bled dry" and needed further streamlining. Dunlop said that producers could no longer compete unless operations were placed back on a viable basis. The Association proposed its own plan for the industry which aimed to reduce minesite costs by 20% and increase export volumes by NSW producers by 5%; a 20% reduction in NSW Government charges, including royalties, port charges and rail freight rates was a key part of the proposal. Association CEO Barry Ritchie argued that if the plan was adopted, retrenchments in the industry could be less than the 3000 that had recently been forecast by the JCB.[213]

The coal industry in NSW received some temporary relief in 1987 when the NSW Government reduced the standard royalty rate to $1.36 per tonne, a cut of 20%; the reduction, however, was for only a 12-month period. The lower rate was then applied for a further 13 months following pressure from the industry. The changes in 1987 also saw the super royalty abolished for underground mines and reduced to $0.50 per tonne for open cut mines. In November 1987 the NSW and Federal Governments agreed on some further assistance to the coal industry, with each to contribute $20 million to reduce freight charges for efficient producers. A few days earlier, Federal Cabinet had agreed to provide $20 million to the industry conditional on NSW Government providing $40 million in assistance, but this proposal was rejected by Premier Barry Unsworth who wanted equal contributions from both Governments. Unsworth said that there would be significant reductions in coal freight rates for those mines which were producing coal efficiently and which were able to help the SRA transport the coal efficiently.[214] The funds were intended to ease the financial cost for the SRA in reducing freight rates, and encouraging investment in more efficient coal loading facilities which would minimise the time to load trains. The assistance from the Federal Government was slow in coming, with half of the $20 million still unpaid in June 1989 when Commonwealth Minister John Kerin announced that his Government was withholding $10 million because of a decision in May that year that rail freight rates would rise by 5%. NSW Transport Minister Bruce Baird accused the Commonwealth of reneging on the agreement and said that the 5% increase had been agreed with the NSW coal industry and came after rates had been reduced by 20% in real terms over the previous three years.[215] The $10 million is believed to have been paid to NSW in due course.

NSW royalty hike in 1989

With a little more optimism in 1989 about coal prices, the Liberal Government raised the royalty rate once again in August; it was back to $1.70 per tonne. Barry Ritchie, NSW Coal Association CEO, slammed the Government over the increase, saying that the industry's experience had been that all governments were greedy and also that "there's a climate which is anti-development — that says 'we are happy enough with what's here, so we won't bother to develop it. We'll tax it heavily and bureaucratic structures will hamper it' ….There is a perception that the coal industry is a sort of cargo cult. Because there are minerals in the ground, the people of NSW must get some benefit and because the resource is a mineral, there must be a great store of wealth."[216]

Premier Nick Greiner argued that the reality was that the industry had been receiving "massive subsidies for years" and that "If the coal industry was serious …they would have said something three months ago…They are now just going through the motions. They are too late".[217]   He later said that the restoration of the royalty rate to $1.70 "should be regarded as representing the end of the extended concessional period rather than an arbitrary increase in royalty" and that the Government's decision "was based on the improved performance of the coal industry during 1988-89 and the prospect that this improvement will be sustained."[218] From the Association's standpoint, the hike in the royalty rate was an opportunistic decision by the Government, made at a time when there was an expectation that the industry was recovering from years of low prices and lack of new mining developments. As for the Premier's claim about "massive subsidies", the Association believed that given the losses made by many companies during the 1980s, the temporary reduction in the royalty rate was hardly a subsidy and nor were reductions in excessive government charges. In fact, with monopoly profits built into rail freight rates for Hunter Valley producers, there was a significant subsidy by the coal industry being paid to the Government.

Company takeovers, comings and goings

Changes in the major companies in the coal industry have been a constant feature of the industry for a number of decades, and certainly since the 1960s when Coal & Allied emerged from the merger between JABAS and Caledonian Collieries, and Utah began to explore for coal and develop its major projects in the Bowen Basin. The 1970s and 1980s were particularly notable for corporate changes such as BHP's purchase of Utah, the surge in interest in the industry following the oil shocks and the belief by oil international companies and local companies such as CSR that coal represented a once in a lifetime opportunity to invest in an industry with major strategic growth and profit potential. Many of the hopes and plans of major investors, however, would prove to be based on shaky foundations.

BHP buys Utah

Utah Development, the Australian coal mining company, was a subsidiary of Utah International which in turn was part of the General Electric group. GE had secured Utah International in 1976, but in 1982 its Chairman, Jack Welsh, decided that he wanted to sell off Utah as it did not fit with his vision for the company. Welsh wrote that "Utah was a highly profitable, first class company that derived its income largely from selling metallurgical coal to the Japanese steel industry. It also had a small U.S. oil company and large proven but undeveloped copper reserves in Chile….To me …it didn't fit the objective of consistent income growth …I felt the cyclical nature of Utah's business made our own goal of consistent earnings impossible. I didn't like the natural resource business, where I felt events were often beyond your control …"[219] Welsh did not find a buyer among US companies, having offered to sell it to Pennzoil and having also talked to other potential buyers. He found little interest in the USA. However, his vice chairman, John Burlingame, "found what he considered the best strategic buyer for Utah" and that was BHP. "The Australian-based natural resources concern seemed the perfect fit."

In September 1982, BHP's board decided to enter into negotiations with GE to buy Utah Development, a decision which would rank as one of the most critical in the company's history. Negotiations between BHP and GE led to a letter of intent being signed in December 1982, by which time GE's price for the sale had been reduced to $US2.6 billion with some of the Utah assets no longer part of the proposed deal. The purchase now had to be agreed by the BHP board, but at its December meeting, it advised GE that it was unable to give approval to proceed with the deal. The price was too high. GE was shocked, but agreed to negotiate and remove some of the assets from the deal, including Utah's oil subsidiary. However, the BHP board decision in December 1982 proved to be just a temporary setback, and in January 1983, GE and BHP announced that they had a tentative agreement that BHP would to buy Utah for $US2.4 billion (around $A2.5 billion).

The proposed takeover raised the ire of NSW Premier Neville Wran who had concerns over BHP's future in the state as a steel producer and over moves by BHP to retrench significant numbers of workers from its Newcastle steelworks. Wran said that he would call for an inventory of BHP's coal mining leases in NSW and would expect BHP to make its intentions clear about continuing as a mining company in NSW.[220] In what can perhaps only be described as unfortunate timing, on the same day that GE and BHP had announced the Utah agreement, BHP told steel industry unions of the retrenchment of 1500 workers from its Newcastle steelworks. These job cuts, however, should not have come as a complete surprise, as not long before BHP had told an Industry Commission inquiry into the steel industry that it would be cutting jobs in Newcastle due to the depressed state of the steel industry and the continuing poor financial performance of BHP's steel division.[221]

The purchase took some time to be completed, with one of the major hurdles the establishment of a financial vehicle which would take ownership of a significant part of the Utah assets. In 1983, the coking coal market was depressed and investors generally were reluctant to invest in a major coal company. To spread the cost of the deal, BHP set up Queensland Coal Trust (QCT) which, with the support of the Utah Mining Australia Limited board, and subsequently its shareholders, took over Central Queensland Coal Associates (CQCA). UMAL was the listed company which owned a share of the Utah Development Bowen Basin mines, with CQCA and other shareholders owning the balance of equity. QCT was then listed on the stock exchange. The complicated financial arrangements and re-jigging of equity holdings saw BHP now owning 35% of the Utah assets (including the Blackwater, Goonyella, Norwich

Park, Peak Downs and Saraji mines); Mitsubishi owning 12%, QCT 21.75% AMP 7.75%, and General Electric initially having 15.5%. BHP held 47% of Gregory, and 58% of Moura, Nebo and Riverside. Mitsui remained a major shareholder in the Moura, Nebo and Riverside mines with a 20% equity. These shareholdings would change over time, for example, with GE selling out and BHP increasing its equity. However, with BHP the dominant shareholder and in charge of operations, it was now king of the Bowen Basin, and set to become king of the internationally traded coking coal trade. Other companies which took up equity in CQCA were Bell Resources (initially 5%, later increased to 10%) and Pancontinental Mining with 3%.

BHP also emerged from the Utah deal with a 30% interest in the Goldsworthy iron ore mine in Western Australia and ownership of major iron ore reserves in the Pilbara. These assets added to BHP's existing investments in iron ore, including its 30% interest in Mt Newman. BHP now also owned coal mines in the USA and another jewel in the crown from the Utah purchase, the Escondida copper mine in Chile.

Utah was possibly a major opportunity missed by Rio Tinto which was also sounded out as a potential buyer by GE; Rio Tinto's CEO Sir Alistair Frame said that he had rejected Utah as too big.[222] However, the Utah deal did open up the opportunity for Rio to expand its copper mining interests. Sir Robert Wilson, who became Rio's CEO and later its chairman, was working on a copper project in Panama in the early 1980s, and had his eyes on the Escondida project which he believed he might be able "to snaffle from under BHP's nose."[223] Wilson thought that GE was placing no value on the Escondida copper reserves because of the huge cost of developing the project. He later met BHP executives but did not succeed in convincing them to sell Escondida to Rio Tinto. However, in 1984 Texaco, which owned a major stake in Escondida, put it up for sale. Wilson and then Rio Tinto CEO, Derek Birkin, negotiated a joint venture with BHP to develop the mine, with Rio Tinto gaining a 30% share of the project.

John Prescott, who at the time of the Utah deal was BHP's general manager of transport, and later became its CEO, gave his managing director, Brian Loton, with much of the credit for pushing ahead with the Utah acquisition: "Brian Loton's vision of the Utah acquisition was that it should be used to expand the company overseas…Although the principal assets GE were offering were the coal mines in Queensland, Brian saw that Utah should be used to diversify BHP out of Australia because BHP was outgrowing its Australian market: steel was much more stagnant than it had previously been, the heydays had gone, there was a Labor party in office in Canberra and after the early experience of a Labor Government in the early '70s nobody was quite sure just how socialist they'd be and how much control they'd try and exercise. It turned out that they didn't, but we didn't know that when we were acquiring Utah. But Brian had this vision and he managed to implement it."[224]

When the purchase was finally completed with the hand-over of a cheque for $US2.268 billion in April 1984, Brian Loton said that he was confident that the excellent quality of the Utah assets would boost BHP's profitability in the short term. Loton also rejected speculation that the deal had been thrown into question in 1983 when BHP was struggling to secure investors for the CQCA joint venture which would own the assets. "I never doubted for a moment that the agreement would take place," Loton said. "We knew Utah wanted it, General Electric wanted it, and BHP wanted it."[225] Loton also pointed to the future, saying that BHP was now entering a new era, as the deal had transformed it into a "moderate sized international company."

The Utah purchase came at a critical time for BHP. It had shown foresight in the 1960s when it teamed with Esso to develop the Bass Strait oilfields. BHP's steel operations, its mainstay since it moved out of Broken Hill, were struggling, with no improvement in sight. But now Bass Strait oil and the Bowen Basin coking coal assets would underpin BHP's business. In 1985, BHP and Shell purchased Woodside Petroleum, and Escondida was another huge project that BHP was able to add to its stable of world class assets. BHP became a moderately sized international company in 1984, and was on the path to becoming the world's largest mining company.

CSR bursts on to the energy scene

In the early 1970s, CSR Limited was a widely diversified company, with interests in sugar refining and marketing, mining, building materials, shipping, quarrying and ready mixed concrete. It was the largest of several sugar refining companies, handling about 20% of the local crop, and responsible for marketing the great bulk of Australia's sugar crop under contract to the Queensland Government. Its mining interests included 20% equity in the Mt Newman iron ore mine, and a major interest in the Gove alumina mine. Originally called Colonial Sugar Refining Company, CSR had a long history in Australia and in Fiji. CSR agreed to sell its Fiji assets to the Fiji Government in 1971.

But with the changing energy scene in the early to mid-1970s, CSR began to envisage a much grander future. In 1973, CSR was interested in the Redcliffs petrochemical project in South Australia, but it withdrew as a potential investor in 1974. In 1973 it began its move into the coal industry, purchasing 50% of Buchanan Borehole Collieries, a NSW company operating the Lemington open cut and underground mines near Singleton and an underground mine near Newcastle. In 1974 it increased its stake in Buchanan Borehole to over 92%. By the 1972-73 year, CSR was earning around 34% of its profits from its major mining investments in iron ore and alumina investments, and this proportion was destined to rise as the company's mining interests expanded to include copper mining and tin mining in Indonesia. In 1975 the company's directors said that the company's coal mines in the Hunter Valley were expanding, and the company was continuing to look for ways to expand its interests in the resources industry.

In 1974 Gollin & Co, with backing from Japanese trading company Toyomenka, commenced construction of a new export coal terminal in Newcastle. However by 1975 Gollin was in severe financial difficulties and in 1976 was placed into the hands of a liquidator. A consortium was formed to take over the project and to complete the construction, with the new company called Port Waratah Coal Services (PWCS). The owners of PWCS were CSR (25.5%), Japanese interests (30%), and a number of

Australia coal producers owning the balance of shares (including Peko Wallsend, Thiess, Coal & Allied, R W Miller and Bloomfield Collieries). CSR also won the right to manage and operate the terminal. The terminal opened in 1977, lifting the port's coal export capacity from around 8 to 20 million tonnes per year, and allowing the Hunter mines to significantly expand exports.

CSR's interests in coal grew over the next few years to include 50% of Western Collieries, the coal miner located in Collie, south of Perth, which had contracts to supply the State Electricity Commission of WA and industrial users including the Worsley alumina plant. In 1977 CSR won the battle with CRA to buy a major interest in the Hail Creek project in the Bowen Basin. The steaming and coking coal resources at Hail Creek had been discovered by Don King, the geologist who had been instrumental in Utah's major early discoveries in the Bowen Basin. King joined AAR after resigning from Utah and his discoveries also included the Yarrabee anthracite deposit. CSR gained control of AAR and its coal interests in 1977.[226]

In 1979, CSR announced an offer to purchase shares in Thiess Holdings, the iconic Queensland company which had pioneered export development in the Bowen Basin and which had grown into a major coal producer. Thiess, as a local Queensland company, with its headquarters in Brisbane, was one of the largest locally owned companies in the state, and one which Queenslanders did not want to see pass into the hands of a company located south of the border. CSR quickly obtained around 16% of Thiess, putting it into a strong position to move to get control of the company.

Queensland Premier Joh Bjelke-Petersen said "We are not happy…I don't like to see control of a firm with coal, engineering and vehicle sales activities going to the south."[227] In fact the CSR bid for Thiess was the second that year for a local company, Henry Jones (IXL) having bid for Provincial Traders earlier in the year, and legislation had been passed requiring the approval of the board of a target company for any bid from an "interstate" company.[228] But despite being unhappy with the takeover, Bjelke-Petersen said that his Government would not be legislating to prevent a takeover.

Sir Leslie Thiess and his fellow directors initially rejected the CSR offer, but in November CSR launched a formal takeover bid for Thiess which valued the company at close to $500 million. The offer included a cash component and a component of CSR shares. By late 1979, the Thiess board's resistance was waning. Sir Leslie said that after CSR had purchased its initial stake in Thiess, serious offers were received from foreign companies at prices which were consistent with the independent valuation obtained by the Thiess board; discussions had also taken place with some local companies.[229] By January 1980 the Thiess directors had to accept that the bid would succeed and recommended that shareholders accept the CSR offer. Sir Leslie said that the Federal Government's foreign investment policy had precluded any possible overseas bid for the company, and with recent increases in CSR's share price, the offer was now more palatable. Thiess then passed into the hands of the Sydney-based CSR, a bitter blow to Sir Leslie, who said that the loss of his company was the worst day of his life.[230]

CSR was now one of the largest public companies in Australia, ranking third behind BHP and CRA in market capitalisation, and just ahead of MIM. It now had control of an impressive stable of new mines and projects. Callide and South Blackwater were owned and managed by Thiess, and Thiess held a 22% stake in Thiess Dampier Mitsui which owned the Moura and Kianga mines and had the rights to develop the Nebo project. The takeover of Thiess also allowed CSR to increase its interest in the Theodore project and in the Rolleston and Yarrabee projects in which CSR has already gained equity when it bought AAR in 1977. In about 7 years, CSR had gone from a standing start to become one of the major coal producers in Australia. And it did not stand still, in 1981 winning NSW Government approval to develop the Drayton open cut mine in the upper Hunter Valley, near Muswellbrook, with a plan to produce 3.5 million tonnes per year by 1990. CSR also brought other companies into its mines, with Shell acquiring a 30% interest in Callide and a 40% interest in Theodore, and AMP a 15% interest in Callide. CSR was also involved in the Winchester South prospect with BP, and in the Wandoan prospect with MIM.

In 1980 the company's plan was to double in size through the expansion of its energy division. CSR's coal mines were producing around 8 million tonnes a year, and plans were being developed to increase production to 28 million tonnes a year by 1985, with expansion of the capacity of the NSW Buchanan Borehole mines and from Callide and South Blackwater in Queensland, and from development of new mines including Drayton in the Hunter Valley, and Yarrabee and Theodore in Queensland.[231] The Drayton mine near Muswellbrook in fact received approval to proceed from the NSW Government in 1981. Drayton, planned to be developed to reach a capacity of 3.5 million tonnes a year, was another interesting story from the 1980s, one we will come back to and which involved Miners' Federation bans and a struggle between members of the union to allow the mine to produce and export.

By the 1980s, CSR could also claim major investment in oil and gas (through the Roma gas fields in Queensland, the Roma to Brisbane pipeline, a pipeline in the Northern Territory, a gas project in central Queensland, and an interest in an oilfield in Indonesia), and aluminium smelting through its investment in the Tomago smelter north of Newcastle. It also had the Julia Creek oil shale project on its books, a potential $5 billion investment, but which Energy Division head Gene Herbert said was unlikely to see large scale production before 1990.[232] But as we now know, the 1980s were to prove a difficult time for minerals and energy companies, and by the mid-1980s CSR was feeling the pain, as were many other coal mining companies which were losing money or making meagre profits. Kelman acknowledged that CSR's investments in coal, oil and gas and aluminium had not produced the returns that had been expected.

By 1985 the restructuring of CSR had begun and it would emerge in a few short years as a much smaller company concentrating on sugar and building materials. In 1985 CSR sold its interest in Mt Newman iron ore and in coal miner Thiess Dampier Mitsui, and an investment in a gold project; in 1986 it restructured the debt which had been a major drag on its Delhi oil and gas assets in the Cooper Basin in South Australia. Like most producers, CSR's coal operations continued to disappoint in terms of their profits, although in 1987 the company's Coal Division chief, David Sawyer, said that the high proportion of domestic sales had cushioned the group's coal revenues from many of the problems experienced by other coal producers in the mid-1980s. However, Sawyer also said that profits continued to be unacceptably low. At that time the company was not looking for a quick exit from coal mining, with Sawyer saying that its objective was to establish its coal business predominantly in the domestic market as a "profitable, competitive, free-standing and largely self-funding activity." He said that this may lead to an eventual public float of the coal operations, but that profitability would have to improve if such a float was to be successful.[233]

In 1987 CSR and its partner Exxon closed the Lemington underground mine in the Hunter Valley. Exxon subsequently purchased the CSR share in the Lemington open cut. The year 1987 also saw the sale of the Delhi oil and gas assets to Exxon (which operated its Australian oil interests as Esso), and in October 1988 CSR sold its remaining oil and gas assets to AGL, including 80 per cent of the Roma gas fields in south Queensland and a major share in a pipeline in the Northern Territory and an interest in an oil field in Indonesia.[234] In 1988 the South Blackwater mine, which had been developed by Thiess in the early 1970s, was sold to John Holland Holdings and Pennant Holdings for the princely sum of $42 million. Gene Herbert, the deputy CEO, said that the sale would be favourable for CSR as the mine had incurred substantial losses in the preceding few years, and that the company was negotiating the sale of its remaining coal assets.[235]

In 1988-89 CSR found a ready buyer in Shell for its remaining key coal assets, which included the Callide mine in Queensland and the Drayton mine in NSW. CSR had now come full circle, back to its core business of sugar and building materials. It had ridden the wave of optimism regarding the coal industry in the 1970s and into the 1980s, only to find that profits were meagre. Selling out of the coal industry between the mid and the late 1980s must have been painful, given the millions that CSR had invested in its coal business. But CSR's exit opened the door for other investors, notably Shell, a company which would also have high hopes of establishing a significant profitable coal business. But Shell would also be gone by 2000, another company that was to be part of the passing parade of coal owners.

The oil companies invest, then sell out

In the 1970s, as the oil shocks led to many companies questioning their future role and their strategies, the major oil companies saw opportunities through investing in coal mining. The major focus of these companies was steaming coal, which came to be seen as a major growth fuel for power generation. While Australia was not the only location for the attention of the oil companies, it became their major focus, along with Colombia and Indonesia. Shell, BP, Exxon and ARCO went on to become significant players in the Australian coal industry in the 1980s and 1990s, although by the early 2000s, they had all waved goodbye to Australia for various reasons. There were also other oil companies which played a lesser role in the coal industry, including Caltex and Agip; they too would depart the scene by the end of the 1990s or soon after.

BP made its first major foray into the local coal industry in January 1977 when it acquired 50% of Clutha with its substantial stable of coal mines located in the NSW Burragorang Valley and its Foybrook underground mine in the Hunter Valley. In July 1978 BP then moved to take full ownership of these assets, this change seeing the end of Daniel Ludwig's period as a significant investor in Australia. BP then developed the Tahmoor mine near Picton and in 1979 it purchased a 49% share of the Clarence colliery near Lithgow. During the 1980s BP's Hunter Valley interests expanded with the Howick and Foybrook open cut and Pikes Gully underground mines. In 1986 BP's Hunter Valley mines included two underground mines (Pikes Valley and Foybrook No. 1) and an open cut operation comprising the Howick and Foybrook leases. Howick was being expanded to supply a new contract it had recently won from the NSW Electricity Commission. BP closed Hazeldene (the Foybrook and

Pikes Gully underground mines) in 1987 and by 1989, only Howick was still operating.

The Clutha mines in the Burragorang Valley proved problematic and several were closed by BP in the early and mid-1980s as coal prices fell and losses mounted. In 1982 it closed two mines (Brimstone 2 and Valley 2) and cut production at Nattai Bulli. In 1983 over 200 workers were retrenched from 3 of BP's remaining 5 Burragorang Valley mines. In 1984 Nattai Bulli and Brimstone 1 were closed, although the process was delayed when the NSW Minister for Mineral Resources Peter Cox initially refused to allow the closures. By 1989, only Tahmoor remained of the southern NSW mines acquired by BP.

Disappointment with its Australian investments however did not deter BP from investing in coal mining in Indonesia and in 1988 BP and CRA announced their Kaltim Prima joint venture in East Kalimantan, an export coal project designed to produce 7 million tonnes a year, making it one of the largest coal mines in the world at that time. But the following year, BP made a corporate decision to sell its minerals division, and BP's exit from the Australian coal industry came with CRA agreeing to buy the Howick, Tahmoor and Western Main mines in NSW and BP's 50% interest in the Winchester South project in Queensland. Oakbridge purchased BP's interest in the Clarence mine. BP's exit from Australia followed years of losses from its coal operations, which overall amounted to several hundred million dollars.

Shell's first major move into Australian coal was in 1976 when, together with Thiess, it announced a plan to develop the Drayton mine in the Hunter Valley commencing in 1977, although this project was then delayed for several years. In 1977 Shell bought 25% of NSW coal miner Austen and Butta Limited (A&B), then a listed company, with mines in the Lithgow and NSW south coast areas, and an interest in the German Creek project in the Bowen Basin. It also made a share placement which saw its equity in A&B increase to around 37%. Just days before the A&B acquisition, Shell said it was also planning to buy MIM's 16.5% share of Queensland miner Thiess.

Austen and Butta's interest in German Creek was held through its 26.7% equity in Capricorn Coal Management (Capcoal). Shell subsequently became the major shareholder in Capcoal when it acquired a 43.3% stake. Shell sold down its direct stake in Capcoal to around 17% by 1983, and subsequently rationalised its Capcoal equity, holding around 38% by 1990. By 1980-81, Shell had acquired a 30% interest in the Callide mine, with Thiess (now a CSR subsidiary) the major shareholder with 55%; Shell also had a 40% interest in the Theodore project, with Thiess holding the other 60%. During the 1980s, Shell increased its equity in A&B to almost 50%, and in 1985 appointed Ian Dunlop as CEO in place of Dick Austen who remained chairman. Dunlop had been Deputy CEO for three years, prior to which he was employed by Shell. The 1980s were difficult years for the company, with significant losses sustained, and some mines closed, but Shell had supported A&B during those years.

In early 1992, Shell made an unfriendly takeover offer of 70 cents per share for the A&B shares it did not own. The price at the time of the offer was $0.53, but Dick Austen, one of the founders of the company along with Angelo Butta, believed that the offer seriously undervalued A&B. However, in June 1992, A&B reported a loss of over $10 million for the eleven months to May, confirming the seriousness of the company's financial position. After a long history of working with Shell as a major investor in A&B, Austen was bitter about the takeover offer; he felt that Shell had turned from a friend into a foe, with the control of the potentially valuable Dartbrook steaming coal project in the Hunter Valley its major goal.

A&B had begun in 1950 as a one truck business carting coal in the Lithgow area, with the business later buying a small local open cut coal mine, and became a public company in the 1970s. Austen told journalist Bruce Hextall that it was the potential of the Dartbrook deposit which turned Shell from a friend into a foe, with Shell "prepared to make a Rambo-style attack when it was obvious that (A&B) was vulnerable."[236] Austen said that he believed the industry was at the very bottom of the cycle and that a recovery was in sight. After a bitter fight involving court battles, Shell gained control of A&B, but the 1990s would, contrary to Dick Austen's optimistic prediction, prove to be another very tough decade.

Exxon was the third major oil company to become a significant producer in Australia, although its involvement was not as extensive as that of BP and Shell. Esso, part of the Exxon group, developed the Bass Strait oilfields with BHP from the 1960s, and the company's investments in coal began in 1977 when it purchased a 25% interest in the Hail Creek project in Queensland. In 1981 the Federal Government approved Exxon's purchase of a 49% interest in the Gloucester steaming coal joint venture in NSW. Exxon's subsequent investments included 50% of the Lemington underground mine in the Hunter Valley which it purchased from CSR in 1986. In 1989 Exxon purchased the coal interests of the White family, including 36% of the Ulan mine near Mudgee in NSW, 55% of the Clermont project in Queensland, and 20% of the United mine in the Hunter Valley.

In 1993 Exxon sold its interest in Clermont to CRA and Arco. The writing was on the wall for Exxon's departure from Australia in 1997 when the company announced that it was testing the market for a sale of its Australian assets which at that time included Ulan, the Lemington open cut, the Mt Thorley coal loader (a rail loading facility for several mines near Singleton) and a share in a zinc and copper mine in WA. No sales of its coal assets eventuated, but in 2000, Exxon announced that it was putting its interests in Ulan and Lemington up for sale. With the market looking a little brighter and Glencore rapidly expanding its coal business, Glencore purchased Ulan in 2001. Glencore's Australian coal CEO, Peter Coates, had been the CEO of Ulan Coal Mines in the 1990s and so knew the asset well. Lemington was purchased by Coal and Allied and integrated into its Hunter Valley Operations group. Shortly after, Exxon also sold out of the huge Cerrejon North coal mine in Colombia.

Some of Arco's experience Australia is covered earlier in the next chapter in relation to the Gordonstone battle. Arco developed that mine in the 1990s, only to close it and sell it to Rio Tinto in 1998-99; Arco also sold out of its other major Australian mine, Curragh, in 2000. Arco also operated coal mines in the USA which were sold in 1998. It later became part of BP.

The saga of the oil majors and coal is an interesting one, and it is easy to be critical of the companies for their decisions to invest large sums into coal mining in Australia and other countries and for the losses which they incurred. However, in the 1970s in particular, steaming coal was correctly seen as playing a major future role in power generation in Asia, with coking coal also having a bright future, although in a market expected to grow more slowly. What the oil companies failed to foresee, as did most other players in the industry, was the impact the oil shocks of the 1970s would have on economic growth in the 1980s, and the significant increases in coal supply capacity in Australia, Indonesia, Colombia and to a lesser extent South Africa which came in response to the increases in coal prices in the 1970s and early 1980s. The growth in the Canadian coking coal industry, supported by the Japanese, was another factor acting to restrain coking coal prices. And there was the inability of the Australian coal industry to become more efficient and move down the cost curve so that it could compete more effectively.

The oil majors – Shell, BP and Exxon – have survived and generally prospered as oil and gas producers and marketers; Exxon merged with Mobil in 1999 and is now ExxonMobil. Coal was only ever a small part of their operations and in and in a sense was a diversion from their principal business.

Peko Wallsend rocks the status quo

Peko Wallsend was a miner with ancestry dating back to the early days of coal mining in Australia. It was formed by the merger of two companies in 1961 - copper miner Peko Mines NL and coal miner Wallsend Holding and Investment Company Limited. Wallsend traced its lineage back to the Newcastle Wallsend Coal Company which commenced production around 1861 and owned two collieries near Cessnock, Pelton and Gretley. By the 1970s, Peko Wallsend had become a diversified miner, with interests which also included mineral sands, scheelite and iron ore in the

Northern Territory. Peko Wallsend moved further into coal, buying 51% of Gollin's Gunnedah colliery in 1976, and a 51% interest in Gollin's coal trading operations. In 1976 Peko was also developing the Ellalong colliery adjacent to Pelton.

It was in the 1980s that Peko gained, depending on one's view, notoriety or fame, when it became a major player in the struggle between unions and management through its involvement in the iron ore industry in the Pilbara region in West Australian. Peko also rose to prominence when it was involved in a battle with the NSW Government after it attempted to purchase the Saxonvale mine in the Hunter Valley from BHP. Peko was a company not well liked by unions or the Labor Governments of the day. Peko bought a 35% interest in the Robe River iron ore mining and shipping operation in December 1983 when it took over Robe River Limited, a listed public company. The operation was managed by a US company which also held a majority of the equity. Peko's chief executive, Charles Copeman, said that his board was told by Peko management that, given its level of production, Robe River employed too many people and had a reputation for buying industrial peace by giving in to union demands too readily, in particular to ensure it was able to meet its shipping program.[237] In 1984 Peko undertook a comparison of employment levels at Robe River with similar Australian operations and overseas operations which, Copeman said, confirmed that there was in fact an excessive number employed throughout the operation. Peko discussed its research with Robe River management who rejected the findings and commissioned independent consultants to carry out another assessment. The consultants confirmed Peko's findings, Copeman said. Peko did not have access to the consultant's report, and Copeman said that the local executives were reluctant to allow Peko to undertake its own investigations or to discuss sensitive industrial relations issues with employees or union officials. This was clearly a frustrating time for Peko, but in January 1986 it gained a majority share of the operation when the US company sold out, with Japanese interests buying 45.5%. Peko also gained management rights to the operation.

When Peko gained control of Robe River, the WA Government

and Japanese shareholders were concerned to ensure that no changes were made to the management or to the project. Copeman says that Peko agreed not to make any immediate changes, but also that it would be unrealistic for this commitment to bind it into the future, and that it stressed that its majority interest gave it the responsibility for management of the project. The iron ore industry in the 1980s saw significant levels of industrial disputes and this gave the major customers cause for concern over the reliability of supply. The full story of Peko and Robe River is a complex one, involving among others, the WA Labor Government, the WA Industrial Commission, the Appeals Court, and the Robe River employees, unions and management. But the essence of the story from Peko's point of view was that it was a company which had made a significant investment in iron ore and was frustrated by its lack of ability to manage its own operations, including making decisions on employment levels and working conditions. It also had a mine where the operations were rife with restrictive work practices and rorts. Peko saw the unions and their officials and delegates on site as effectively taking over much of what would be regarded in other industries as the prerogative of management, with adverse consequences for efficiency and the cost structure of the whole operation. Peko eventually won the right to manage the operation in practice, and between July 1986 and January 1987 was able to reduce the workforce from 1680 to around 1250. Copeman said that productivity had doubled as a result of its changes, and huge savings were made in capital expenditure and in operating costs. But Peko emerged from the battle carrying a stigma in the eyes of many in the union movement.

Saxonvale was an open cut mine which was developed by BHP from 1981 and which officially opened in 1983. BHP was reported as intending to use Saxonvale's coal to feed its blast furnaces in Newcastle, but the company found that the coal was unsuitable and then obtained NSW Government approval for the mine to become an export operation. After three years of losses, however, BHP decided to sell the mine, telling employees of its decision in November 1986. Peko offered to purchase Saxonvale from BHP in 1987 for about 40% of what BHP had spent to develop the mine. However, the NSW Government's Mines Minister, Ken Gabb, inserted a condition in the lease that required that a minimum 236 people be employed at the mine. This followed a CIT decision that required Peko to maintain the existing conditions for the employees, and which rejected its bid to overhaul work practices.[238]

Peko was planning to expand its coal interests in the full knowledge of the prevailing difficult market conditions, but on the basis that it could make Saxonvale's cost structure competitive at a time when coal prices were low and losses were common. Peko took the unusual step for the time of an advertisement in the media, headlined "NO MINISTER" which questioned Gabb's demand on employment and said that Peko was prepared to employ the required number of people provided it was allowed to "eliminate the inefficient and wasteful work practices" which it said were "largely responsible for Saxonvale's current plight."[239] Peko refused to sign the lease on the Government's terms and the deal fell through. While Peko and the NSW Government were holding discussions, BHP had signalled that if Peko did not take over the mine, it would have to make cuts to the workforce, and once Peko had walked away from the deal, it wasted little time in retrenching over 130 Saxonvale employees.[240] Gabb's decision had deterred Peko from gaining control of Saxonvale, but at a significant cost to jobs.

But the Saxonvale saga had further to run. Peko, which also owned the Bulga prospect adjacent to Saxonvale, and which it had described as "the most attractive coal area awaiting development for export sale in Australia", was taken over by North Broken Hill Limited in March 1988. NBH was not particularly interested in Peko's coal assets, and found a willing buyer in Elders Resources which had purchased the Saxonvale mine from BHP. Elders now owned the Newcastle Wallsend underground mines, Saxonvale, the Bulga prospect, half of the Hebden joint venture with Coal & Allied, and interests in the Newcastle export terminals. Elders also held an option over 20% of coal miner Oakbridge , which owned the Clarence and Baal Bone collieries in the Lithgow area. The time was right for Elders and Oakbridge to consider rationalising their coal assets and the companies did hold talks in 1989. In 1990 Elders launched a takeover bid for Oakbridge but lost out to shipping and transport group McIlwraith McEacharn and its two Japanese partners Toymenka and Nippon Oil. McIlwraith now had the management of Oakbridge and its stable of mines.

McIlwraith had a long link to the coal industry as one of the major shipping companies in Australia in the late 1800s and early 1900s which controlled the interstate shipping of coal and which were involved in the famous Vend case in the 1910 to 1912 period. But McIlwraith's re-entry into the coal business in the 1990s did not last for long. In 1993 US company Cyprus Minerals bought a major share in Oakbridge from some of its major local shareholders, including TNT. McIwraith chief executive Tony Lawrance said that the Bulga mine needed additional capital expenditure of $200 million over the next few years to develop it to its full potential, and that TNT was a reluctant investor in areas such as coal which it saw as non-core operations. In 1994 Cyprus moved to increase its stake in Oakbridge and gained control of the management of the company's mines. Cyprus then had a relatively brief period in the industry before selling out to Glencore and Centennial.

Peko was one of the passing parade of coal investors in the 1980s, but it was an important one, having played a key role in challenging the status quo in Australia's industrial relations system which prevailed at that time. Under Copeman's leadership, the WA iron ore industry was never the same, and the changes to Robe River reverberated far beyond the remote Pilbara region.

CRA, the Blair Athol saga and foreign investment policy

The possibility of the large scale development of the Blair Athol deposit in the late 1940s was one of the bright spots in the industry in Queensland, with British interests looking at developing a 3 million tonnes per year mine, with a rail link to the coast. This proposal did not proceed, and mining on a small scale continued until the two companies then mining at Blair Athol were taken over by a joint venture between CRA and Clutha in 1968. Clutha had earlier purchased 25% of the two companies operating mines at Blair Athol, and agreed to take a 40% interest in the new joint venture with CRA. Clutha wanted to develop Blair Athol as an export mine linked to a deep water port, and was looking to a deal to supply the Californian Gaslight Company with coal for power generation.[241] However, the investment required for the project and the lack of an assured market for its steaming coal saw no progress made for the next ten years.

In March 1974 CRA's chairman, Sir Maurice Mawby, told the company's annual general meeting that with the dramatic changes in the world energy market, the company was reappraising its approach to investment in uranium and coal. CRA already had interests in those sectors, although its coal assets were still relatively modest. Later that year it was reported that CRA's plans for Blair Athol were for the project to be developed into a mine producing up to 10 million tonnes per year, largely for export. CRA was also reported to have signed a letter of intent with a group of Japanese companies headed by EPDC, and including Nippon Steel, for the supply of 8 million tonnes per year. However those plans hinged on the removal of Japan's embargo on the importing of steaming coal and on approval by the Australian Government.[242] The lifting of the Japanese embargo was announced in January 1974 but it was only a partial lifting, with the tonnages allowed still very limited. And in 1974 the Whitlam Government's foreign investment policy of promoting the maximum level of local equity in major resource projects would have precluded CRA from proceeding with Blair Athol.

In the second half of the 1970s, with CRA looking to expand its interests in Australia, the company, although listed on the Australian stock exchanges, was hamstrung by the fact that it was a foreign owned company, part of the London – based RTZ (Rio Tinto-Zinc Corporation) group. CRA was interested not only in developing Blair Athol, but buying into the Hail Creek project, and taking a major stake in the NSW coal producer Coal & Allied. In 1976 CRA shelved its plans to develop Blair

Athol on the basis of a lack of demand from Japan, the coal export levy introduced in 1975, and the less attractive tax regime in place after the Fitzgerald report and changes to the Tax Act. But the company's interest in Blair Athol was dormant, not dead, and would soon be revived.

CRA was certainly feeling unloved by the Government in the mid-1970s. In May 1975 the RTZ chairman, Sir Val Duncan, told shareholders in London that the atmosphere in Canberra towards companies like CRA, which had until recent years been regarded as British, but now as foreign, were not exactly assisting local management. This was despite the fact that CRA's management and staff in Australia were almost all Australian, and that CRA had almost 80,000 shareholders with Australian addresses. Duncan said that he hoped that the CRA operations would one day be judged by their contribution to Australia's welfare rather than the passports of their shareholders.[243]

In July 1977 CRA joined with Howard Smith to make a bid for all the shares of Coal & Allied, with Peko Wallsend making a counter bid which was later withdrawn when Peko realised that CRA and Howard Smith already held over 50% of Coal & Allied. The joint bid had been approved by the Commonwealth Government on the basis that CRA would agree to become a majority Australian company over time, and CRA amended its offer to shareholders in Coal & Allied to give them the opportunity to take up shares in CRA. CRA chief executive Roderick Carnegie (soon to be Sir Roderick) said that the share offer was part of a plan to make CRA more Australian. But the NSW Government then threw a large spanner into the works with its decision to award 51% of a mining lease for the large Warkworth deposit in the Hunter Valley to the NSW Electricity Commission. Coal & Allied was invited to take the other 49% of the project, but had expected to be granted 100% ownership of the lease and was planning to spend millions to prove up the deposit. Coal & Allied was no longer quite as attractive as a takeover target, and CRA now had to contend with a NSW Government which was not keen on foreign investors taking the majority share of major new projects.

Following discussions with the NSW Government, CRA formally

withdrew its bid for Coal & Allied in August 1978; it was clear that CRA's foreign ownership did not sit well with the NSW Government's policy of keeping the coalfields in Australian hands.[244] Howard Smith had withdrawn its bid in May, although by that time it had already increased its stake in C&A to around 50%. CRA would have to bide its time to swallow Coal & Allied; that would not come for another 13 years. And Coal & Allied would soon be awarded another major lease which would see it developing the Hunter Valley No.1 open cut, one of the largest mines in the Hunter.

The CRA/ Howard Smith bid for Coal & Allied and the conditions for the Government's approval were followed by a significant change in the national foreign investment guidelines in June 1978, when the Government introduced a novel concept of "naturalisation". Companies which had at least 25% Australian ownership, with a board composing a majority of Australian citizens, and which had undertaken to move to at least 51% Australian ownership could now be regarded as "naturalising" and therefore eligible to participate in new projects before they reached the 51% local ownership required under the previous policy. This policy change was directed clearly at the mining and resource industries which had been held back under the Labor Government's strict foreign investment policies, and the modest changes made by the Coalition Government in 1976. CRA, along with MIM and Consolidated Goldfields of Australia, all with a minority of local shareholders, were the major mining companies seen as having the potential to become naturalised.

But CRA did not simply accept the status quo in terms of foreign investment policy. CRA in fact had been lobbying the Government to change its foreign investment guidelines which it said did not adequately recognise the positon of foreign-owned companies which had increased their local ownership. CRA said that its proposals would increase local participation in foreign-owned companies and it proposed the concept of a naturalising company through the creation of three categories of companies: those with 25% or more foreign ownership to be regarded as foreign; those with 50 to 75% foreign ownership, but with an intention to progress to 51% local ownership, to be classified as naturalising; and those with over 50% of local ownership to be classified as Australian. CRA also proposed a new concept of control, whereby companies with 25% local ownership and a majority of directors who were local residents would be classified as Australian.[245]

The Treasurer, John Howard, rejected the CRA proposals on a number of grounds, including that naturalised companies (with 51% or more local ownership) would be regarded as 100% Australian, and would be able to undertake any new mineral project or take over other companies without restrictions, leading to the possible situation where new projects could have significantly less local equity than the required 50%. Howard also argued that CRA's new concept of control could lead to situations where there was still effective foreign control of companies. CRA's lobbying had not resulted in the full range of changes it was seeking, but, nevertheless, it was an impressive win for the company and for its determination to achieve a change in a critical area of public policy. CRA and its parent company RTZ agreed with the Government that CRA would progressively move to majority Australian ownership over time, with CRA becoming the first company to take advantage of the naturalisation policy. CRA had its eyes on some large coal projects, having been somewhat slow to that time in gaining a major stake in the coal business.

By the late 1970s, CRA's plans for Blair Athol were to develop a massive operation producing 8-10 million tonnes per year, but there was as yet no definite timetable for development. Prices for export steaming coal were around US$20 a tonne, and the costs of developing new infrastructure were seen as high; and Australian producers would need firm long term contracts to justify the investment required in new mines, rail links and ports. Don Carruthers, head of CRA's coal division, said in April 1978 that it was difficult to see steaming prices increasing sufficiently to support the infrastructure costs of new projects.[246]

Blair Athol's development into a major mine would not be confirmed for another two years, and in that time ownership changes involving Clutha, BP and Arco would play out before the final shareholders in

Blair Athol would be decided. Daniel Ludwig sold out of his Australian coal interests in 1977 and 1978, with BP buying 50% of the company's NSW Clutha assets in January 1977, and the remaining 50% in 1978. By that time Ludwig's financial interest in Clutha had been transferred into the Ludwig Cancer Research Institute based in Switzerland, with one newspaper article announcing the sale to BP saying that the Clutha story had a happy ending, and that "… at least the sale of New South Wales's richest coal company could have a major benefit to mankind in the fight against cancer."[247]

By this time American oil giant ARCO (Atlantic Richfield) was another company which had decided to get into the coal business in Australia, and with Japan expected to become a significant market for steaming coal, ARCO obtained Federal Government approval to buy Ludwig's 38% stake in the Blair Athol project held by S&M Fox Pty Limited. The Government's decision was seen as a rebuff for CRA which had been denied approval to buy Clutha's stake. However, the ARCO purchase got the green light on the basis that it would not lead to any increase in foreign ownership, but was simply a transfer of ownership from one foreign company to another. The project was, however, expected to increase its local ownership once development got under way.

In 1979 ARCO purchased Clutha's interest in the project, and EPDC of Japan signed a letter of intent with CRA to take an equity interest in the project and to enter into a contract for supply of coal to the Japanese power utilities. The basis for the project to finally proceed was set in February 1980 when CRA and EPDC signed contracts for the supply of 5 million tonnes of coal per year for 15 years, with EPDC having an option to take equity in the project. However, there was another hiccough two months later, when the FIRB rejected EPDC's bid to take a 19% equity in the project. Doug Anthony said that the Japanese request did not pay sufficient attention to the need to secure Australian equity in the project. Paul Keating, then Opposition resources and energy spokesman, argued that ARCO should never have been given the green light to participate in the project and that, to maintain Australian equity of at least 50%, there had to be a drop in ARCO's stake to accommodate the Japanese. At that time, CRA held 62% and ARCO 38%.[248] This was a major problem as the project was depending on contracts from Japanese power utilities which would flow once Japanese equity had been locked in. The shareholding structure was then re-jigged, with CRA reducing its stake to just over 50%, ARCO reducing its to just over 15%, new partners AMP and Bundaberg Sugar taking just over 12% each and the Japanese (EPDC and JCD) a combined 10%.

CRA then committed to the investment in the Blair Athol project in January 1981, and export sales contracts were signed in July for 72 million tonnes to be supplied over 15 years. Exports of coal commenced in 1984. It had taken almost 40 years to achieve the hopes expressed in the 1940s for a large scale development of this valuable deposit.

CRA and CSR fight for Hail Creek

The Hail Creek deposit, in the Bowen Basin south west of Mackay, was discovered in 1968. In 1969 Mines Administration Pty Limited (85% owned by AAR and 15% by a CRA subsidiary) was awarded a licence to prospect in the Hail Creek area. The company soon located thick seams of coking coal and went on to undertake a detailed evaluation of the deposit. From 1971 other companies joined the consortium, including Western Mining Corporation (WMC), Marubeni and Sumitomo Shoji Kaisha. A sale agreement was signed in 1974 with Japanese steel companies for the supply of 66 million tonnes of coking coal over 15 years commencing in 1978, but the agreement lapsed when development stalled. WMC withdrew from the project in 1975, and CRA increased its interest. In October 1977, the Government announced that it had rejected CRA's bid to increase its stake in Hail Creek to 30%. Doug Anthony, Deputy Prime Minister, and Phillip Lunch, Treasurer, said that they recognised the important role which CRA played in the Australian mining industry, the substantial level of Australian participation in the company, and its long association with Hail Creek. However, CRA's proposal would have taken the local ownership of the project to well below 50%, and this was unacceptable to the Government.[249] But the way was now clear for CSR to become the major player in Hail Creek and it was able to buy AAR, the company with the major stake in the project. Esso (a subsidiary of Exxon), now also keen to become a significant coal producer, received approval to purchase 25% of Hail Creek in August and Japanese companies 15 %.

CRA's chairman, Rod Carnegie, and managing director, Russell Madigan, issued a statement following the Government's rejection. They said that CRA had been a partner with AAR in Queensland for 16 years, the partnership having discovered the Hail Creek deposit, and that CRA had earlier been given permission by the Coalition Government to purchase, under certain conditions, an additional interest to bring its overall interest in the project to 43%. Carnegie and Madigan went on to point out that, in the proposal that the Government had just rejected, CRA had offered to actually reduce its interest to 30% and to increase the number of CRA shares held by Australians. CRA had spent significant funds as its share of the exploration costs of the project and had managed the project from April 1976 to August 1977, completing mining and engineering design work, negotiations for marketing and finance and undertaking negotiations with the Queensland Government. In what must have been one of the under-statements of the decade, Carnegie and Madigan said: "We are therefore naturally disappointed."[250]

Hail Creek was a major setback for CRA and a win for CSR. CSR, as a major Australian owned company, had been opposed to the relaxation of the foreign investment guidelines to allow companies like CRA to increase their involvement in Australian resource projects. CSR's chief executive, Gordon Jackson, was opposed to the naturalisation decision, and supported the foreign investment policy that then Treasurer Lynch had detailed in April 1976, which required a minimum 50% local ownership of projects and which was not substantially different to the policy of the previous Labor Government. CSR argued that the previous policy was working as intended and had not inhibited foreign investment.[251] When the Government announced the new foreign investment policy, Jackson slammed the decision, and said that ways should be found to make a company's promise to naturalise "irrevocable and enforceable".[252] CSR had won the Hail Creek battle with CRA, but lost the argument on foreign investment policy, and was now on the way to becoming a major player in the coal industry. CSR's move in 1977 was well timed, and by 1979 Hail Creek's partners were planning for the mine to be in production by 1984.

In January 1979 CRA announced that it was now formally moving to naturalised status, with the necessary changes to its articles of association to be put to the upcoming annual general meeting. RTZ and CRA had held discussions with the FIRB and reached the necessary understandings in relation to moving to reducing RTZ's shareholding to 49% of a period of time.[253] The path for CRA to become much more involved in Australian resource development was now clear.

CRA maintained its interest in Hail Creek, and in 1980 the stand-off between CSR and CRA ended when CRA agreed to sell its shares in AAR for an increase in its stake in the Hail Creek project to 25%. The deal between CSR and CRA also saw CRA sell its stakes in several other Queensland projects to CSR, including the Theodore steaming coal project, the Taroom steaming coal area, and the Yarrabee anthracite prospect. The new ownership structure of Hail Creek had been approved by the Federal Treasurer John Howard under the Government's foreign investment guidelines. Howard said that the Hail Creek project would, after allowing for the naturalising status of CRA, have an Australian equity of just over 56%, including CSR's 44%. Hail Creek was all set to go, but it would be another 20 years before construction began, by which time CSR would be long gone from the coal industry and Rio Tinto would be in control of Hail Creek.

Demands for a coal marketing authority

The demand by unions for a government authority to take responsibility for the marketing of coal has a long history in the Australian coal industry, but it was in the 1980s that these demands arguably reached a peak. The unions, and in particular the Miners' Federation, wanted to see the Federal Government take responsibility for coal marketing away from the producers and invest responsibility in a statutory authority. At times the demand was also for the authority to have the power to control the rate of development of the industry. The desire for a coal authority evolved over time out of the Federation's traditional policy of seeking the nationalisation of the coal industry. In 1972, the Federation put a proposal to the Federal Government for a National Fuel Board which would control the mining and marketing of energy resources including coal.[254] Nothing came of this proposal, and the election of the Labor Government in 1972 and the imposition of export controls by Whitlam and his Resources and Energy Minister Rex Connor saw arguments for a national authority fade away. But in 1979, with the Coalition Government in power, the National Conference of the Labor Party took a tentative step forward by resolving to accept the concept of a national coal authority as a long term goal.[255]

In the 1980s, many coal producers, while not necessarily supporting a government authority, also had major concerns over the way the Japanese went about contract negotiations, and the overly rosy forecasts of coal demand that the Japanese had made. In 1982, for example, Australian companies at the annual Australia Japan Business Cooperation Committee meeting in Kyoto were reported to have "lashed out at Japanese resources buyers for knowingly encouraging an over-supply of iron ore and coal."[256] The Australian companies "complained about their weak bargaining position against the cartelised Japanese steelmakers in resources negotiations and said Australian companies should be more coordinated – either by the industry itself or the Federal Government."

The Miners' Federation's desire for a coal authority came to the fore at various times in the 1980s, including in March 1983. Around 300 mine workers marched into the Sydney city offices of the JCB on 29 March and staged a sit-in in protest against recent retrenchments. The Federation's general president, Bob Kelly, told the media that there had been more than 2,000 job losses since May 1982 and he laid some of the blame on the JCB which he said had "sabotaged the industry."[257] The unions, employers and the NSW and Queensland Government representatives were meeting the Federal Government the following day and the unions were pushing for a moratorium on retrenchments and the establishment of a national coal board. Whether a national coal authority could have made a better fist of negotiating coal contracts with the Japanese is debatable, but these words certainly resonated with many employees in the industry, as well as with many in the general community. The Federation believed that the JCB should use its powers to control the development of the industry in order to balance the development of open cut mining with the much more labour intensive underground mining sector. Kelly said that "If the Joint Coal Board is allowed to proceed as it has been, then we could see the system of underground mining disappear completely to be replaced in the short-term with open cut mining, and the skills of the underground mining industry lost."[258] The NSW Coal Association's CEO, Barry Ritchie, rejected the Federation's proposals, arguing that reducing government charges would be the most significant factor in assisting to minimise retrenchments. Ritchie said that the biggest single cost burden on the industry was the NSW Government's freight rates and royalties which were worth around $300 million per year to the Government. Ritchie also warned that while a reduction in charges would provide some protection for employment, it would not save further retrenchments from being implemented.[259]

Following the meeting in Canberra on 30 march 1983, the Government did make some concessions to the Federation when Deputy Prime Minister Nigel Bowen, along with Resources Minister Peter Walsh and Employment Minister Ralph Willis, issued a statement promising a better coordinated approach to coal export sales, and support for the coal industry through tighter export controls. The meeting had agreed to form a new body – a coal industry consultative council – to review the industry's problems. Bowen, Walsh and Willis announced a five-point plan involving a review and strengthening of export control procedures for coal; sponsorship of export missions that would represent all facets of the industry to Australia's main overseas markets and to emerging markets; promotion of education in coal technology in less developed countries, through technical seminars and training programs in Australia, to enable these countries to turn confidently to coal; research to identify market opportunities and to assess the likely size of markets open to

Australia; and strengthening of bilateral relationships with consumer governments, promoting recognition of the mutual benefits available from long-term stable relationships. At a media conference later in the day, Walsh also advised that a new coal industry specialist position would be established in the Australian Embassy in Tokyo.[260] Bowen said that the Government recognised that Australian coal producers had marketed their coal fairly and rejected the idea of a central marketing authority.

In November 1983, the day before a meeting of the Australian Coal Consultative Council, Peter Walsh secured Cabinet approval for his recommendation that the Government should reject the Miners' Federation proposal for a national coal marketing authority.[261] This decision clearly allowed Walsh to go into that ACCC meeting with a firm Cabinet decision and settled any possibility that the Federation would win the day at the meeting. While that Cabinet decision made the Government's policy clear, it did not put the debate to rest or cause the Federation to drop its policy. The debate over a coal marketing authority waxed and waned during the 1980s with the ups and downs of the coal market and the Government's policy at any one time on export controls. Export controls had been in force since early 1973 just after the Whitlam Government was elected, and had been eased somewhat in the 1970s by the Coalition Government under Malcolm Fraser as Prime Minister and Doug Anthony as Resources Minister.

Some companies, of course, were strongly opposed, not only to a coal marketing authority, but to the Government's involvement in the market through export controls. CRA was one the most prominent, and in 1985 at a coal industry conference its finance director, John Carden, said that export controls had actually led to some loss of sales. Carden referred to a statement the previous year from a senior officer of the Department of Trade, the department responsible for administering those controls, who said that export controls were a key instrument in ensuring that Australian producers received a fair and reasonable prices in line with market trends.[262] Carden said that while Australia was a significant mineral exporter, it was "in no position to control the market or dictate to any country the prices it must pay….Guidelines which aim to enhance ability to hold set positions in the market or protect the most vulnerable producers are neither realistic nor in the national interest… When markets were firm the guidelines became a ceiling because it became known, but when they were soft Australian sellers had to meet it with reduced prices…The only winners from the guidelines are usually Australia's competitors who can adjust their prices accordingly and gain tonnage at Australia's expense."

But the Government was not ready to do away with export controls, and in October 1985 John Dawkins, Minister for Trade, said that the coal market was not ready for deregulation, despite the relaxation of controls on other resources announced that week. The Prime Minister had just announced that the Government would continue its review of coal export controls and Dawkins said that he would "not be in favour of reducing controls on coal in the absence of a fairly broad acceptance of that in the Australian industry. There is no such broad agreement at present, although they also want greater flexibility in the system." Dawkins said that the Government would need to be satisfied that deregulation would lead to increases in quantities sufficient to offset a decline in price, but that "I don't think anyone is totally convinced of that at the moment."[263] However, two months later the Government agreed to undertake an independent review of the export control system, and funds were allocated for a review; this decision was reversed later in 1986, and a revised policy on coal export controls was agreed.

The Cabinet documents related to these decisions provide some insight into the positions of the Australian Coal Association, the NSW Government and the major Federal departments. The ACA supported the eventual removal of export controls through a process of deregulation, a stance which reflected the desire of some companies to abolish the controls as soon as possible, while other companies wanted to retain the controls in the belief that this would help support higher prices and a more orderly market. The NSW Government was concerned that removal of export controls would affect the $1 billion it had invested in upgrading the state's rail and port infrastructure.

The inquiry agreed in December 1985 was delayed according to the submission by Trade Minister Dawkins (which was considered by Cabinet on 1 September 1986) because of the major productivity based log of claims case underway before the CIT at that time. Dawkins said that the cooperation of the unions was required for the independent inquiry to proceed, but this was not possible while the productivity case was still running. The "necessary delay" had therefore, according to Dawkins, reduced the value of the inquiry, and as an alternative to the inquiry Dawkins put forward three options for Cabinet to consider. Option one was the complete abandonment of export controls on coal: this was complete deregulation, with the Government having no further role. Option two was for "automatic approval of exports on the basis of monitoring": coal would remain subject to controls but export "permits would be freely available." This option would remove the Government from involvement in marketing, but "provide a means for ensuring information on coal contracting, pricing and tonnage commitments as a basis for addressing larger trade policy concerns." Option three was for "removal of the administrative guidelines" – the "requirement to obtain approval of negotiating parameters in advance of negotiations" – while considering "final contract settlements in the context of the policy of maximising benefits to Australia through maximising export returns." This third option would "represent a considerable administrative simplification and help to overcome a number of administrative difficulties of dealing with the problems of increasing complexities in world markets."[264]

The tortured bureaucratic language involved in these options perhaps reflected the almost impossible task borne by the Federal officials who were responsible for administering the export control system. Dawkins' submission referred to the fact that export controls had become too difficult to administer; they had, he said, become "too complex to be continued." The reason lay in the way the coal market had developed and the Cabinet submission made it clear that the Government needed to acknowledge the realities of the market. "With market developments since last year, including increasing competition and changes in the oil market, concern over union opposition to change in export controls should be subordinated to the need to facilitate increased market penetration through encouraging the more efficient parts of the industry. In these circumstances, delay in instituting change, inherent in holding an inquiry, is no longer appropriate."

The Treasury supported option one. Option two was supported by the Departments of Prime Minister and Cabinet, Industry and Finance, although PM&C supported the ultimate abolition of export controls. Option three was supported by the Department of Resources and Energy as it said that option would maintain a role for the Government in marketing; this Department also noted that CEOs of coal companies were aware that evidence of severe undercutting of the market would result in selective re-imposition of administrative guidelines and an overall review of the situation in six months. Resources and Energy was saying that if option three was chosen, the coal producers, having been on notice to act responsibly, would quickly see tighter controls once again if they indulged in unreasonable price cutting.

The submission by Dawkins referred to the extremely competitive international coal market in 1986. Coal prices expressed in US dollars had been falling, and South Africa had increasing its sales into Australia's key Asian markets. In fact, 1986 was the year in which trade sanctions were imposed on South Africa by a number of countries in response to its Apartheid policy, with that country's coal exports into Europe hard hit. South Africa responded in arguably the only way it could by pushing into Asia and offering coal at a significant discount to the prevailing prices. Up until the mid-1980s, South Africa exported the bulk of its coal to the member countries of the European Union (then the EC); in 1985 26 million tonnes went to the EC out of total exports of 42.5 million tonnes. Exports to the EC then dropped significantly, falling to 22.7 million tonnes in 1986, and to only 19 million tonnes in 1988, although the tonnage rose again by 1990 to 23.8 million tonnes. South Africa was able to maintain a reasonably level overall export volume by attacking the key Asian markets in Hong Kong, Taiwan and South Korea. Japan reduced its imports from South Africa after 1985, however with increases in the three other Asian markets just mentioned, South Africa's overall exports to Asia rose from 11 million tonnes in 1985 to 17.5 million tonnes by 1990.

Sanctions were imposed at different times by different countries, but by late February 1986, under the threat of sanctions and the competition in the coal market, with coal supplies readily available from all the major exporters, South African steaming coal was selling on the spot market for around $US25 per tonne, compared with $US32.50 in early 1985.[265] The spot price for Australian steaming coal had fallen from around $US33.50 to $US29.50 (or around $A47.85 to $A42.15) over the same period, and Australian exporters were in the process of attempting to negotiate contracts with the Japanese for the new Japanese financial year commencing in April. Australian exporters were reported to have given up any hope of achieving any price increase and were happy if they could achieve the same price as in JFY 1985.

The negotiations by the steaming coal exporters for JFY 1985 ran until June, with the Australian companies securing a deal that, given the market conditions, was reasonable. CSR achieved a price of $A48.72 compared with $A43.40 for JFY 1984; with the Australian dollar worth around US67 cents, that equated to around $US32.60 per tonne. Price increases were also achieved by R W Miller, Warkworth and Coalex. The Australian companies were not successful in having their contracts changed from Australian dollars to US dollars, but they were successful in reducing the differential in price between their coals and the price which Blair Athol was receiving from a high of around $A8 per tonne to around $A1. These negotiations were also significant in terms of the role which had been played by the Australian Government which led to a cessation of supply in May after an edict from the Government that it would not approve any contract prices below the equivalent of $US36 per tonne. That edict apparently forced the Japanese to make concessions, with the Government also moderating its position and giving the companies a negotiating price range which proved achievable.[266]

At the Cabinet meeting on 1 September 1986, option 3 was approved, and on 17 September the Government announced that it would relax export controls on coal and also on oil, alumina, bauxite and tungsten. Under the new arrangements for these commodities, companies were now not required to submit their negotiating guidelines to the Government for approval, but would need to gain approval for prices before being able to finalise contracts. John Kerin, then acting Trade Minister, said that export approval would be readily forthcoming where export prices were fair and reasonable in relation to the market situation, and provided longer-term contracts included adequate provisions for review of prices.[267] The Miners' Federation was not happy with the changed guidelines, nor was the NSW Government which claimed that they could lead to the closure of some mines. The concern was that the easing of export controls would allow lower prices to be negotiated, with BHP's Utah reported to have said that it could sell an additional $1 billion of coal on the export market if prices were lower. But while that may have been positive for BHP, it would have had the potential to damage higher cost producers in NSW.[268] John Maitland, Miners' Federation general president, warned the Prime Minister that "Whether Hawke likes it or not …we will not allow him to jeopardise the (south coast coal) producers and our jobs. We will use whatever means we have, including industrial action, to preserve those operations."[269]

BHP's Utah had supported the easing of export controls, with its spokesman, Ian Dymock, referring to the frustrating approval system, saying that up to then "we have had to report back at every stage of negotiations."[270] Bill Myles, general manager of NSW underground miner Oakbridge, was cautious about what might happen, saying that "The important thing overall is to get the right price for Australia's goods…One big producer could sell at a false price and that would then become the international market price." It did not take too much imagination to assume that Myles had Utah in mind when referring to "one big producer". The coal industry unions and their members did not take the new guidelines lying down. National 24-hour strikes were called to protest about the guidelines and the low prices in the export market.

Only weeks after the new export control guidelines were announced,

Dawkins and Gareth Evans, the Minister for Resources, had some particularly strong words to say to the Japanese. With the Japanese pressing for lower prices and cut-backs in tonnage, and following the reduction in prices in 1985 of 10%, the Ministers did not hold back. The Ministers suggested that moves to cut back Australian exports in favour of US and Canadian supplies were a response to the political problems caused by Japan's huge trade surplus with the US and pressures from the Japanese coal industry: "Japanese pleas for understanding from Australian suppliers would carry more weight if they were couched in economically rational terms…Australian coals are already competitively priced and many of the Japanese difficulties could be addressed by increased reliance on Australian coals rather than continuing to purchase high priced coals elsewhere…For Japanese buyers to expect Australian exporters to accept substantial price cuts while they maintain their purchases of non-competitive coals is tantamount to asking Australia to subsidise other inefficient sources – notably Japanese domestic output and some Canadian mines and the United States."[271] The Ministers were raising the issue that had existed for some years, with the Japanese steel mills continuing to source a significant proportion of their coking coal from higher cost producers in the USA and Canada. This was an understandable strategy for the Japanese in terms of diversifying their sources of supply, and not being overly-reliant on one country, but it meant higher costs for the Japanese steel industry and lower tonnages for Australian coal exporters.

A test for the new export control guidelines soon came in November, when Dawkins advised a meeting of the Australian Coal Consultative Council that he had just rejected two contracts with Japanese buyers, saying that the prices were too low and not in the national interest. One contract was for lower quality coking coal and one for steaming coal. Dawkins said that "Both submitted prices significantly below levels that have been and are being achieved by other Australian exporters." Within a matter of days, both companies had resubmitted their contracts for approval. Maitland said he would continue to press the Federal Government to accept a national coal authority to control sales and a watchdog body, to include union representatives, to monitor the new system.

In 1987, the coal market was, if anything, tougher than in 1986. Speaking at a coal conference in March, a senior Australian Government speaker said that South Africa would shift around 6 to 8 million tonnes of steaming coal from its traditional European markets onto the Asian market, and also warned that spot prices could hit a low of $US20 per tonne.[272] Other speakers at the conference noted that China was planning to expand its coal exports from just under 10 million tonnes in 1986 to around 30 million tonnes by 1990. A large mine in China developed by US company Occidental and Chinese partners was about to commence production, with 8-9 million tonnes of its total output of 15 million tonnes expected to be exported, adding to the already over-supplied market.

In July 1987, with the coal market still depressed, the Miners' Federation gave the Hawke Government an ultimatum, with John Maitland warning that Hawke had one week to agree to developing a long term strategy for the industry, or the union would call mass meetings of members with a view to industrial action. Barry Swan, the Federation's national secretary, was optimistic that the Government would agree to establish a national coal authority, claiming that Senator Evans had said that he strongly supported such an initiative.[273] Swan also claimed that the government had given a commitment before the election that a national coal strategy was a "very real possibility" if it was re-elected.

Swan's claim about Evans seems to have related to Evans' statement in Parliament in November 1986 when he said that the Government was "considering, as it has been for some time since it is a matter of long-standing Party policy, the feasibility and desirability of establishing a national coal authority. It is far too early to say what the outcome of those deliberations will be. I point out only that there are a multitude of different models one could choose between if one wanted to go down that track. There is a very big difference between a national coal marketing authority performing some function along the lines of the Australian

Wheat Board at one end of the spectrum and, at the other end of the spectrum, by way of contrast, a body established, perhaps, along the lines of the Steel Industry Authority existing essentially to monitor the direction of the industry and to give advice to the relevant government as to policy changes and so on that might appropriately be implemented. There are a number of models in between, perhaps involving some variation in the kinds of powers and functions that are exercised by the Commonwealth-New South Wales Joint Coal Board that might also reasonably be taken into account in such an evaluation." Evans went on to say that the Government was aware of strong differences of view about such an authority, with the coal producers "by and large solidly against it" and the mining unions very much in favour.[274]

Senator Warwick Parer, an ex-Utah senior executive, had asked Evans whether the Government was looking at the feasibility of a national coal authority, this question producing Evans' statement quoted above. In his response to Evans' statement, Parer did not hold back, saying that "The thought of even introducing a bureaucratic monster such as a national coal authority - Senator Gareth Evans painted certain scenarios as to what that could be - is enough to damage the mining industry, of which the coal industry is part. We cannot afford to let this happen in this country today with our disastrous balance of payments situation."[275] Perhaps Evans went into too much detail given the fact that establishment of an authority was highly unlikely. However, his statement indicates the seriousness of the issue at that time, and the pressure the Government was under from the unions and a number of coal electorate MPs.

The Federal Labor Caucus met in July 1987 and adopted a resolution calling for the urgent establishment of a national coal authority and the development of strategies to support the coal industry. Stephen Martin, the Member for the southern Sydney electorate of Macarthur, had put the motion to the meeting, pointing out to colleagues that the Labor Party's Platform included a commitment to set up such a body.[276] The following month John Maitland put the case for a national coal authority in an article in the Sydney Morning Herald, saying that "Excess capacity is the result of overstated forecasts by Japan and the International

Energy Agency in the past, coupled with too many individual companies expanding capacity to meet the forecasts…These investment decisions were rational on an enterprise level, but not rational at the industry level …The manifestation of excess capacity is downward pressure on coal prices, as individual companies adopt the policy of price-cutting to gain market penetration… Price competition between Australian producers has been prevalent since the requirement for prior Department of Trade approval for prices was removed in September 1986 by the Hawke Government. Only central industry planning through a National Coal Board, and export controls, can ameliorate the effects of excess capacity."[277] A national coal marketing authority which the Federation had supported in the 1970s was now proposed as an authority with powers much wider than just marketing of coal. The Federation wanted it to have the power to control production, much as the Joint Coal Board could do in NSW, but with the authority's powers covering other states.

The Hawke Government's policy regarding a national coal authority had been pretty clear for some time, but in September 1987, Hawke accused the Miners' Federation leaders of misleading their members on this question, saying: "A grave disservice is being done to people employed in the coal industry by those who would delude them, that by establishing an authority, jobs would be saved… Imagine that Australia, by setting up an authority, is going to be able to determine the price we can get. All the evidence is that this is a buyers' market and will continue to be as far as we can see into the future."[278] Hawke ruled out any possibility of establishing a coal authority, telling a media conference in Gladstone that it would be a "bureaucratic irrelevancy."

In March 1988 the NSW south coast coking coal producers - Clutha, Kembla Coal & Coke, Bellambi and BP – won an increase of $US2.90 per tonne in the price negotiations with the JSM. This price was below the $US5 per tonne that they had been seeking. With Japanese steel production having recovered in JFY 1988 to around 103 million tonnes, the $US2.90 price rise was seen as unreasonable; and with the value of the Australian dollar rising from around 64 cents in early 1987 to around 74 cents, Australian producers were actually worse off in local dollar terms. Canadian exporters had earlier accepted the same price increase, although with changes in the classification of some of its coal, their price increase was believed to be closer to $US6 per tonne.[279]

John Kerin, the Minister responsible for export controls on coal, deferred approval for the deal on the grounds that it might not be in the national interest. JCB chairman Jack Wilcox came out publicly advocating establishment of a national coal authority, telling the Business Sunday program on Channel 9 that "Individually, the companies negotiate as well as they can. But the question is can the best efforts by individual companies necessarily lead to best results for the countries as a whole?.... Individual negotiations don't lead to the best results for Australia." Wilcox warned that the proposed new coal contracts with Japanese consumers could plunge Australia's troubled coal industry even deeper into the red, causing further mine closures and putting more miners out of work.[280]

The stance of the biggest coal exporter, BHP, on the issue was crystal clear. Speaking at the industry's biennial Australian Coal Conference in April, BHP Utah's general manager, Gavin McDonald, told the conference that negotiations by individual companies were fundamental to the survival of the coal industry. McDonald referred to the popular view that a "coal supermarket" should be established, with a standard price and with coals attractively displayed.[281] McDonald cautioned the Government to resist the temptation to make export controls even stricter or to establish a coal marketing authority, and pointed out that "We don't have a monopoly in coal and one supplying country alone cannot determine for us unilaterally." He went on to advise the Government that it should concentrate on the macro aspects of world trade, establish the climate for inter-country trade and untangle the "web of tariffs and protection measures that keep Australian producers out of a number of markets overseas".

Kerin spoke at the conference later that day, advising the industry executives present that he had grudgingly approved the deal between the Australian exporters and the JSM, saying that he was "less than satisfied", but cautioned that the approval would be subject to assurances from the companies involved about employment levels in their mines. Kerin however did acknowledge that the nominal price increase of $US2.90 per tonne for hard coking coal was not as disappointing as it may have seemed as it "significantly (understated) the effective price increase inherent in the majority of cases....The difference occurs because of quality upgrading and tonnage increases."[282] Kerin also told the industry that he would use his export control powers to refuse to approve contracts involving unrealistic prices that meant that jobs in the industry would be endangered. ACA executive director Barry Ritchie fired back at Kerin, saying that threats to veto contracts would not prevent the closure of mines or loss of jobs as the rationalisation of the industry would continue, with closures and job losses inevitable.

Relations between the coal companies and the Minister were going through a difficult period, but relations between the Government and the Miners' Federation were arguably worse and hit rock bottom in 1988. Following the Government's outright rejection of the Federation's demands for a national coal authority, the coal committee of the Federal Labor Caucus met in April to consider a motion from Bob Brown, the Newcastle MP, that "the Government establish a National Coal Authority." Brown's motion was adjourned. That same month John Kerin advised the Australian Coal Conference that he did not support a coal authority and gave several reasons – it would deter investment; he doubted it would be effective; it would not stop new mines being developed; and he doubted that the State Governments would hand over their power to issue mining leases (a power an authority would need if it wanted to control coal production).[283]

In early August, the Government released details of a strategy to support the coal industry; notably absent was a national coal authority. The new plan for the industry was effectively a clayton's plan, offering "a more centralised and coordinated approach to coal marketing and strengthening of export controls", additional funding for Kerin's Department to coordinate marketing by coal companies and to promote trade, more detailed research on the coal market by ABARE (the Bureau of Agricultural and Resource Economics), and a restructuring of the

Australian Coal Consultative Council (ACCC). The new Australian Coal Industry Development Council would have a wider membership, and would advise the Minister on marketing and other issues.[284] John Maitland said that the plan to overhaul the ACCC would fail: "We're not going to be involved in it. It's a waste of time and effort and precious union funds."[285] To add salt to the Federation's wounds, the National Conference of the Labor Party had also agreed to scrap any plans for such an authority. Before the Conference, the Party's energy platform committee decided to delete reference to a coal authority from the party's policy. The issue was then debated at the Conference and a resolution was passed supporting the Government's decision to set up a new coal council.

The Federation then severed its links with the Australian Labor Party. In August, the union's policy-making body, its central council, unanimously decided to break away from the ALP and took that decision to its members. The decision was taken despite personal interventions by John Kerin, then Minister for Primary Industries and Energy, and Bob Hogg, the ALP national secretary.[286] In Queensland, meetings at 8 locations voted unanimously to disaffiliate, and the Federation's Queensland president, Andrew Vickers, threatened to stand as an independent candidate in the next Federal election in the seat of Oxley held by Bill Hayden, although he did not follow through on this threat.

In announcing the strategy for the industry, Kerin had said that "Full consideration was given to the unions' call for establishment of a national coal authority but the Government remained convinced that the appointment of a bureaucracy to handle a highly commercial operation was not a solution to the marketing problems of the industry."[287] This was not the last word on the national coal authority. With price cuts forced on Australian exporters in early 1994, and the Federation (now the CFMEU) taking industrial action, a major inquiry was launched by the Government headed by Rae Taylor which would again have a national coal authority as one of its major issues. We will come back to this inquiry in the next chapter.

Coal industrial relations in the 1980s

Industrial relations in the coal industry did not get off to a positive start in the 1980s. In 1980 the coal mining industry lost 710,000 working days to industrial disputes and accounted for over 21% of days lost by all Australian industries. The great majority of these disputes were in NSW and Queensland, and for an industry employing only around 25,000 workers in these two States, that was not a record to be proud of. The number of days lost to industrial disputes in the coal industry did not reach the 1980 level again in the 1980s, but was still high and saw coal continue as the leading industry. In 1981 318,000 days were lost in the coal industry and this rose to 525,000 in 1982; the disputation rate eased back in 1983 and 1984 before rising again and peaking in 1988. In 1988, 471,000 days were lost, equivalent to 15.5 days for each and every employee in the industry; across all Australian industries the comparable figure in 1988 was less than 0.3 days lost.[288]

There were many reasons for industrial disputes in the 1980s including minesite issues, wage claims, other major cases before the CIT (such as the work practices decision in September 1988) and strikes on what could be termed political issues such as coal prices and calls for a national coal authority. We will come back to the 1988 CIT case shortly. One major dispute in 1982-83 was quite different and involved the Federation and other unions placing a ban on the Drayton coal mine in the Upper Hunter Valley. Drayton was an open cut steaming coal mine and was developed by CSR, Shell and Japanese and Korean companies for export, but mining was subject to a ban by the unions in early 1982 because of retrenchments in the industry. The unions objected to job losses in the underground sector and saw new open cut mines as a threat to job security.

NSW Premier Neville Wran, who was also Minister for Mineral Resources at the time, convened a meeting of the parties involved in October 1982 in an attempt to broker a settlement. Wran had visited Japan and Korea only the month before and had told Drayton's partners that he thought that the bans would soon be lifted.[289] The union ban continued into 1983 despite the mine beginning to lose export contracts. By October, two shipments which were to have been supplied by Drayton were supplied by another Hunter Valley mine. However, Drayton's general manager, Darcy Wentworth, said that this was part of only one of six export contracts under which the company was to supply almost 2 million tonnes per year to Japan, Europe and Korea. He claimed that the loss of Drayton's contracts would not benefit other Hunter Valley producers, but competitors located in Queensland and overseas, and that with the project at a critical stage "if we don't get operating soon, we are not going to supply our other contracts, which start coming in from January next year."[290] At the annual Australia Japan Business Cooperation Committee meeting in Tokyo which was being held at the same time as the Premier's meeting, the chairman of the Japanese Coal Association made a plea for the bans to end. Mr Shingo Ariyoshi, who also was chairman of Mitsui Mining (one of the shareholders in the Drayton project), said that all the partners in the project were embarrassed by the delays and warned that the impact on Japanese customers, who had entered into purchase agreements with Drayton, could not be ignored. Ariyoshi also warned that "frequent strikes and big wage increases are gradually undermining the competitive position of Australian coal and, at the same time, making the coal users extremely nervous." He said that he "sincerely (hoped) both employers and labour will realise that labour problems in Australia cannot be confined at home and that they will deal with it with such a comprehension."

The bans had been imposed by the Northern Liaison Committee of the Combined Mining Unions, members of which were the Federation, the FEDFA and three other unions. The FEDFA however had defied the Committee since September 1982 with 27 of its own members working at the mine, initially operating machinery to strip the overburden. By January 1983 these FEDFA members began to actually extract the coal, but the lost European contracts now totalled 350,000 tonnes and shipments to Japan and Korea were due to commence shortly after.[291] With the bans still in force in early May 1983, a public meeting was held in Muswellbrook attracting a reported 2500 people; the meeting was addressed by Darcy Wentworth and the FEDFA's John Thorley and called for the mine to re-open. Telegrams demanding the re-opening were sent after the meeting to the Prime Minister and to the ACTU president, with a copy to Bob Kelly, general president of the Miners' Federation.[292] The major fight within the unions became even clearer when around 1500 miners from the open cut sector met in Singleton and decided to fight their own unions over the bans. The open cut workers saw the bans as strangling the development of the whole open cut sector and agreed to establish their own committee to oppose the policies of the combined unions. The open cut sector meeting agreed on the formation of its own open cut liaison committee which would have "the power to coordinate strikes and bans among all workers in the 14 Upper Hunter open cut mines."[293] The meeting also agreed that open cut workers would operate the Drayton mine by Friday of that week despite there being a threat from the maritime unions that all coal exports through Newcastle would be stopped if any unauthorised workers attempted to carry out mining.

With this major split in the unions, the end of the dispute was to come later in May when the NSW Government brokered a deal with the unions which was endorsed by a meeting of 4,000 Hunter Valley mine workers. The deal allowed Drayton to operate on a limited basis, supported an increase in severance pay for retrenched workers in the northern district who had been denied this increase by the CIT because of the bans, and included the reinstatement of over 80 workers at Preston colliery near Gunnedah who for the previous two months had been underground staging a sit-in. Other points of the deal included retrenched workers to be given priority for employment, Government action to ensure that the industry conformed with the award which provided for a 35-hour week, and future retrenchments only to be implemented if all options had been tried, including involving the CIT and the State and Federal Governments.[294] Another key element of the plan was for all future developments to ensure that there was a 70/30 ratio between underground and open cut production.[295] The NSW Government then met with the NSW Coal Association and a few days later both were able to announce the agreement to allow Drayton to re-open and the reinstatement of some of the Preston colliery workers. By the end of May Drayton was back in operation and was commencing the hiring of the 140 new workers it planned to employ over the following two weeks. The mine was now able to proceed after a delay of around a year.

Australian Industrial relations – the precursor to a new era

Major industrial relations reforms were implemented in Australia in the 1990s, with the Keating Labor Government's changes to the legislation in 1993 which put the focus on enterprise agreements, and the Howard Coalition Government's further changes to the legislation following its election in 1996 which narrowed the scope of awards and introduced individual workplace agreements. However, we also need to look at the 1980s for the major changes in thinking which created the climate for the 1990s reforms. The series of Accords between the Labor Government under Prime Minister Hawke and the labour movement under ACTU secretary Bill Kelty were a major factor in the way the economy developed in the 1980s and in the way the pressure on wages was eased by trade-offs which provided a higher "social wage" for families and households.

While the coal industry employers had generally supported the CIT up to the 1970s, the 1980s saw a major change as a result of the severe economic pressures on the industry, and a realisation that it was time for the industry to become part of the mainstream in terms of industrial relations and regulation generally. The Federal Government in 1983 appointed a three-man committee to review the nation's industrial relations framework; the panel comprised Professor Sir Keith Hancock (vice chancellor of Flinders University), Charlie Fitzgibbon (senior vice president of the ACTU) and George Polites (previously an executive director of the Confederation of Australian Industry). The Hancock committee as it was known reported to the Government in July 1985 and recommended some major changes to the system, including a new body – the Australian Industrial Relations Commission (AIRC) – to replace the existing Conciliation and Arbitration Commission. Other changes recommended included abolishing the penal sanctions in the Act and strengthening the deregistration provisions, abolishing the industrial arm of the Federal Court and replacing it with an Australian Labour Court, and requiring unions to have at least 1,000 members (with existing unions having five years to comply). The Hancock committee also recommended that the specialist industrial tribunals should be abolished and their functions transferred to the new commission. Those tribunals included the Coal Industry Tribunal, the Academic Salaries Tribunal and the Flight Officers Industrial Tribunal.

In early 1986 there was major national strike action over a log of claims submitted by the coal industry unions, the log based on productivity increases in the industry. The case was heard by CIT chairman David Duncan who brought down his decision in June, agreeing to some of the claims which had been agreed by the producers and the unions, and in addition agreeing to a travelling allowance of around $21 per week which was incorporated into wage rates and an increase in annual leave of 5 days. The CIT also passed on the national 3% wage increase awarded by the Arbitration Commission.

However, the release of the Hancock report, industrial action by the unions and the parlous financial state of many coal companies had seen a hardening of the attitude of the Federal Government towards the coal mining unions. Resources and Energy Minister Gareth Evans came out publicly criticising the CIT in July 1986 for its decision, a move not often seen from a minister with a major responsibility for the coal industry. Evans told an industry seminar that the decision "tied another millstone around the necks of the coal producers" and he went on to caution that the Government was giving close consideration to the policy implications of the decision, with that consideration including Cabinet discussions on the Hancock recommendations regarding incorporating the CIT into the Arbitration Commission. Evans recognised that the decision was going to be costly, initially raising open cut costs by around 50 cents per tonne and underground costs by around $1 per tonne, and when fully implemented by July 1987, raising average mine workers' earnings by around 8.7%. Evans also put the unions and their members on notice, saying that their expectations had been unrealistically high, "never more so than in the recent national dispute", and advising them that they would have to drastically revise expectations if the industry was to have a chance of survival in its current form.[296]

The CFMEU demanded that the CIT be retained, with its national secretary Barry Swan threatening that his and the other mining unions would "stop the coal industry" if the Government attempted to abolish it.[297] Swan at that time was responding to a report that Employment and Industrial Relations Minister Ralph Willis had accepted that the CIT should go. Swan argued that the CIT had been designed to serve the unique interests of the coal industry since it was established in 1947, and added that "There is a need to maintain an industrial relations atmosphere through an intricate knowledge of the industry by all parties for resolving disputes quickly and settling health issues."[298]

In March 1987, the Australian Coal Association, representing the coal producers in NSW and Queensland, made a submission to the Federal Government arguing for the abolition of the CIT. The ACA's submission was aimed at influencing the details of legislation then being developed for the new AIRC. While not directly criticising David Duncan, who had been the one person Tribunal since 1975, it was nevertheless a document strongly critical of the Tribunal and its operation, saying that "The policy of appeasement inherent in the terms of reference of the tribunal may indeed be a source of industrial friction in today' coal industry."[299] This was a reference to the CIT's charter set out in the 1947 Coal Industry Acts and which specified that the purpose of the Acts (and of the JCB and the Tribunal) was the "securing and maintaining adequate supplies of coal to meet the needs for that commodity throughout Australia and in trade with other countries and for providing for the regulation and improvement of the coal industry in the State of New South Wales and for other matters relating to the production, supply and distribution of coal…"

The position of the coal producers on the future of the CIT, or at least the majority of the producers, had changed dramatically since the

1970s. In 1977, then Employment and Industrial Relations Minister, Tony Street, made a submission to Cabinet on the CIT recommending its retention. That submission followed a review of the JCB and CIT in 1976 by Cabinet's Administrative Review Committee, chaired by Henry Bland, which had recommended no changes to the Coal Industry Acts.[300] Street noted that he had consulted with the major stakeholders on the question of the CIT's future, and the coal producers in both NSW and Queensland "were firmly committed to the retention of the Tribunal" although they did wish to see a mechanism introduced which would allow appeals to the full bench of the Conciliation and Arbitration Commission. Street reported to Cabinet that the unions did not wish to see any significant changes at all to the CIT, and that both the employer groups and the unions were satisfied with the performance of the CIT and its "contribution to the generally sound industrial relations within the industry." Street said that the CIT had "worked remarkably well in recent years" and also noted that even if change was sought, there would be major practical difficulties, in particular with the need to obtain the support of the NSW Government for any change to the Coal Industry Acts. Street said he would consider at a later date the need for an appeal mechanism as sought by the employers, but was not convinced of the need for such a change. The CIT continued to operate as a self-contained industrial tribunal for the coal industry throughout the 1980s.[301]

The ACA's 1987 submission argued that the CIT had not reduced the incidence of strikes or provided a superior mechanism for wage fixing to that provided by the Conciliation and Arbitration Commission. It also charged the CIT with allowing wage increases which were over and above and inconsistent with national increases. The ACA said that the continuation of the CIT was no longer consistent with the needs of a major export industry, and that there was no longer any rationale for its continued existence. In an unusual step, David Duncan, who was hearing a major wage case, made an opening statement saying that industrial relations in the industry were being destabilised, although he had "no quarrel with the endeavours of employer associations to seek to abolish or alter the tribunal." However, Duncan warned the ACA against its "continuing use of the media as a means of putting pressure on the Tribunal."[302] Duncan also said that he was "unaware of any specific plans to change the operations of the Tribunal in any way, despite the continuing speculation which abounds in the present media activity."

Coal Industry employers, however, would be disappointed, but not surprised, when, in May 1987 the Federal Government announced the details of a new industrial relations Act. While the other specialist tribunals were indeed to be absorbed into a new Commission, the Coal Industry Tribunal was spared. Despite the lobbying from the ACA, changes to the CIT and the Coal Industry Acts, of course, required the support of the NSW Government, and it was understood that the Miners' Federation had been given private assurances by the NSW Labor Premier, Barry Unsworth, that he would not agree to any request from the Commonwealth to abolish the CIT.[303]

Ralph Willis clarified his position, if not necessarily that of the Government, later in 1987 in Parliament when he was asked to confirm a media report that he was considering abolishing the CIT, and if so, whether he would publicly seek to obtain the agreement of the NSW Government in doing so. "The report that was in the paper about the Government proposing in the strategy paper for the coal industry that the Coal Industry Tribunal should be abolished was wrong" Willis said. "That is not what is said in the strategy paper. It simply raises, as an issue for consideration, the future of the Tribunal and suggests that it may be appropriate that it no longer exist, but does not propose its abolition as a strategy for the industry at this stage. What is relevant is that the Tribunal is established by joint legislation between ourselves and the New South Wales Government. Its abolition by us unilaterally would still leave the Tribunal operating in New South Wales and would only be relevant, therefore, to Queensland. What would happen in that State would then depend on whether the unions involved and the employers sought to come under a Queensland State award or seek a Federal award. The Tribunal is a bit of an anachronism in the sense that with the proposed rationalisation of tribunals in the Industrial Relations Bill which we brought forward earlier this year, we left the Coal Industry

Tribunal to one side because of the difficulty with New South Wales and the fact that New South Wales did not, at that stage, support its abolition…I think a rational system of industrial relations in this country would see that Tribunal no longer in existence. However, I think we need to discuss that further with the New South Wales Government and have agreement between the two governments before any step can be taken in that direction."[304] The mining unions had obviously won the day, no doubt with strong support from the Parliamentarians who represented coal electorates in areas such as the Hunter and Illawarra, and with the NSW Labor Government unwilling to make a change. However, Willis' statement made it clear that the CIT's days were numbered and that many in the Government would support its abolition. But it would be another 7 years before the CIT would be abolished by the Keating Labor Government in the face of strong opposition from the CFMEU (the union formed from the merger of the Miners' Federation and other unions) and Labor Parliamentarians.

However, the industrial relations system was changing in the 1980s, even if the coal industry was left to continue to operate with its own tribunal outside the mainstream system. The changes came in part from the Hancock committee and the Hawke Government, although those changes were evolutionary rather than revolutionary. Changes were also being driven from the new thinking being promoted by the Federal Opposition under John Howard as Leader and Neil Brown as Deputy and shadow minister for industrial relations, and particularly from the Business Council of Australia (BCA). In 1984 the Federal Opposition's industrial relations policy provided for decentralised bargaining between employers and employees, with provision for parties to elect to opt out of the award system. In 1986 the Opposition released a new industrial relations policy, arguing for a system with greater flexibility, with the ability for voluntary workplace agreements to be made by businesses employing less than 50. These agreements would not be made within the Arbitration Commission system, but would have the same status as awards. The policy also proposed allowing claims for damages to be made through common law actions against unions involved in industrial disputes. Brown said the policy was "the most significant reform ever made to our industrial relations system since it was established in 1904".[305] The Opposition's policy would retain the Arbitration Commission, but the Commission in its national wage case decisions would need to take into account the economic climate and an industry's capacity to pay. The Opposition's policy was supported by the major employer groups, including the BCA and the Confederation of Australian Industry (CAI).

There were also some landmark industrial disputes in the mid-1980s which broke new ground in terms of the ability of companies to take action against unions and to secure awards of damages. The two major cases were Dollar Sweets and Mudginberri Abattoir. Dollar Sweets was a small Melbourne firm which was picketed by the Federated Confectioners' Association during a dispute. The case went to the Victorian Supreme Court which ruled against the union, exercising common law jurisdiction and bringing the dispute to an end. A young lawyer, future Treasurer Peter Costello, made a name for himself representing the company. Costello said that prior to this case "there was considerable doubt about whether State Supreme Courts would exercise common law jurisdiction to grant injunctive relief against unions involved in industrial disputes. It was argued that since the Commonwealth Parliament had set up a legislative scheme for dealing with industrial disputes (through compulsory arbitration), the Courts would not (and ought not) exercise general reserve common law jurisdiction. Indeed this was put on behalf of the unions in the case. The argument was rejected and the Court granted the orders sought."[306] The company, in a settlement between itself and the union, also received $175,000 to compensate for losses it suffered from the picketing.

Mudginberri Abattoir in the Northern Territory signed individual contracts in 1985 with its employees without the involvement of the union, the Australasian Meat Industry Employees Union (AMIEU). Union members picketed the abattoir for four months, with the union concerned that its members were not being paid award wages and other benefits. Commonwealth Government meat inspectors refused to cross the picket line, and without the necessary certification the company was unable to export its beef. Jay Pendavis, the Abattoir owner, took action under section 45D of the Trade Practices Act, resulting in significant fines being imposed on the AMIEU for its refusal to remove the picket lines. The National Farmers Federation and the employer organisation, the Meat and Allied Trades Federation, supported Pendavis in his legal action. The dispute was settled in 1986, with the agreement of the Abattoir that it would notify the union of any contract arrangements with employees and that conditions agreed for the Northern Territory would not set a precedent for other parts of Australia.

In the 1980s, the BCA was one of several major employer organisations including the Confederation of Australian Industry and the Australian Chamber of Commerce, which were prominent in debates on taxation and other issues central to the economic direction for the country. The BCA was the body which represented the largest companies and it arguably became the most influential organisation in the mid to late 1980s and the 1990s in terms of industrial relations; its industrial relations agenda was also the most radical of the major employer groups. In 1986, at a time when Sweden was seen by the Labour movement as a model for Australia in terms of industrial relations, the BCA sent its own mission to that country. The ACTU looked favourably on the Swedish model involving tripartite agreements between unions, management and government. The BCA mission was led by Peter McLaughlin, the BCA's deputy CEO, and reported back saying that Sweden had little to offer as a model for Australia. The mission found that the Swedish policy of lifting the wages of the lowest paid workers had led to a reduction in the relativities; shortages of skills had emerged, together with pressures for wages to increase unreasonably and damaging strikes. The mission also saw some positives in Sweden, including a lack of interference by national union officials and a greater commitment by management to good relations with employees at the enterprise level. These positives had contributed to raising productivity and job satisfaction.[307]

In March 1987 Sir Roderick Carnegie, the BCA president, and Amcor CEO Stan Wallis, chairman of the Council's industrial relations committee, launched an industrial relations policy entitled "A Better Way of Working". Carnegie said that industrial relations for too long had been ruled by the adversarial approach and that "The Council rejects categorically that enterprises and their employees need to be in conflict. In fact, the interests of both parties are best served in an environment of mutual trust and support, with a strong focus on common goals… Management had to place less reliance on barristers advocating in front of tribunals." The BCA policy advocated a system of enterprise agreements, negotiated directly between management and employees and unions, but with workers in each enterprise represented by only one union, with the BCA advocating a move towards industry-based unions. Wallis said that the Council would appoint an expert group to study the strategy and assist to develop a process of managing the change to the new system which it said should happen as part of a gradual process, rather than taking place overnight.[308]

The BCA's expert group – its Study Commission - reported in October 1989, but in the meantime, the Hawke Government had repealed the Conciliation and Arbitration Act and replaced it with the Industrial Relations Act of 1988. The new Act provided for a new option for employees and companies – certified agreements – and also streamlined the process for union amalgamations. The BCA Study Commission was chaired by Fred Hilmer, who in 1992 and 1993 chaired the review into national competition policy for the Federal Government. Perhaps the most fundamental change recommended by the Study Commission was the creation of two streams of regulation, one for employers wishing to remain part of the current award system, and one for more innovative employers. The second stream was designed to allow employers and employees at the enterprise level to enter into fixed term agreements which would have the force of Federal awards, with twelve months the minimum term. The agreements would contain procedures for preventing and settling disputes, and the parties involved would need to be able to satisfy the AIRC that they had entered into the agreements freely. The Commission report acknowledged that cooperation between employers, unions and the government had improved, but said that Australia had to move beyond the prevailing situation, jettisoning "the industrial relations mindset within our enterprises where it still rests on the outmoded assumption of conflict and move to 'employee relations' ... (which) assumes that employers and employees have much more in common than they have differences... An employee relations world does not imply a 'free-for-all' or a return to 'the law of the jungle', but it does imply a major shift in Australian thinking about the regulation of relations at work."[309]

The Study Commission said that the biggest single impediment to more efficient workplaces was the antiquated structure of the trade union movement, with the complex, multi-union representation in most workplaces hindering improvements in work practices. Ideally it wanted to see one bargaining unit at each workplace, and that unit could be a branch or division of a large union, and it wanted to see a move away from the prevailing structure of unions which were mainly craft or occupation based, reflecting the way that work was organised 50 years before.

The 1988 Act's provisions were well short of the revolution the BCA was seeking, but it did go some way along the BCA path with its enterprise bargaining provisions and provisions for rationalising the structure of unions. However, the late 1980s also saw the Conciliation and Arbitration Commission and its successor AIRC beginning to shift towards more flexibility in award decisions. The former introduced a two-tier wage decision in March 1987, with tier one awarding a flat national wage increase of $10 per week for all workers; the tier two increase of 4% was conditional on an industry or enterprise agreement to improve work practices. In August 1988, the AIRC then introduced the Structural Efficiency Principle, the objective of which was to "improve the efficiency of industry and provide workers with more fulfilling and better paid jobs". Employers and unions were offered a bargaining agenda to follow which was designed to achieve a range of outcomes including skill related career paths which provided an incentive for workers to continue to improve their training and skills, elimination of impediments to multi skilling, and creating appropriate wage relativities between different categories of workers.

In 1989 the AIRC in its national wage case decision, and following the ACTU's "Blueprint for Changing Awards and Agreements", increased minimum wage rates and provided for general wage increases in exchange for reforms to work classifications, removal of demarcation barriers which prevented more flexible work organisation within enterprises, and the creation of career paths to create incentives for workers to acquire skills. The decision also expanded the areas for bargaining to work practices and conditions including working hours, penalty rates and annual leave. The decision showed that the industrial relations system was evolving and attempting to push employers and unions to accept more flexibility and negotiate accordingly.

The coal industry was also caught up in this push in 1989 through a major case before the CIT. In September 1988, David Duncan brought down his decision on a coal industry work practices case in a detailed 160 page judgement, hailed at the time as bringing in the most significant changes in the industry for decades. Key elements of the decision included an increase in the standard shift from 7 to 8 hours, but with the standard working week remaining at 35 hours; longer shift lengths would be possible if negotiated on a minesite basis; production at underground mines would be possible 6 days per week (versus 5 days); production would be possible 52 weeks per year, eliminating the traditional shut-down over Christmas and Easter; rosters would become more flexible and would be determined by employers; and annual leave would be determined by employers. The employers were generally pleased with the decision, although they had asked for production to be possible 7 days a week, and for shift lengths to be 9 hours.

Duncan, having sat through several weeks of hearings, acknowledged that industrial unrest was likely as mine workers fought to retain conditions that had been in place for some time. Duncan was correct, with wildcat strikes at several mines occurring almost as soon as the decision was brought down. Coalcliff and Tower on the NSW south coast, and Collinsville and Peak Downs in Queensland were among the mines hit by these strikes. The Miners' Federation leaders reluctantly accepted the CIT decision and mass meetings of workers were then arranged in the major districts. The meetings in the western district of NSW at Kandos, Lithgow and Mudgee rejected the decision by 237 to 136.[310] The NSW northern district also voted to reject the decision by 1366 to 1230 and the meetings were critical of the union's leadership, particularly national president, John Maitland, for the CIT's decision. In the NSW south coast district around 2000 miners voted narrowly to accept the changes. The results of mass meetings in Queensland were not initially disclosed by the union, but from private indications given by union and company sources, there was a clear vote in favour of accepting the changes proposed by the CIT.[311] The Queensland vote was the decider, resulting in the changes being accepted on the basis of the aggregate vote in all the districts.[312] The challenge now for producers was to get the changes accepted and implemented at the minesite level.

Pit ponies become union members

It was in February 1990 when the last of the pit ponies which were employed in underground coal mines in Australia were "retired from service" at MIM's Collinsville mine in Queensland. Wharrier and Mr Ed were the last in a long line of horses used extensively in Queensland, NSW and Victoria to carry out much of the hard work underground; their memory now lives on in Collinsville where there is a full size bronze statue of a pit pony in the town's centre.

With mechanisation throughout the industry doing away with the need for horse power, most mines had pensioned off their horses by the 1960s and 1970s. At Collinsville, the mine management's attempt to pension off their last two horses, however, led to a major dispute with the mine's workers. The CFMEU, in an attempt to block management's decision, had enrolled the two horses as honorary union members, a move which allowed the union to claim that their new honorary members were now protected by the seniority rule, under which the longest serving members were the last to be retrenched. The case went to the CIT and was subsequently resolved; Wharrier and Mr Ed then left the employ of the mine, entrusted to the care of Bill Hoffmann, who had been their handler in the mine. Wharrier and Mr Ed presumably let their membership of the CFMEU lapse. While this dispute would have been the cause of considerable mirth in the industry at the time, it perhaps was emblematic of the tendency in the industry to hang on to its past – and of the need for change. That change would come in the 1990s and industrial relations in the industry would never be the same again.

Japan's coal industry death spiral continues

Japan's once large coal mining industry began a process of rationalisation in the 1950s, although its peak production was reached in 1961. Through various Government-led coal policies and programs, and due to competition, particularly from oil, the industry rapidly and painfully shrank in the 1960s and 1970s to a fraction of its former size. Imported coking coal from Australia, Canada, the USA and other suppliers took over from the domestic industry and fuelled the dramatic growth in Japan's steel industry. In the 1970s imports also began to play an increasing, although still modest, role in supplying steaming coal for Japan's electricity generators. By 1980, the Japanese coal mining industry's employment was down to less than 20,000, or only around 5% of the number of workers in the industry at its peak during World War Two.

However the Japanese Government, with its understandable concern for energy security and its total reliance on imported oil, as well as on imported uranium for its growing nuclear power sector, was determined to maintain some level of domestic coal production. In 1981 the Government introduced another plan for the coal industry, this time setting a production target of 20 million tonnes per year, with approximately equal contributions from coking and steaming coal. Production at the 20 million tonne level was expected to continue through to 1995.[313] Government subsidies of around 60 billion yen in 1980 were still provided to the local coal producers to encourage production and to assist with modernisation of the industry. The previous embargo on the importing of steaming coal, partially lifted in 1974, was finally fully lifted in 1977, and imports of steaming coal grew to a modest 1 million tonnes by 1979, but were expected to exceed 50 million tonnes by 1990.[314] The Japanese Government was also actively supporting the construction of coal handling facilities in its ports, as well as providing financial assistance for the development of coal mines in other countries (this assistance totalled 4.2 billion yen in the 1980 financial year).

By the mid-1980s the Japanese steel mills, after several years of pressure from the global downturn of the early 1980s, were refusing to pay the higher prices required by the local coal mines and announced that they would only pay the import price for domestic coal. The electricity utilities were also pushing back, demanding price reductions from the local coal producers. This was also the time when the imbalance in trade between the USA and Japan became a major political issue between the two countries, with the USA pressuring Japan to buy more from the USA, and with commodities including coal and beef seen by the USA as prime targets for higher sales to Japan.

Acting on the recommendations of the Coal Mining Council, an advisory body made up of representatives of the coal producers, steel mills, power utilities, cement companies and unions, the Japanese Government implemented another coal plan in 1986, the eighth in a series since 1962. Under the new plan, domestic production would fall from over 17 million tonnes in 1986 to 10 million tonnes by 1991; coking coal production was to be completely phased out, and steaming coal would remain the only type of coal produced to supply only a part of the rapidly growing demand from the Japanese power utilities. By 1986 there were around 26 operating coal mines, of which around 10 or 11 could be regarded as of reasonable size. These mines employed around 20,000 workers, but employment would fall to less than 9,000 by 1991. And by that year, only 6 of the larger mines were in operation.

The power utilities, which had been planning to reduce their use of local coal, were pressured by MITI into agreeing to purchase 8.5 million tonnes over the period 1987 to 1991.[315] While this agreement gave the local coal industry a temporary lifeline, it was merely postponing the eventual phase-out of domestically produced coal. On the day that the

Government announced the new plan, Mitsubishi closed its Takashima mine with the loss of almost 1,000 jobs. That mine had been the longest running coal mine in the country, first developed in the late 1860s with British involvement, and purchased by Mitsubishi in 1881. The mine was located below Takashima Island, a small island near the port city of Nagasaki, and now perhaps better known as the location for part of the action in the James Bond movie Skyfall. Among other major mine closures which were to follow were Mitsubishi's Minami Oyubari Yubari mine in Hokkaido in 1985 (an explosion there killed 62 miners deep underground) and Mitsui's Sunagawa mine also in Hokkaido in 1987 where just over 700 workers were employed.[316] Following a major explosion at the Hokutan Yubari mine late in 1981 which killed 93 miners, that mine was closed in 1982.

The island of Hokkaido in Japan's north was one of Japan's major coal mining regions, with the city of Yubari its major mining centre. Coal mining was largely dominated by two companies - Mitsubishi and the Hokkaido Colliery and Steamship Company. Yubari had a population of over 120,000 at the peak of mining in the 1950s and 1960s, but with mine closures that fell to only around 21,000 by 1990. The city went bankrupt in 2007 as its declining population was unable to support its need for services. It now has a population of less than 10,000. Yubari is arguably the most dramatic example of the impact that the death of the Japanese coal industry had on one particular area.

The Japanese coal industry limped on through the 1990s and by 2001 production had fallen to only 3 million tonnes. Mitsui's Miike had seen another major disaster in 1984 when a fire in the mine and poisonous gases killed 83 workers; more than 600 other workers were fortunate to escape. The Miike mine finally closed in 1997. With the closure of another mine in 2001, and restructuring of another, national production was down to 0.7 million tonnes in 2002. In 2003 the IEA reported that only one mine was still operating.[317] The BP Statistical Review of World Energy for 2019 reported Japanese coal production of 0.6 million tonnes in 2018. From a peak of around 51 million tonnes of production in 1961, Japan's industry progressively declined and now supplies only a tiny proportion of the country's energy needs. The winners, of course, have been exporting countries, particularly Australia, which have not only replaced Japanese coal production, but captured all of the growth in the Japanese market since the 1960s.

3

Inquiries, Reforms, and Industrial Battles

Queensland development resumes

After the virtual hiatus in major developments in the second half of the 1980s, there was a little more optimism in the early 1990s and a new determination on the part of the Queensland Government to get some new projects under way. A change in the Federal Government's foreign investment policy was also critical. The Gordonstone mine, located about 45 kilometres from Emerald in central Queensland and developed by US oil company ARCO (Atlantic Richfield), created something of a first in terms of foreign investment policy when, in August 1900, Paul Keating as Treasurer brushed aside the prevailing policy requiring a minimum 50% Australian equity and approved the development of the mine on the basis of up to 95% foreign ownership. At that time ARCO had 65.5% ownership of the project, Suncorp 29.5% and Lend Lease 5%. The approval was conditional on ARCO reducing the foreign share of investment to 50% within 7 years of completion of construction of the mine. Keating said that the policy still allowed a project that did not meet the Government's guidelines to proceed provided it was not contrary to the national interest, and provided the lack of local equity would not unduly delay development.[318]

A $500 million development, Gordonstone was planned to be the largest Australian underground coal mine up to that time, with an output of 3 million tonnes of coking coal and 1.2 million tonnes of steaming coal, and was touted as one of the most technically advanced coal mines in the world. And it was a project to which the Federal and Queensland

Governments clearly wanted to give the green light. The coal industry had entered the 1990s with international markets still extremely competitive, although prices had improved in 1989-90 from the depressed levels of the late 1980s. Exports from Queensland mines were increasing only slowly, and apart from the Jellinbah mine, which began exporting in 1989-90, there had been no other major new project developed for several years.

Following Keating's approval, development of the mine commenced and the first export shipment took place in December 1992, with longwall operations commencing in April 1993 and with a second longwall commissioned in 1994. By 1993 employment had grown to around 450. In August 1991 ARCO purchased Suncorp's equity, taking its stake to the permitted 95%. Later that year Mitsui bought 15% of the project, reducing Arco's share to 80%. ARCO was also expanding its interests in Australia, in 1993 buying Exxon's 55% stake in the Clermont project in conjunction with CRA. By 1994-95 ARCO also owned around 31% of Blair Athol and had boosted its share in its other major investment, the big Curragh mine, to 87%. By 1995-96 Gordonstone was producing 4.5 million tonnes of coal, with plans to expand to 6 million tonnes per year, and had commenced simultaneous operation of both longwalls. Employment at Gordonstone had increased to around 500 and the future for ARCO in Australia looked promising, although conditions in the international coal market were still difficult. A major battle with the company's employees and the CFMEU, however, was just around the corner. More of that story later in this chapter.

Japanese trading company Idemitsu Kosan bought a 22.5% stake in the Ensham project, a large steaming coal deposit around 40 kilometres from Emerald, in January 1984. Other partners in the joint venture were Bligh Coal 15%, Allied Queensland Coalfields 15%, CRA's Pacific Coal 15%, Italian energy company Agip 15%, Korean company Lucky Goldstar (LG) 5% and German coal company Rheinbraun 5%. In February Queensland Mines Minister Ivan Gibbs announced in principle approval for the project and the granting of an authority to prospect, adding that Ensham was one of the best steaming coal deposits in the State. The joint venture partners were now able to proceed with a feasibility study for a new port north of Rockhampton and electrification of the rail line. Production was expected to commence in 1987, but development was delayed by the reluctance of the companies to proceed at a time of over-capacity and low prices. Idemitsu wanted to push ahead to develop the mine and in February 1988 made an announcement in Tokyo that it had decided to invest in its development, although it admitted that it had yet to reach any agreement with the Queensland Government on rail freight rates. Idemitsu said that the mine would commence production in 1990, with initial output of 1 million tonnes a year, rising to 3.6 million tonnes after 3 years; Idemitsu and LG would take all the initial production which would be sold to power and industrial users in Korea and Japan, including the Chubu and Hokuriku power stations and cement producers. This announcement, an apparent attempt to stir its partners into agreeing to an early start to the project, was arguably a little premature. Later that year CRA, keen to increase its stake in the project, made a bid for Bligh Coal (whose major asset was its 22.5% interest in Ensham) and quickly secured around 3% of Bligh Coal's shares. Idemitsu and Agip also entered the bidding for Bligh, and in October CRA upped its offer, but also signalled that it was not prepared to go higher. Idemitsu won the battle, emerging with ownership of Bligh Coal and a total share of Ensham of 45%, with CRA now holding 19% of the project.

Idemitsu increased its stake in Ensham to 57% in 1989 and that year also saw the Queensland Government meeting with the joint venture partners to try to get them to commit to a firm development plan for the mine. Idemitsu was keen to proceed, but CRA and Agip were not, believing that the market conditions made the timing not yet appropriate.[319] Shortly after the Labor Party gained power in Queensland in December 1989, the new Resources Minister, Ken Vaughan, wrote to the partners regarding development of the mine. Only Idemtsu and LG were apparently prepared to give the Government the firm commitment it was seeking to proceed with the project, and in May 1990 the Queensland Cabinet excluded CRA and Agip from the project by renewing the authority to prospect only for Idemitsu and LG. This was a decision which would not only upset CRA and Agip, but also the mining industry generally. However, the announcement of the ejection of CRA and Agip by the Government was made before the Resources Department had notified the companies that they had been sidelined. The head of the Department, Kevin Wolfe, who had been retained by the new Government in his position, soon found his job being advertised. The Queensland Chamber of Mines protested against the decision to cut CRA and Agip out of the project, and in a meeting with Premier Wayne Goss sought assurances about security of mining tenure. Goss was keen to allay concerns and said that the decision was a one off.[320] Terry O'Reilly, head of CRA's Pacific Coal, said his company was extremely concerned about the Government's "unprecedented action", with CRA having been involved in the project since 1983, and with some millions of dollars spent on the project in the intervening years.[321]

The development of the Ensham mine commenced in 1992-93 and its first export shipment was made in October 1993. But CRA and Agip did not depart the scene quietly, launching legal action against Idemitsu in the Queensland Supreme Court. In February 1993, Justice Ryan found Idemitsu guilty of "misrepresentation and deceptive conduct" and ordered the company to pay CRA and Agip $29.5 million each. Justice Ryan said that Idemitsu had produced letters of intent from Japanese customers who were interested in buying up to 1.6 million tonnes of coal from Ensham. However, the judge said that Idemitsu also had side letters, which it did not reveal, and which in his opinion "rendered them valueless as evidencing any genuine intention to purchase the coal." Ryan found that Idemitsu's representations to the Government "were clearly derogatory of the conduct of Pacific Coal and Agip" and breached Idemitsu's duty to its partners.[322] Justice Ryan was also very critical of CRA, finding that it "abused the process of the court" and had threatened to ensure that the project was "put on ice" if it did not gain control. He also said that CRA and Agip had at one time breached the joint venture agreement. However, an appeal by Idemitsu to the Queensland Court of Appeal saw CRA and Agip ordered to repay a total of $27.5 million to Idemitsu, although the Court still found that key parts of the Supreme Court decision were correct. The president of the

Appeal Court, Justice Tony Fitzgerald, referred to Idemitsu's "continuing misrepresentations" about potential customers for the mine's coal. He also referred to delay and obstruction by CRA and Agip which breached the original joint venture agreement, with these actions partly responsible for the Government's decision to exclude them from the project.[323] So while CRA and Agip emerged winners from the court battles, they lost the war and were still no longer part of the project.

When the mine's development was finally approved, the partners in the joint venture were Idemitsu with 95% and LG with 5%. Subsequently, Japanese company EPDC bought into the mine, purchasing 10% from Idemitsu. Ensham had also been a beneficiary of Keating's change in foreign investment policy in 1990 that saw Gordonstone able to proceed with 95% foreign ownership. Ensham began production in 1993-94 with an output of 500,000 tonnes and was producing almost 4 million tonnes a year by 1998-99. Idemitsu also owned the Ebenezer in the West Moreton, gaining full control in 1987. In 1989 it bought the Muswellbrook open cut mine in NSW, and in 1992 increased its share of the Boggabri mine in NSW to 100%, buying out BHP and Agip.

North Goonyella, a high quality coking coal deposit, located around 150 kilometres west of Mackay, was another major development which kicked off in the early 1990s. The mining lease was granted in October 1991 and development of the underground mine commenced in April 1992, with the longwall commencing production in 1993. This project was a joint venture between White Mining (51%) and Sumitomo Corporation subsidiary Sumisho (49%). Sumitomo had bought into the project in 1989 with an initial 10% stake. Unlike Ensham and Gordonstone, North Goonyella had an uneventful birth, with the time between the granting of the lease to initial coal production only around 2 years. The mine's production reached 2.2 million tonnes in 1994-95 and 3 million tonnes in 1997-98. Japan was the mine's major customer in 1997-98 taking 1 million tonnes, but other significant customers included India, Taiwan, UK, Belgium and Germany. North Goonyella became a fully owned subsidiary of Sumitomo in 1997-98 and was then purchased by German coal mining company RAG and Thiess in 2000-01. It changed hands again in 2003 when Peabody became the new owner.

As the 1990s progressed, other new developments also got underway, with coal companies and investors recognising that the outlook was for strong growth in coal demand, even if competition for markets could be expected to remain fierce. The Queensland Government also commenced to more aggressively promote the State's potential for steaming coal developments in 1994 with the commencement of the QTherm initiative. Until that time Queensland was rightly seen as the premier coking coal centre, with NSW, and in particular the Hunter Valley, recognised more for its steaming coal resources as well as for lower quality coking coal. In January 1994 Savage Resources, then with significant coal operations in the Hunter Valley, won the tender issued by the Queensland Government for the Togara North project, a steaming coal deposit in the Bowen Basin. Savage partnered with Mitsui and Korean companies to win this tender, but it was subsequently taken over by Pasminco and the project was on-sold to Glencore in 1999. Despite the initial interest in the project, it remains undeveloped. Another steaming coal project which did get underway in the 1990s was Wilkie Creek, located on the Darling Downs near the town of Macalister. Wilkie Creek was developed by AQC (Allied Queensland Coalfields), with construction beginning in 1993-94 and with its first coal railed to customers in February 1995. The major market for the mine has been Japan, with some sales made to local power stations. Peabody Energy bought the mine in August 2002.

Coking coal projects dominated the developments in Queensland in the 1990s, with 1995-96 the year which saw five new developments commenced, as well as expansion at Blair Athol and Ensham, and a second longwall installed at Gordonstone. The new mines were Burton (Portman Mining 95%), Moranbah North (Shell 87%), South Walker Creek (BHP 80%; Mitsui 20%) Newlands open cut (MIM 75%; Itochu 25%) and Oaky North (MIM 75%; Sumitomo 15%; Itochu 10%); all but Newlands were coking coal mines. QCT's Kenmare underground mine, part of the South Blackwater mines, commenced production from its longwall in November 1996. BHP's Crinum underground mine (part of its Gregory/ Crinum operation) commenced production in June 1997, with development beginning in July 1994. Foxleigh, a PCI mine in the Bowen Basin, commenced development at the end of the decade, and started producing in January 2000.

Queensland's overall coal production grew strongly in the 1990s, up from 78.4 million tonnes in 1990-91 to 124.3 million tonnes in 1999-2000, an increase of around 59%. Coking coal production was up 55% and steaming coal 63%. Exports grew from 62 million tonnes in 1990-91 to 105 million tonnes in 1999-2000, an increase of 69%. In terms of both production and exports, the growth in the coal industry in Queensland easily outpaced that in NSW.

NSW mining developments also get underway

The 1990s in NSW saw a number of major new mines and some smaller mines developed, but the decade was also marked by many mine closures due to market conditions, lack of profitability and resource depletion. Closures occurred in all the major coal districts. In the West, Western Main underground and open cut mines were closed in 1994, although the open cut re-opened under different ownership in 1998. Other closures in the West included Canyon (1997), Clarence (it closed in 1998, and re-opened later that year) and Invincible. In the southern district, South Bulli closed in 1997 (it re-opened as Bellambi West in 1997) and Oakdale in 1999.

Camberwell Coal, located 10 kilometres north west of Singleton, was a joint venture between civil engineering company Henry Walker (50%), Toyota Tsusho Mining (40%) and Dia Coal Mining Australia (10%). The joint venture was formed in 1989, with Henry Walker buying into the project in March 1990. Construction of the mine infrastructure began in April 1990 and the mine was officially opened by Premier Nick Greiner and Mineral Resources Minister Neil Pickard in May 1991. With little new development in the mining industry having occurred under his premiership, and with an election looming later in the month, Greiner was obviously keen to be associated with a significant new mining development. Camberwell's development had taken only 15 months, with the company claiming that time a record for the industry.[324] The initial plan for this open cut operation was for annual production of around 1.3 tonnes of steaming and soft coking coal, but production grew to 2.2 million tonnes by 1998. Camberwell later became part of the Vale group, the Brazilian mining company previously known as CVRD, and was integrated with the Glennies Creek joint venture.

Shell Coal's Dartbook underground mine near Muswellbrook (with other shareholders Marubeni, SsangYong Resources and Showa Coal) commenced construction in June 1993, and production commenced in October 1994, its longwall producing coal in October 1996. The mine was a challenging one, with steep grades and above average levels of gas, although the results from the mine's first longwall block were good and by 1998-99 Dartbrook was producing 2.9 million tonnes of coal.[325] Shell's experience with new enterprise agreements at Dartbrook is discussed later in this chapter.

The Bengalla mine was developed in the 1990s by Peabody, with joint venture partners Wesfarmers, Kepco, Mitsui and Taipower. Located close to Muswellbrook, the steaming coal project kicked off in February 1990 with the NSW Government calling for proposals from interested companies. It would take another 9 years before the new mine would start to produce coal. In what could be seen as a foretaste of community reaction and opposition to new mines in the Hunter Valley in more recent years, a court challenge was launched in December 1994. The challenge to the decision of Commissioner Kevin Cleland was by the Oatley family's Rosemount Wines which had vineyards close to the proposed mine, with the Oatleys claiming that dust from the mine would cause damage to their vineyards. Cleland had headed the Commission of Inquiry into the mine's proposed development and recommended it be approved. In January 1995 the Land and Environment Court upheld the challenge, finding that Cleland's recommendation to approve the mine was "manifestly unreasonable" and that material submitted to the Inquiry demonstrated that the mine would intrude visually into large parts of the surrounding countryside.[326] The Court referred the matter back to the Commissioner, and a second public inquiry began in April 1995.

The NSW Government, under pressure to ensure that major new projects in the State were not unduly blocked, gazetted a new state environment planning policy (SEPP 45) which overrode a council's local environment plans. Planning Minister Craig Knowles then approved the Bengalla project under the new policy. The Oatleys launched another challenge in the Court, and were successful, with the Court finding SEPP 45 unreasonable. The Government appealed to the Court of Appeal, but before a decision was handed down, it legislated to validate SEPP 45. Knowles said that the Bengalla project had stalled for more than three years under the previous Coalition Government, but that within 6 months the new Government had approved the mine and had clarified the rules for approving new mines. Knowles also said that Bengalla would be subject to some of the most stringent environmental criteria ever imposed on a mine: "Thirty conditions containing more than 90 controls will be monitored by an independent environmental audit. In addition, a consultative committee will also be set up to provide direct and ongoing input from local residents. These conditions have set new benchmarks for the management of the environmental impact of mining activities on a national basis."[327]

The mining lease for Bengalla was issued in June 1996 and development of the mine commenced. The mine despatched its first coal in April 1999 and by 2001-02 it was producing over 5 million tonnes per year. The ownership of Bengalla changed in 2001 when Peabody sold its NSW coal assets to Rio Tinto subsidiary Coal & Allied, and changed again in 2016 when Coal & Allied sold its share of the mine to Queensland mining company New Hope. In 2018 Wesfarmers completed its exit from the coal industry, selling its interest in Bengalla to New Hope, having announced the sale of its Curragh mine in Queensland in December 2017 to Coronado Group.

Springvale colliery near Lithgow in the western district was developed in the early 1990s to supply the new Mt Piper power station; Springvale had won a contact with the NSW Electricity Commission for 2 million tonnes per year for 20 years, with an overland conveyor running from the mine to Mt Piper. Clutha Ltd developed the mine in a joint venture with Samsung of South Korea and planned to export around 600,000 tonnes of its production. The power station contract called for coal to be supplied from 1995, but by the start of that year Clutha was in financial trouble and was placed into administration. By May the administrator had also closed Clutha's Oakdale and Brimstone mines in the Burragorang Valley. Springvale is now operated by Centennial Coal which has a 50% share of the mine, with the other 50% owned by S K Kores Australia.

The Mt Owen mine, a steaming and semi soft coking coal deposit between Singleton and Muswellbrook, began operating in 1993 under the management of Hunter Valley Coal Corporation Pty Limited. BHP acquired the mine in 1995. Thiess was responsible for constructing the mine and has operated it under contract since 1995. In 1998, Mt Owen was purchased by Glencore, then in the early stages of its coal mine acquisition program, the acquisition adding over 3Mt to the Glencore stable. BHP's acquisition of Mt Owen was a strange move, as by the end of the 1980s, the company had sold off its Saxonvale mine near Singleton and its Macquarie group of coal mines in the Newcastle area and was concentrating on its Illawara group of hard coking coal mines for supply to its local steel works and for export. The sale of Mt Owen in 1998, however, did not mean a permanent exit for BHP from the Hunter Valley as the merger with Billiton in 2001 brought the large Mt Arthur North mining operation (now Mt Arthur Coal) and the Coal Operations Australia mines into the new company. By 2001, BHP was back in the steaming coal business.

NSW also saw a number of mines close in the 1990s, only to re-open under new ownership or after a reassessment of viability of their operations. One period of only 14 months (the period between February 1998 and June 1999) saw 7 permanent mine closures. The Gretley mine was closed in July 1998 by Oakbridge (then part of Cyprus), but re-opened in May 1999 as New Wallsend No.2 under new owners New Wallsend Coal. The Pelton/ Ellalong underground mine was closed in May 1998 by Oakbridge and re-opened in August 1998 as Southland under new owners Southland Coal. The Lake Macquarie mines, Chain Valley and Moonee and Chain Valley, closed in 1994 and 1999 respectively.

But there were also closures which were permanent, including Oakdale in 1999, the last mine in the Burragorang Valley. CRA's Vickery mine in the Gunnedah area closed in 1998 following the prolonged strike which we discuss in this chapter. The Preston Extended mine near Gunnedah was closed by Centennial in 1998; the Preston mine dated back to 1970 and was the first mine purchased by Centennial when the company was established in 1989. In the western district, CRA's Western Main open cut and underground mines closed in 1994, with the open cut opened and closed more than once under Centennial's Springvale Coal ownership, before it finally closed again in 1999. The Western Main open cut minesite is now the location for Springvale mine's processing facilities. Coalpac's Canyon mine closed in 1997. Cyprus' Clarence mine closed in February 1998 following an industrial dispute, re-opening in September that year under new owners Centennial. In the southern district South Bulli closed in August 1997, re-opening the same month as Bellambi West. The Metropolitan mine, first developed in the 1880s, closed in November 1998 and re-opened under new ownership of Helensburgh Coal in May 1999; it is now owned by Peabody Energy. BHP's Elouera mine opened in 1993, amalgamating the Nebo and Wongawilli mines.

Billiton enters Australian coal industry

One of the major companies to enter the industry in the 1990s was South African miner Billiton, a subsidiary of Shell; Billiton's Australian assets in the early 1990s included major interests in the Worsley bauxite and alumina operations and Boddington gold mine, both in Western Australia. South African company Gencor made a bid for all of Shell's international metal mining businesses in 1993, and by 1994 it had secured the Billiton assets in Australia. Gencor at that stage did not seek to buy Shell's Australian coal assets. Gencor had considered bidding for McIlwraith McEacharn in late 1992 to gain control of McIlwraith's Oakbridge coal assets, but was unable to proceed due to South African foreign exchange restrictions. Cyprus Mining was subsequently successful in gaining control of McIlwraith and Oakbridge.

Gencor's Ingwe Coal then made the move into the Australian coal industry and its target was Coal Operations Australia Ltd (COAL), the company established by Tony Haraldson in 1994 with the backing of AMP and which had purchased the Coal & Allied underground mines following CRA's takeover of Coal & Allied. Haraldson, the ex CEO of Coal & Allied, knew the assets well, and although these mines were not generating great profits, he understood their potential. Haraldson had assembled an impressive board which included John Kerin, a senior minister in the Hawke and Keating governments, and Toby Rose, an ex-director general of the NSW Department of Mineral Resources. By the start of 1995, Haraldson and his board were gearing up to float COAL and were within weeks of submitting a prospectus to ASIC for approval when the Ingwe bid for the company emerged. Ingwe was successful in taking control of COAL, keeping Haraldson on as CEO, and appointing David Murray and Mick Davis to the board. Billiton was divested from Gencor in 1997 and listed on the London Stock Exchange. Ingwe Coal became wholly owned by Billiton in 1998.

COAL won the right to develop the big Mt Arthur North deposit in the Hunter Valley in 1998, undercutting bids from a number of other companies. Critical to its win was the price it offered for the supply of coal to the state-owned Bayswater power station, with the Government requiring the successful developer of Mt Arthur North to supply Bayswater with 5 Mt of coal per year for 5 years. COAL offered a price of 29.5 cents per gigajoule, which equated to around $4 - $5 dollars per tonne of coal. While the coal quality specified for Bayswater was not high, the low price in the bid proved to be a cheap way to gain the right to a lucrative deposit and develop a mine largely for the export market. COAL also won the right to develop the Wyong Area leases, which are now owned by Korean company Kores. COAL was also able to gain control of the Bayswater No.2 and Bayswater No.3 mines in the Hunter Valley when Caltex sold out of Bayswater Colliery Company.

Another project owned by COAL was Foxleigh located north of Blackwater in the Bowen Basin. COAL had done some initial exploration on the Foxleigh deposit, but Ingwe's chief coal geologist did not see any great potential for this deposit. On the geologist's recommendation, COAL's board decided not to renew its exploration lease with the Queensland Government. However, eyeing a potentially attractive project, John Thorsen and other colleagues from COAL then won the company's approval to ask the Government to transfer the lease to them. The Government agreed to the transfer, the new mine was developed quickly with POSCO and Itochu as joint venture partners and began producing PCI coal for export in January 2000. Anglo Coal bought a 70% interest in Foxleigh for $A700 million in 2007, delivering the ex-COAL executives a handsome profit. Unlike Ingwe's negative view of Foxleigh, Anglo was very positive, with its CEO Cynthia Carroll saying that Foxleigh's operations and undeveloped assets were "located in one of the world's largest and best-developed coal provinces", an area with which Anglo was very familiar. Carroll saw Foxleigh representing "a valuable strategic and complementary addition to our portfolio of coal assets in Australia."[328] Anglo sold its interest in Foxleigh in 2016 to Middlemount South Pty Ltd.

Coal prices, strikes and the Taylor Review

The 1994 year was memorable in more ways than one. Major national coal strikes rocked the industry in the early months of the year, with mine workers protesting against cuts to prices accepted by Australian coal exporters. The miners marched on Canberra. Prime Minister Paul Keating announced that the Coal Industry Tribunal was to be abolished. With the CFMEU once more calling for a national coal authority, the Government appointed respected senior bureaucrat Rae Taylor to undertake a review of the industry.

In January the bad news about coking coal was announced – the negotiations between BHP and the JSM had resulted in cuts in prices of $US3.30 to $US3.85, or around 8%, with cuts to tonnages of around 10%. This was devastating news for the industry in both Queensland and NSW, as the price cuts applied for both higher quality coking coal and soft coking coal. The CEO of NSW underground miner Clutha,

John Doherty, called this the worst blow to the industry since World War Two.[329] The announcement was followed by a series of 24 hour strikes in the various coal districts and in late February around 20,000 mine workers began an indefinite strike which shut down mines in NSW and Queensland. Mine workers had walked off the job on Saturday night 26 February in advance of mass meetings to be held the following Monday to protest at the price and tonnage cuts. A large contingent of miners was also planning to travel to Canberra on Tuesday for a rally outside Parliament House to increase pressure on the Government for changes in the way the industry was run.

On 1 April the media reported on the negotiations between Shell, Ulan and MIM and Chubu Electric Power Company which resulted in a cut in the benchmark price for steaming coal of $US2, or around 5.5%. The contract price was now down to $US34.35, a price which Shell said it regarded as "a good result in the light of much greater price falls for both oil and natural gas which compete with coal in the power station market…The Japanese price for thermal coal is the highest available to Australian suppliers and it is essential that we continue to maximise our position within this market." MIM said that after seven rounds of negotiations over several months that were "very long, very detailed, and thorough… we regard it as an equitable result." MIM also pointed to the fact that Chinese coal exports to Japan had suffered a slightly greater price cut of around US$3.50.[330]

One of the significant outcomes of the 1994 steaming coal negotiations was the severing of the traditional link between coking coal and steaming coal prices. The nexus between coking coal and steaming coal is lower quality coking coal which is produced largely in the Hunter Valley. As the IEA noted in a 1997 publication: "There are some grounds for expecting prices to be linked, since some substitution between coking and steam coal is possible. Coking and steam coal prices in the Japanese market have been linked by reference to a price for 'semi-soft' coking coal, which can be used as a low grade coking coal, or as steam coal." The IEA went on to say that "The use of (PCI) in blast furnaces has also enhanced the substitutability of steam and coking coal", while also noting that the link had been "stronger than would be expected from normal market considerations."[331]

The coking coal negotiations in the 1980s and 1990s tended to be between Utah, and later BHP, and Nippon Steel, or one of the other major steel mills, acting on behalf of the mills. The steaming coal negotiations involved different Australian exporters and one of the major Japanese power utilities, acting on behalf of the other utilities. The break in the nexus was a sign that the coking coal and steaming coal markets were developing in different directions, and that the major coal customers – the steel mills and the power utilities – were subject to very different economic forces.

Around 1500 mine workers descended on Canberra on 1 March 1994 to protest at the price cuts and demanded that the Government set up an independent inquiry into coal pricing. The miners held a rally in Canberra and voted to remain on strike until the end of that week. The CFMEU's John Maitland said that he believed that an inquiry was the only way the problems confronting the industry could be resolved, and said that a person such as former Primary Industry and Energy Minister John Kerin would be capable of addressing the industry's problems.[332] Maitland said a national coal authority should be established to "oversee the sale of all Australian coal" but admitted that the union "was not naïve enough to believe that the Government would allow the establishment of such a board." The mine workers who travelled to Canberra had come from various areas in NSW and Queensland, many from the Bowen Basin towns, and towns such as Rockhampton and Mackay and areas further north. The rally outside Parliament House was addressed by Parliamentarians from coal electorates, with all the speakers reported to have supported the union's demand for a national coal authority.[333] A tent embassy called the "Boot End Colliery" was set up near Parliament House.

The Federal Minister for Resources, David Beddall, announced on 14 March that an independent review of the industry would be undertaken, and on 24 March named Rae Taylor, an experienced Federal bureaucrat, to head the inquiry. Taylor had been the Secretary of several Federal Government departments and was the CEO of Australia Post from 1989 to 1993. Beddall said that the review would be "commissioned" by the Australian Coal Industry Council (ACIC), the tripartite body representing the coal producers, the CFMEU and the Federal, NSW and Queensland governments. In reality, although the ACIC was nominally given oversight of the inquiry, the inquiry was undertaken by Taylor, supported by a secretariat comprising staff from the Federal Department of Resources. The inquiry's terms of reference were relatively brief, although they allowed for it to canvass a wide range of issues: "Recognising the size of the NSW and Queensland black coal industry and the important place it occupies in the economy as a major export earner, generator of returns to the community, employer of labour and the regional importance of mines in each state...examine and make recommendations for the Council's consideration on appropriate action by governments, companies and unions on impediments to, and opportunities for, optimal coal industry, coal market and mine development in an environment of significant change and increasing market potential including: (a) the introduction of world's best practice in key aspects of coal production and export and (b) the expansion of Australia's coal trade and the capturing of new markets."

While the terms of reference noticeably avoided any reference to a coal marketing authority, the CFMEU welcomed the review, with Maitland saying that the union was pleased because the marketing of Australian coal was a subject that had been identified specifically as requiring examination. He also said the union intended to vigorously pursue its arguments with the inquiry, and would also be approaching coal producers to seek a moratorium on retrenchments while the inquiry was in progress.[334] The coal producers were less enthusiastic, some seeing it as an imposition and a waste of time, while others recognised it as a necessary part of the political process, with the potential, however, for positive or negative outcomes.

However, in a move which the union would soon live to regret, on Monday 18 March a national 48 hour coal strike began involving around 20,000 mine workers, in protest at the outcome of the contract negotiations and the refusal of the Federal Government to intervene in the market. The Government had gone a long way to accommodate the CFMEU's demands with the independent inquiry, but now the union's members were striking. At the union's Central Council meeting on March 21 to 25 a resolution was adopted warning the Government that the Taylor inquiry had to result in the outcomes the unions was seeking: "Council notes that the establishment of the study is the first step in dealing with the coal industry's problems and will not be judged a success unless its recommendations include measures to address coal marketing, job security and mine development. Council warns that failure to deal with the above matters will lead to further industrial action."[335]

At midnight on 17 April another national coal strike commenced in protest against the lack of action by companies and the Federal Government on the cuts to coal prices. The union had also been seeking a moratorium on retrenchments in the industry and Maitland said that this, together with a lack of a coordinated response on jobs, were other factors causing the strike action.[336] The latest strike action, however, was the final trigger for the Government to take what can only be described as a courageous move. On 19 April Employment Minister Laurie Brereton announced that one of the industry's sacred cows, the Coal Industry Tribunal, would be abolished, and would be absorbed into the Australian Industrial Relations Commission. Arguably, the union had shot itself in the foot, but now would fight to overturn the decision by Brereton and his cabinet colleagues. The coal producers however, or the great majority of them, were equally determined to see the end of the CIT. We will come back to the demise of the CIT shortly.

The 1994 coal price settlements

At times the constant public questioning by the unions and some media of the competence of the Australian coal company executives involved in price negotiations (and in fact of the system of price negotiation) has tended to mask the commercial realities and the complexities of the international coal market. Therefore, before looking at the Taylor inquiry and its outcome, some background on the price settlements agreed between the Japanese and the Australian exporters may assist to put the 1994 market situation into perspective.

The Australian Coal Association's summary booklet of its Winning Coal submission to the Taylor inquiry contained a brief, but nevertheless informative, background to the negotiations between Nippon Steel and BHP in relation to the coking coal negotiations, and between Chubu Electric Power Company and Shell, Ulan Coal and MIM in relation to the steaming coal negotiations. The following extracts reproduce material from that summary document.[337] The background to the coking coal settlements was that "Participants in the JFY 1994 negotiations were aware of the Japanese mills' heavy losses in steel making, the then (late 1993) current pessimism about the outlook for recovery in the Japanese economy and the corresponding likelihood of downward pressure on the benchmark price. The background encompassed an anticipated decline in Japanese blast furnace steel production in JFY 1994, slow growth in steel output in other metallurgical coal importing countries, and strong competition from other suppliers of coals.

"The round was brought to a head when Canadian coal producers, in mid-December 1993, offered Nippon Steel, the lead Japanese negotiator, to cut hard coking coal prices by about $US4/tonne off the 1993 level. According to Fording, the dominant Canadian exporter to Japan, the mills first determined the maximum price cut acceptable to the Australians and then went to the Canadians, who 'indicated their willingness to the competitive.' The Canadians, who had become relatively high cost producers, suffering heavy losses in recent years, were determined to win back tonnage after some years of declining market share. Canadian mines have undertaken major structural reforms and productivity has increased markedly. Those without long term contracts were prepared to be very aggressive. With the depreciation of the Canadian dollar and some rail freight concessions, the Canadians were reportedly able to offer the Japanese even larger price cuts, perhaps as high as $US6/tonne off the 1993 price, and still increase their revenue (in Canadian dollars).

The Canadian offer was instrumental in determining the international benchmark price for Australian producers.

"Faced with a deal with the Canadians, Australian coking coal sellers had only two options: either to accept the Canadian price as the benchmark for their price-quantity negotiations or reject it and continue bargaining. With other world markets unlikely to offer better terms and certainly incapable of taking up the additional tonnage, the tough alternative for the Australian producers would have involved tonnage cuts and risked future losses of market share. (It was the Australian attempt to force the Japanese to offer higher coal prices in the 1970s that prompted Japan to invest in setting up the Canadian coal industry). It was Hobson's choice.

"Nevertheless, having accepted the Canadian price cut as the basis of the 1994 price benchmark, the Australian negotiators managed to improve on the Canadians by securing the price cut of $US3.85/ tonne for hard coking coal and $US3.30/ tonne for soft coking coal with only a small cut in tonnage (though sizeable tonnages were reclassified to lower price categories). The negotiations produced an estimated average price of $US44.90/ tonne for Australian coking coals in JFY 1994, an eight per cent reduction on the JFY 1993 level.

"In April 1994, US coking coal supplier Pittston finalised negotiations with the JSMs for JFY 1994 deliveries, accepting a price reduction similar to the Australian and Canadian settlements. Modifications to specification were reported to have partly compensated for the price reduction, in recognition that market conditions had firmed since December. The difference between long tons and metric tonnes is said to have given Pittston a further small benefit. Announcing the settlement, Pittston claimed success in negotiating new categorisations. The mills denied the changes offset the benchmark price cut. No publicity was given to coking coal settlements by other North American exporters to Japan."

The background to the steaming coal settlements was as follows: "Subsequent to the JSM settlements with Canadians and the Australians, the coking coal benchmark price became the key reference price for the JPUs, led by Chubu Electric Power Company. Chubu took the line that, given the high degree of substitutability between semi-soft (coking coal) and thermal coals, Australian producers should match the achieved cut for semi-soft coal with that for thermal coal. This was despite stronger demand for electricity generation in Japan, strong profits made by the Japanese power utilities and the tightening supply of thermal coal. Spot market prices had firmed from $US21/ tonne (FOB) in the second half of 1993 to $US 28-30/ tonne (FOB) during the early months of 1994. Also, the 1994 China-Japan agreement, which preceded the Australian negotiations, resulted in Chubu and China Coal Export Import Corporation settling the JFY 1994 price cut at the year's semi-soft coal price cut of $US3.30/ tonne. (In exchange, Chinese suppliers gained additional 0.8-1.0 million tonnes of contracted thermal coal for JFY 1994).

"Early in 1994, Australian coal production and exports declined as a result of industrial unrest, flooding in Queensland and technical mining difficulties in New South Wales. Coal stocks fell by 14 per cent in the first quarter of 1994 (to the lowest level since August 1989). In the weeks following the somewhat premature coking coal settlements, the outlook for the Japanese economy turned significantly for the better.

"Accordingly, the Australian negotiators, led by Shell, Ulan Coal Mines and MIM Holdings, rejected the $US3.30/ tonne proposed price cut. Chubu proposed a two-tier system which would have involved a price cut of $US1.50/ tonne in the first six months of the new contract year, rising to a $US2.50/ tonne in the subsequent six months. Finally a flat rate reduction of $US2.00/ tonne (FOB for 6700 kcal/kg) was agreed, with the NSW benchmark coal prices for JFY 1994 set at $US34.35/ tonne.

"The JFY 1994 thermal coal settlement broke the nexus between thermal and metallurgical contract-based coal prices, narrowing the gap between the semi-soft coking and thermal coals to $US1.95/ tonne. Although the 1994 price cut had been particularly painful because it coincided with the appreciation of the $A relative to the $US, breaking the nexus with coking coal represents a considerable achievement for the Australian producers.

"In the Japanese industrial market, the very steep discount to power utility contract prices, which was a feature of JFY 1993, has narrowed by several dollars in the first half of calendar 1994. The further narrowing margin between the Japanese electricity generating sector's contract-based prices and industrial sector's spot prices is likely to continue in the second half of 1994. This should benefit thermal coal exporters."

The Taylor inquiry

The Taylor inquiry was treated as a potential watershed for the industry by both the coal producers and the CFMEU. The Australian Coal Association, as the representative of the NSW and Queensland companies, commissioned ACIL to prepare a major submission, with input from the companies and the two State associations, the Queensland Mining Council and the NSW Coal Association. The inquiry commissioned some research, including an analysis of enterprise agreements by ACIRRT (the Australian Centre for Industrial Relations Research and Teaching) but relied largely on the submissions it received, its own research and meetings with a range of major companies, unions and the employer groups.

The CFMEU submission, *Out of the Red into the Black*, was prepared by its national research officer, Peter Colley, and laid out the union's case for intervention in the coal market. The submission was notable for its apparent recognition that the decades-old strategy of pushing for a national coal authority was unrealistic. The submission focussed on the "collusive buying practices" by Japanese "cartels" which accounted for almost half of all coal sold by Australia, with the cartels using "their collective purchasing power to set a ceiling price for coal which flows through into sales into other markets."[338] The CFMEU argued that coal consumers and trading companies investing in the industry were doing so in order to drive down prices, and were accepting losses on mining investments which were more than recouped through lower coal prices. It also said that customers were using minority shareholdings in order to gain inside information on costs of production. Interestingly, the CFMEU's submission did not propose a national coal authority, and said that Australia needed to "learn to play as a team in its coal trade if the industry is to ever make a contribution to Australis that its tremendous natural advantages should make it capable of."[339] To achieve this, it recommended "the creation of an effective countervailing ability to match the collective practices we face", with the creation of a high-profile, well-resourced unit by the Australian Government to administer the export permit system. A range of other measures was proposed including "the establishment of a peak industry association dealing explicitly with price formation and marketing issues."

The CFMEU stated that economic modelling undertaken for the union indicated that at least 5,000 and possibly up to 20,000 jobs had been foregone as a result of "artificially depressed" coal prices, with these jobs being in coal mining and in a range of other industries. The modelling report, an appendix to the submission, showed the results of two sets of simulations. The first simulations were based on a higher price for Australian coal assuming that the price was based on a sharing of the benefit of the appreciation of the Japanese currency since 1982 between the Japanese coal customers and the Australian exporters. The second set of simulations was based on assuming a narrowing of the differential between the prices paid by Japan for Australian coal and coal from other sources. The economic modelling report did in fact show that these numbers of jobs could be created throughout the economy, although for the higher jobs figure, the increase in the price of Australian coal to Japan was 60%. Importantly, the consultants pointed out that their report did not address the issue of whether the coal price increase assumed in its simulations could be achieved by changing the marketing arrangements for Australian coal. For Australia to have been able to obtain significant increases in prices for its coal, without any loss of market share, is of course unrealistic. Not only did the 1980s and early 1990s see the emergence of major competitors in the international coal market, including South Africa, Indonesia, China and Colombia, these years were also notable for the intense pressures on Japan's economy from the appreciation of its currency, with that appreciation reducing the cost of imported materials such as coal and iron ore, but also reducing the competitiveness of its major export industries such as steel and motor vehicles which were major consumers of the same raw materials. The CFMEU recommended that fair and genuine competition in the international coal trade required action by Australian governments, and that all forms of collective bargaining by sellers and buyers should be illegal. How the Japanese would be forced to change their approach was not clear.

The ACA/ ACIL submission was a detailed one examining various aspects of the industry, with the consulting team led by John Daley, who had also worked for the Federal Government in various senior roles (including in the Coal Branch of the Department of Resources and Energy) and had a good understanding of the coal industry. The ACA already clearly understood the CFMEU arguments in relation to the market, arguments which the union had been espousing for some time, including those by John Maitland in a presentation to the National Press Club in 1994. The ACA submission therefore took these arguments head on and, in relation to the coal market, and, in particular, Japan's role as the major buyer and price setter, emphasised that the Japanese did have a raw materials procurement strategy which worked effectively to provide the quality and dependability of coal supplies that its major consuming companies demanded. That strategy also provided Australian exporters with the highest prices, with those exporters queueing up after the annual benchmark price settlements to sell as much as possible at the benchmark price. The Japanese system paid a premium price to secure quality, dependability and diversity of supply, with their contracts guaranteeing priority when supply was tight. Of course, the Japanese had not developed their purchasing strategy to be generous to Australian exporters, however, it was a rational strategy which satisfied their need for quality, dependability and priority in supply. The strategy also aimed to minimise costs over the long term, subject to ensuring acceptable quality and tonnages.

The ACA submission also argued that intervention by the Government in the market would jeopardise the "dependability premium" which

Australian exporters received, and would almost certainly induce customers to diversify their purchases to Australia's detriment. The Japanese were already well aware of their heavy reliance on Australia as their major source of supply for both coking and steaming coal and would be expected to retaliate in the event of government interference. The parties who would then suffer would be the industry and its workforce.

Taylor's report was released publicly in December 1994 and covered a range of issues, but the key section of the report was its conclusions on coal marketing, and specifically on the CFMEU's proposed alternative. Taylor concluded that: "The Study considered that the CFMEU approach of setting up a countervailing force to Japanese joint negotiating arrangements in the price formation arrangements is a high risk strategy and is not appropriate."[340] The report stated that several factors led the inquiry to that conclusion, including the fact that Japanese buyers paid the highest prices and were also our biggest customers. Those customers were believed to have the potential to react adversely either individually or collectively to such a market signal by paying higher premiums to other suppliers, making medium to long term investments in coal in other countries, reducing the contract tonnages for Australian suppliers, and looking to technology to reduce coal demand, and to alternative fuel strategies.[341] The report also questioned the CFMEU's strategy which implied that Australian companies would be willing to withhold supplies from the Japanese market, a situation which would lead to financial losses which would need to be shared by others in the industry, and noted that "Other supplier countries, rather than cooperate with Australia in restraining supply, may see it as a marketing opportunity." The report also stated that it was not clear that the marketing practices of the Japanese had been detrimental, with the Australian coal industry having achieved significant growth and a high share of the Japanese market.

Taylor, while opposed to the CFMEU's proposals relating to coal marketing, still argued for export controls to remain in place, saying that "there had not been sufficient change in circumstances since the 1992 Government review to warrant their abolition."[342] The abolition of export controls was to come in another 3 years and occurred after the Howard Government won power. However, Taylor did make some recommendations to improve price transparency and provide greater confidence in the marketing of Australian coal. His key recommendations were that the Minister should meet annually with companies to review the operation of export controls and discuss market information; that the Department distribute on a regular basis a statistical series on prices for Australian coals and coals from other exporters; and that ABARE should regularly publish information on world coal trade.

The Taylor report was the subject of a meeting of the Australian Coal Industry Council in March 1995 chaired by David Beddall. Beddall told the media after the meeting that it had resulted in unprecedented agreement between trade unions and employers, and that the industry had turned the corner towards better productivity and marketability. He said that the meeting had agreed to improve the industry through better industrial relations, market diversification and attention to greenhouse gas emissions. "For the first time, there's a cooperative spirit between companies and unions, with a common goal." Mr Beddall said.[343] He said that the unions had accepted that Taylor's recommendations to improve price transparency could be implemented, and that he would take a package of measures to Cabinet, one of which would be for the Australian Bureau of Agricultural and Resource Economics (ABARE) to receive special funding to provide regular information on movements in coal prices, a measure that would boost the confidence of the coal industry's workforce in coal prices. Beddall also said that the climate was right for industrial reform, with an increase in cooperative spirit and projections of significant price rises for the year ahead. Another key outcome of the meeting was agreement on the need for market diversification, with the potential for trade missions to India, Thailand and southern China.

Beddall took a package to Cabinet later in the year and won approval for some modest measures in August 1995. The major initiative, announced as part of the 1995-96 Budget was a $12 million Coal Australia Promotion Program; another program involved ABARE examining the possibility of a public price series for coal which could involve other coal producing countries, and providing ABARE with the resources to improve its capacity to forecast world coal supply and demand. Industrial relations also got a mention, but that component of the program drew on existing programs in the Department of Industrial Relations, with work to focus on developing performance indicators of world's best practice, study tours for mine operators and workers to inspect top operations in other countries and training programs to help the industry adopt a stronger enterprise focus. The announcement of the measures was rounded off with the Government urging the industry "to develop a stronger enterprise focus in the implementation of change" with this requiring "unions to be flexible and willing to embrace change" and management "to be consultative, and demonstrate a willingness to allow genuine work force participation in the decision-making."[344]

The coal promotion program did result in some trade promotion missions to The Philippines and to Thailand, but overall, the package of measures announced by Beddall, while providing some projects for his Department to work on, made no significant change to the industry. For the coal producers the outcome was positive in that further government involvement in the key aspects of the industry was minimised. For the CFMEU, there was little support in the report for its major concerns, but having taken part in the inquiry, it now had to essentially live with its findings.

Japan and Australian coal trade

By 1990 the coal trade between Australia and Japan was just over 30 years old, dating from the first significant coking coal contracts signed with NSW South coast producers in 1958. The steaming coal trade was less mature, with exports becoming significant in the 1970s. The 1990s saw the role of government in the industry reduce, with the 1994 Taylor review of the Australian coal industry opposing intervention in the market, and with the formal abolition of export controls by the Federal Government in 1997 (although by that time the controls had been administered in a light handed manner for several years). But the 1990s also saw a greater focus on the coal trade between the two countries by analysts, academics and journalists, with the emphasis on the question of Japan's negotiating strategies, its encouragement of excess capacity in Australia and other countries over a number of years, the role of the Japanese trading houses, and the lack of coordination in negotiations by Australian exporters. The general tenor of a number of articles, submissions and books on the question was that Australian exporters were achieving lower prices than were justified, and were failing to coordinate their negotiations. The end result, according to some of these writers was that Australia had failed to capture all the value from the coal trade with Japan, with the Japanese steel mills and power utilities benefiting at the expense of Australian producers and the Australian community. The policy proposed to counter this situation was often the strengthening of export controls by the Australian Government or the establishment of a national marketing authority.

The Australian Government and the JCB started to raise concerns about the prices being received by Australian coking coal exporters from the late 1960s. In its 1967-68 annual report the JCB said that the prices for Australian coals exported to Japan "…were unduly low compared with prices paid to United States, Canada and other suppliers, even when quality allowances were made."[345]   The JCB's 1971-72 annual report referred to the then current export contracts being based on a cost plus a "reasonable profit" formula, this practice carrying over from the 1950s and 1960s when the export market was being developed. The JCB warned that it was now "…essential that the coals be sold on their merits, reflecting the true value to the consumer."[346] Its next annual report called for coordination between NSW and Queensland, and criticised the "unduly low" export prices being received. The JCB warned exporters against being "invited to participate in competitive price reductions", saying that "The response to any such invitation must be a flat refusal."[347] The views of the Federal Department of National Development on the coal market were clear from a Cabinet submission in April 1972 which went into detail about a number of issues confronting the industry, including the trade between Australia and Japan. The submission said that, allowing for quality factors, there was "strong evidence to suggest that our coals are being undersold on world markets" and that the Japanese buyers acted as "monolithic organisation" and were adept at "playing one supplier against another."[348]

One prominent book on the resources trade between Australia and major Asian countries, particularly Japan, was published in 1994 by Australian Financial Review journalist Michael Byrnes; it was extremely critical of Australian iron ore and coal exporters.[349] Byrnes described the annual negotiation process that took place in the 1970s and 1980s which involved around ten Australian soft coking coal exporters and the seven Japanese steel mills: "This involved the Australians ritually travelling to Tokyo each year, promising each other they would form a common Australian front. The Japanese also formed a united front, by appointing a 'lead negotiator' who was backed up by a committee of senior executives from the main Japanese mills. However, the Japanese united front always held, while the Australian front typically broke down… Each year the Japanese steel mills breached Australia's negotiating walls by offering increased sales tonnages to individual Australian companies, in return for them agreeing to prices on offer. At a certain stage it became too tempting for an individual coal company to refuse the mills' offer, as long as the individual company thought only of its own returns. ..Once one company has accepted Japan's price the remaining Australian companies must also accept, or find themselves cut out of the market." Byrnes also said that foreign companies in the coal and iron ore industries were the leaders in breaking with the Australian front to accept the Japanese steel mills' pricing demands, with Utah (foreign owned until its purchase by BHP in 1984) often the first to give in.[350]

Byrnes went on to look at the export controls applied by the Australian Government in the 1970s and 1980s and what he claims were the Government's attempts to "enforce negotiating unity among the Australian coal companies by refusing to allow exports to take place if prices were judged too low." He says that the policy failed for three reasons. First, Japan was able to buy coal (and iron ore) from other countries. Second, having achieved virtual control of the international coal trade, the Japanese had information about the costs and prices of the industry which was not available to the Australian producers or the Government. Third, it was in the interests of individual coal companies to control their own organisations, rather than ceding control to the Government. The Government was also frightened, he said, to enforce unity among the Australian exporters for fear of being accused of being socialist or anti-business.[351] Byrnes also referred to the fact that Rex Connor as Minister for Resources and Energy was able to force the Japanese to pay higher prices as evidence that the increase in price reflected the loss that Australia was suffering as a result of the lack of unity by our exporters. Byrnes concluded that "If by forcing unity on the Australian coal negotiators Mr Connor was able to produce a price rise of up to US$20 per tonne, that US$20 a tonne would seem to be one clear economic cost to Australia of its systems when pitted against those of Japan."[352] In fact the claimed $US20 price increases achieved by Connor did not mean that Australian exporters had achieved a long term gain relative to their competitors. Connor did force up prices to a certain extent before he lost his job in 1975 in the Whitlam Ministry, but one has only to look at the data on prices received by the major coking coal exporters in the 1970s to see that Australia continued to receive lower prices from the JSM than the USA and Canada. And in fact all the major exporters saw significant increases in prices for coal sold to Japan in the 1970s.

In attempting to make an assessment of the fairness of the coal trade between Australia and Japan, there are many factors to consider. One of these is market share. When Australia entered the coking coal trade with Japan from 1958, the USA was Japan's dominant supplier. However, within a relatively short period, Australia was challenging the dominance of the USA, and by the 1980s had become the leading supplier to the JSM. It was not long before the USA was consigned to the role of "swing supplier", a term used in the trade to describe a supplier which swings into the market when times are good and prices are high, and out again when times are tough. A second factor is the composition of coal traded, and in particular the relative shares of high quality coking coal and lower quality soft coking coal. A significant component of the Australian coking coal exports to Japan in the 1970s and 1980s was the lower grade coking coal which brought a lower price than the hard coking coal from the Bowen Basin and the NSW southern district. Australia's exports of soft coking coal have continued to be significant, with production mainly coming out of the NSW Hunter Valley and the Gunnedah coalfield. Lower priced PCI coals, also used in steel making, have also grown in importance since the 1990s. These are factors which need to always be taken into account when comparing prices for coking coal exported by various countries, with the USA and Canada exporting mostly high priced hard coking coal.

In a submission to Cabinet in 1972 the Minister for National Development included an analysis of Japanese coking coal imports in 1970 and 1971 from Australia and other countries broken down into 4 categories – hard coking coal less than 8% ash; hard coking more than 8% ash; semi hard and soft coking coal less than 8% ash; semi hard and soft coking coal more than 8% ash. The average value per ton received by Australia was well below the average for the USA in all 4 categories, the difference in the second category in 1971 for example being $A14.43 for Australian exports compared with $A24.92 for US exports and $A16.83 for Canadian exports (all these values were c.i.f. values ie delivered prices into Japan). This was certainly evidence of the Japanese giving Australia a much lower price than it gave to the USA and Canada.

In a "perfect" market, and with all coals being of similar quality (which of course is never the case), one would expect that the f.o.b. prices (ie the price as loaded in the port of export) would diverge by the difference in transport costs from the point of export to Japan, resulting in the c.i.f. prices (the price including freight etc cost at the destination port) being equal. On this basis, and because of lower transport costs, the f.o.b. price of Australian coals should be greater than the price of US or Canadian coals.

Comparing major brands of hard coking coal from Australia, the USA and Canada, with all prices in $US and on an f.o.b. basis, it is clear that in the early 1970s, the price received by Utah for Australian Goonyella coal was well below the price for US Pittston coal. In JFY 1971 Goonyella coal's price was $US12.99, Pittston's $US20.80 and Luscar's $US11.86. In JFY 1974 the Goonyella price rose from $US11.79 to $US26.48 – an increase of $US14.69. The Pittston price in JFY 1974 saw an even greater increase of $US25.90 – up from $US22.30 to $US48.20. In JFY 1975, the Goonyella price surged again, up from $US26.48 to $US 46.67, while Pittston's increase was more modest – from $US48.20 to $US59.00. So over the two year period, Goonyella coal price was up by $US34.88, compared with an increase of $US36.70 for Pittston; there was still a major differential in favour of Pittston of $US12.33 in JFY 1976.[353] Prices continued to rise in the late 1970s and by JFY 1980 the differential in favour of Pittston was a little higher - $US13.83. However, with Australia becoming more dominant as a supplier to the JSM, a major change was soon to occur. By JFY 1983, Utah finally achieved a higher price for its Goonyella's coal (US$60.69) than Pittston ($US59.05). From JFY 1974 Canadian coals were priced in Canadian dollars; the price for Luscar coal in JFY 1983 was $C69.83, or around $US61.25.

Richard Koerner has written extensively on the Japan Australia coal trade, and made detailed submissions to the Productivity Commission's 1997-98 inquiry into the Australian coal industry. Koerner had worked for coal companies and had gained significant experience in the coal industry, and had a good understanding of the quality attributes of different coals and their use in steel making. Koerner's February 1998 submission noted the Utah submission to a Senate Committee in 1972 in which Utah stated that "Utah's prices have at times been compared unfavorably with other producers' prices by uninformed commentators. Such comparisons either ignore the facts or fail to comprehend the significance of major quality differences between coals from different sources. Utah's coking coal prices have been in line with market values."[354] Koerner also quoted the report of the 1991 Industry Commission Inquiry into the Australian Mining and Minerals Processing Industries that in relation to the coal trade: "In the Commission's view, distorted purchasing arrangements do not exist or are insufficient to justify use of export controls".

Koerner's conclusion is quite clear: "Contrary to the conventional wisdom, the findings of this chapter suggest that the persistent differentials in cif cost between Australian, US and Canadian hard coking coals cannot be adequately explained by the quality differences which, from a priori technical expectations, should impact each coal's value in coke blending and ironmaking. Quality differences do not seem to be significant relative to price discrimination, which is the major contributing factor."[355]

The Federal Government's research agency, ABARE, published a report in 1997 which analysed the prices Australian coal exporters received in the years from 1989 to 1996 for exports to a range of countries. The analysis used the Government's coal export controls database and covered hundreds of contracts for hard coking coal, soft coking coal and thermal coal. The report concluded that there were not large differences in the prices (adjusted for quality) of Australia's metallurgical coal exports to Japan and other regions. "Excluding west Asia, quality adjusted prices for Japan tend to be slightly higher for hard coking coal, but slightly lower for semisoft and PCI coal. Quality adjusted prices for hard coking coal exports to west Asia were similar to those for Japan in a number of years, and prices to west Asia fluctuated around the Japanese price in other years."[356]

The ABARE report found that for steaming coal, the quality adjusted prices for exports to Japan tended to exceed prices for exports to other markets, although the price gap had narrowed significantly since JFY 1993. The report, however, was not designed to answer the question of whether Japanese buying practices had acted so as to depress the prices Australian exporters received. The report simply noted that "...whether Japan's dominant market position influences the absolute price level of different coal categories in the regional coal market remains an issue."

On the question of the alleged failure of Australian exporters to successfully cooperate to achieve higher coal prices, an argument forcefully put by writers such as Byrnes, there were of course a range of major factors which confronted the exporters. Foremost perhaps were the US anti-trust laws under which there were (and continue to be) significant penalties for companies which act to restrain trade, collude in restraint of trade, form cartels etc. As we saw in chapter 1, the Australian coal exporters actually disbanded their export committee in 1978 following receipt of legal advice in relation to the US anti-trust legislation and its potential impact on companies operating in Australia. While a new export committee was set up shortly after at the request of Doug Anthony, it was careful to steer clear of being involved in price negotiations or any actions that could be seen as collusion. And in the annual (later biennial) conference between the Australian coal producers and the Japanese steel mills and power utilities (the Australia Japan Coal Conference), the Australian side was always careful to avoid pricing issues, apart from regularly questioning the Japanese on the reasons for favouring Canadian exporters with higher prices, or other general market-related questions.

There is also the question of how effective stronger collusive practices (even if legal) or stronger cooperation between exporters might be under different market conditions. After all, weren't the OPEC nations successful in dramatically forcing up the price of oil in the 1970s? Conditions in the 1970s oil market of course allowed the producers to force up the price: supply was tight, demand was strong and the OPEC producers accounted for a high proportion of world oil production. However, when demand fell in the 1980s and new oil suppliers entered the market, oil prices fell and OPEC's previous power evaporated. In the case of coal, strong collusion between Australian exporters, or strong coordination, may have been successful in forcing up prices in the short term if the right market conditions had been present. As we saw in chapter 2, favourable market conditions did exist in 1981 when the Australian soft coking coal exporters were able to win the day and force the JSM to accept higher prices than the mills had been expecting to pay. Those negotiations in 1981 had followed a strong outcome for the steaming coal negotiations which in turn reflected the pressures in the market following the second oil shock in 1979. However, at most times in the coal market, and particularly in the 1980s and 1990s, there is potentially much more coal that can be produced and supplied to the international market. And when this is the case, any price rises may be short lived as customers will look to alternative suppliers and to cut back on orders from countries or companies seen to be acting to force prices above what the customers believe is reasonable.

Steaming coal market becomes more competitive

Whatever may be said about the coal market in the 1980s and 1990s, and the claims by unions and others that Australian coal producers failed to secure the best prices, there can be no argument about the state of the market - it was extremely competitive and tough for exporters. In chapter 2 we looked at the pressures in the 1980s as early predictions about the boom in demand proved to be over-hyped, and as new suppliers entered the market and competition intensified. The 1990s were also tough years, but with new factors also emerging to increase pressure on prices. Camberwell Coal, a relatively modest size operation established in the early 1990s, was one of many companies struggling to be viable and its submission to the Productivity Commission in 1997 provides a useful summary of the changes in the steaming coal market in the mid to late 1990s from the point of view of a smaller producer. Those changes centred on the growth of export contracts for shorter periods, with a significant impact on prices:

"The coal market is currently undergoing significant change. In the past, prices to the Japanese Steel Mills (JSM) and the Japanese Power Utilities (JPU) were decided by negotiations carried out between a 'champion negotiator' representing the buyer and the seller. The 'champion negotiator' from each side was usually the biggest buyer/supplier. For example, Chubu Electric has been the negotiator on behalf of the JPU, and Drayton, Ulan and MIM as the biggest suppliers to Chubu were the lead negotiators from the Australian side. The prices decided at these negotiations then flowed on to other Australian suppliers. This system is referred to as the 'Benchmark Pricing System'. The result of this system was that Japan was paying the highest prices for Australian coal.

"Two years ago, the JSM scrapped the benchmark system, and each steel mill now conducts an individual price negotiation with each of its suppliers. Camberwell supplies all six major steel mills and has a different price for each mill, whereas 2 years ago our price was the same for each mill. Tonnage supplied to some mills has increased and dropped to others.

"About the same time, Taiwan Power Company moved to a system of awarding long term thermal contracts on the basis of the lowest price bid for year one of the contract. Subsequent years were priced at 'Japanese Benchmark'. Through this system, Taipower was able to secure coal in the first year of a 5 - 7 year contract at discounts of up to US$ 15/tonne in the first year. Effectively the discount spread out over the life of the contract was equivalent to a discount of about US$3/tonne from the Japanese Benchmark. Taipower also moved to a purchasing system which combined long term contracts supplying about 80% (70-90%) of their requirements (20 Mtpa) and a 3 month open tender system supplying about 20% (10-30%) of their requirements. The 3 month tenders are conducted on a world-wide basis - effectively, Taipower puts about one million tonnes up for tender every 3 months, which gives a good indication of the spot price prevailing at the time of the tender. The combination of Taipower's system of awarding long term contracts and 3 month spot tenders, means that they have been purchasing coal at a significant discount to the JPU.

"The Japanese power utilities, threatened by the development of IPPs (Independent Power Utilities) in Japan, have realised that they are paying too much for their coal compared to other power utilities in Asia. They are therefore moving rapidly towards use of the spot tender system, and in 1996 about 5% of their coal was purchased at open or closed spot tender. This has increased to about 50% in 1997 and for the first time Chubu Electric will use a formal tender system to decide the award of long term contracts commending in April 1998.

"The introduction of the tender system for deciding the award of long term contracts and the use of 3 monthly spot tenders to purchase up to 30% of coal requirements has put Australian thermal coal under intense competition from low cost producers such as Indonesia, China and South Africa. In the last 10 years exports of thermal coal from Indonesia have risen from about 1 Mtpa to 36 Mtpa and they are expected to further increase to 50 - 60 Mtpa by the year 2005."[357]

As the Camberwell submission shows, a major coal buyer like Taipower was able to take advantage of the state of the thermal coal market in the 1990s to extract lower prices, and the Japanese power utilities, themselves under pressure, also responded. The impact was harsh for coal producers, but there was little they could do; the market was operating effectively for coal buyers and producers were forced to raise their efficiency or perish. The arguments about coal pricing for both thermal coal and coking coal and the role of the Japanese would continue for several more years, but with the changes in the market in the 2000s, when imports by China and India grew rapidly, those arguments tended to die away as Japan was no longer the all-powerful or dominant coal buyer. However, the changes in the market in the 1990s, including the move to short term purchasing of thermal coal, and the change of government in 1996 would see the export control system for coal in Australia finally disappear.

Export controls on coal abolished

Export controls on coal were implemented by the Whitlam Government in February 1973 and were abolished by the Howard Coalition Government in 1997. In November 1986, at a time when the Miner's Federation and others were pushing for the establishment of a national coal marketing authority, Senator Warwick Parer made his position on export controls very clear. Parer had been a senior executive with Utah before being elected to the Senate in the 1980s and so had a detailed knowledge of the coal industry and the Japanese coal purchasing practices; he had also chaired the Australian Coal Association's export committee during the 1970s. Parer said in Parliament that "One of the problems for the Department of Trade that was introduced in the Whitlam-Connor era was the introduction of export controls. I know the reasoning behind it. The thinking was, quite incorrectly, that somehow the introduction of these export controls would artificially lift or maintain the prices of our export products so that we did not somehow slip below the rest of the market. Having been involved in this, I can tell honourable senators that it was an utter failure. In no way did the introduction of export controls have any effect on the sort of prices we got for our mineral exports."[358] But was there evidence that export controls had improved the prices Australian coal exporters received?

ABARE's 1993 report on export controls noted that "One of the main reasons for extending the use of export controls to all of Australia's mineral exports in 1973 was in order to be able to counter the possible effect of collective buying practices in international trade. Although the prevalence of collective buying has not diminished, export controls for which the primary purpose was to achieve 'reasonable' prices have now been removed for other Australian mineral exports which are sold in similar markets, such as iron ore. The question therefore arises whether export controls on coal have a continuing economic role to play."[359] ABARE noted that the major customers for our coal, the steel mills and power utilities, particularly in Japan, continued to use collusive buying practices, but it concluded that "However, when changes in Australia's pattern of coal trade since the early 1970s are taken into account, along with a number of theoretical and practical aspects of the use of export controls, it is doubtful whether export controls are helpful in the long run in increasing the prices obtained for Australian coal."[360]

ABARE also concluded that "From an examination of the Australia-Japan price outcomes for coal over the 1980s, it does not appear that Japanese buyers have been dominating the bilateral trade relationship. After allowance for quality differentials, the sharing of the gains from transport cost differences between markets has been roughly equal throughout much of the 1980s, as would be expected if firms in both countries are keen to develop and maintain a mutually beneficial trading relationship."[361]

Coal market growth in the 1990s

The 1990s was a remarkable decade for a number of reasons, not the least of which was the growth in the Australian coal industry. While profits for many companies were poor, production rose from around 160 Mt in 1989-90 to 237 Mt in 1999-2000, with the bulk of that growth in Queensland. Coal exports grew by 70% - from 104 Mt to 177 Mt; coking coal exports totalled 97Mt in 1999-2000 and steaming coal exports 80 Mt. Australia entrenched its position as the dominant coking coal exporter in the 1990s, capturing 55% of the seaborne export market by 2000, ahead of the USA, with around 17%, and Canada around 16%. In the more competitive steaming coal market, Australia had 25% of the global seaborne market in 2000, ahead of South Africa with 19% and China and Indonesia, both with around 14%. Australia, however, would lose the number one position in the steaming coal market to Indonesia by the early 2000s.

In the critical Japanese coking coal market, by 2000 Australia had captured 58%, with Canada a lowly 2% and the USA just 4%. The other major supplier to Japan in 2000 was China with 17% of imports. In 1991, Australia's share of Japanese coking coal imports was only 46%. While Japan increased its imports of steaming coal in the 1990s, Australia lost significant market share, largely to the rapidly growing Indonesian industry. Australia's share of Japanese steaming coal imports fell from 70% in 1991 to only 64% in 2000.

The competition in the coal market, rapid growth in production from a number of countries and shocks such as the Asian financial crisis in 1998, saw coal prices hit unsustainably low levels by the late 1990s. For high quality Australian coking coal sold into Japan, the fob benchmark price for Goonyella coal was $US52.80 in 1990, but this fell to around $US42 in 1999 and $US40 in 2000. Australian steaming coal sold to Japan averaged $US40 in 1990, but only around $US30 in 1999 and just under $US29 in 2000.

Price discrimination and excess supply

As we saw in chapter 2, the Japanese invested in metallurgical coal mines in Canada in the 1970s and 1980s and wrote major supply contracts with those mines, assisting to create a major competitor to Australia and the USA in terms of supply to the JSM. The new Quintette and Bullmoose mines developed in the early 1980s in Canada were underwritten with prices far in excess of what the JSM were prepared to pay for Australian or US coal. However, by the 1990s those prices were unsustainable and the two mines eventually closed once prices were reduced to levels equivalent to prevailing market prices. This support for the new Canadian mines also followed the period in the 1960s and 1970s when Australian coking coals clearly suffered price discrimination at the hands of the JSM.

However, on the question of whether the JSM had a deliberate policy of encouraging excess supply of coking coal, we need to remember that following the rapid growth of the Japanese steel industry in the 1950s and 1960s, the industry hit a brick wall in the early 1970s. Japanese steel production peaked in 1973 at 119 million tonnes; it fell to only 102 million tonnes in 1975. Production recovered somewhat later in the decade, but hopes of a stronger industry in the 1980s were dashed by the second oil shock of 1979 and global economic troubles in the early 1980s. Japanese steel production in 1990 was still only 110 million tonnes.

One ex senior coal company executive told the author that a Japanese trading company executive had once admitted to him that the Japanese did have a strategy of encouraging excess supply of steaming coal through investment in mine capacity, the rationale being that for Japan it made economic sense to achieve lower coal prices even if the returns from coal mining investments were low or negative. Coal is a major cost for power stations; lower coal prices help to make power utilities more profitable. On the other hand, one ex CEO's response to the question "did the Japanese deliberately encourage an excess supply of coal?" was simply that he did not know, but he did acknowledge that the Japanese always paid the best prices.

The 1980s had been a tough time for coal exporters, and the 1990s proved to be equally tough. Fortunately, by the early 2000s China's massive growth would begin to have a major impact on coal markets, as its imports started to climb and as it scaled back its previously significant exports. Better times and profits were ahead, although few companies were optimistic. The period when the Japanese dominated the international coal trade for coking coal, and had a major influence on the steaming coal trade would come to an end. Although Japan in the post 2000 period would continue as a major force in the coal trade, it would no longer be the dominant force. The controversies regarding unfair or collusive buying practices, price discrimination, encouragement of excess supply and lack of coordination by Australian exporters would soon fade away. Australia would emerge in the new century as the dominant force in the global coking coal trade and the number two thermal coal supplier behind Indonesia.

The Joint Coal Board's powers reduced

The future of both the JCB and the CIT became one of the issues the Coalition government was considering after it regained power in 1975. In 1976, Cabinet's Administrative Review Committee presented a paper to Cabinet advising against abolishing the JCB. The Committee recognised that there would be a violent reaction by the unions to any such decision, and concluded that the modest outlay by the Commonwealth in supporting the JCB's operations was a fairly cheap price for continuing the "political free character" of the NSW coal mining industry.[362] However, Ministers Peter Nixon and Tony Street, obviously not happy with the Administrative Review Committee's recommendations, then followed up with their own submission to Cabinet which recommended that the Commonwealth should withdraw from the joint arrangements, advise the NSW Government of its decision, and negotiate with NSW on the details of the withdrawal.[363] However, with a Labor Government elected to power in NSW in 1976, any desire by the Commonwealth to withdraw from the arrangements was bound to go nowhere. In 1977, National

Resources Minister Doug Anthony told a meeting of the Australian Coal Association's Export Division that it was his Government's intention to preserve the JCB, but to reform it. He said that he had discussed the matter with NSW Mines and Energy Minister Pat Hills who was receptive, but that there had been no positive proposals back from Hills. Anthony said that he would like to see the JCB enlarged to provide broader representation, including having an employer representative on the board.[364]

In 1983 the Commonwealth and NSW Governments discussed the future of the JCB and this led to a change in the composition of the board of the JCB the following year.[365] The Commonwealth and NSW Governments, now both Labor governments, had agreed that one of the Members (ie directors) would be appointed to represent the employees, one to represent the employers and the third to be an independent to represent the governments. This was a change that the unions had been pushing for, and from that time the employee Member would be drawn from the ranks of Miners' Federation officials. Later that year, Jack Wilcox was appointed the new JCB chairman to succeed George Tredinnick who had retired in July. Bill Chapman, the Federation's northern district president, was also appointed to the new board. Wilcox had been a senior official of the Commonwealth Department of Trade, and came to the position with a detailed knowledge of the industry, particularly in relation to export markets.

In 1984, with pressure on the Commonwealth Government to do something about the declining prospects of the coal industry and the pain local exporters were taking as the Japanese cut prices, the unions were again proposing that the Commonwealth intervene by establishing a national coal marketing authority, or by extending the powers of the Joint Coal Board to include Queensland. The Commonwealth Government, and not surprisingly the Queensland Government, rejected the option of extending the JCB to include Queensland, and senior Ministers were not in favour of radical solutions.[366] As we saw in chapter 2, the push for a national coal marketing authority had been rejected by the Cabinet in November 1983. There were many arguments against such an authority, one of which would have been the concern that Japanese customers had about creating such a body or extending the powers of the JCB. Visiting Japanese missions had told the Government on a number of occasions that a situation where the Government decided which mines would get new contracts was unacceptable.

Employment and Industrial Relations Minister Ralph Willis, campaigning in a by-election for the Sydney seat of Hughes, was reported to have said in early 1984 that "Through the Joint Coal Board the Commonwealth has substantial powers. But what you have to understand is most of the big open cut mines are in Queensland and Queensland is not covered by the Joint Coal Board... The prospects of us getting a joint coal board including Queensland are not high because we don't expect to get their cooperation."[367] Given the way in which the Queensland Government under Bjelke-Petersen viewed Commonwealth–State relations and the jealous guarding of its powers, Willis' words about the prospects for an extended Joint Coal Board were certainly one of the under-statements of the campaign.

By 1989, the JCB had been operating for over 40 years, with the Coal Industry Acts which governed its operations largely unchanged. The JCB had evolved over the years as the industry developed and by the late 1980s was a less dominant player in the industry than in its early years. Nevertheless it still had in place a range of Orders (which carried the same legal force as regulations) which required companies to seek its approval for the opening and closing of mines and mine expansions, to supply details of export contracts, to have training schemes in place for employees, to provide fortnightly statistics on production, sales, employment, wages etc. Another Order provided the JCB with the power to put a limit on the level of a mine's production, and in 1990 the Board brought in a new Order requiring mines to submit plans for approval for the installation or removal of longwalls. Since 1948 the JCB had also operated the workers' compensation scheme for the NSW coal industry which mandated that all producers had to insure their employees through the JCB's subsidiary company Coal Mines Insurance Pty Ltd.

While the producers had generally supported the JCB over the years, during the 1980s the NSW companies, through the NSW Coal Association, changed their stance, arguing that it was time for the organisation to be abolished. But to make any changes to the Coal Industry Acts or the JCB's various Orders, both the NSW and Federal Governments would have to agree, and Labor had held power in Canberra since 1983 and Labor in NSW had been in power since 1976. But the Association saw a glimmer of hope in March 1988 when the Liberal/ National Coalition won power in NSW, taking government with a policy of abolishing the JCB.

In Canberra, however, the Hawke Labor Government, although implementing major reforms across the economy, was not prepared to upset its voters in key Labor electorates in the Hunter Valley and Illawarra by reducing the powers of the JCB or abolishing it, although the Government did approach the NSW Labor Government in 1986-87 to seek its support for abolishing the Coal Industry Tribunal.

In 1989, with the new Liberal Government in power in NSW, the NSW coal companies believed that an opportunity existed to make some changes. When the term of Matt Blackham, the employers' representative on the board of the JCB, expired in February 1989, the NSW Coal Association declined to nominate another person to take his place. The JCB continued to operate, but with a board of only two – the union nominee Barry Swan, and the JCB CEO and chairman Jack Wilcox. This was a difficult time for the JCB, with the industry association and most member companies opposed to its continued existence, and with the NSW Department of Minerals and Energy resenting the JCB's powers and the duplication of responsibilities between the two organisations. There was little communication between the NSWCA and the JCB or the between the Department and the JCB.

The decision of the NSW Coal Association not to nominate a replacement for Matt Blackham was taken for a number of reasons. "The most pressing of these was the political context of the time which led Association members to believe that the Board would soon be disbanded and that a nomination may have sent the wrong signals to both Governments. In addition, members believed as a matter of principle that to nominate a replacement would have been inconsistent with the Association's policy in favour of the industry being able to stand on its own feet without the need for a regulatory body."[368] The appointment of Meredith Hellicar, the Association's first female executive director (CEO), in late 1989 did not change the Association's opposition to the JCB. At their first meeting, Wilcox asked Hellicar what her priorities would be in her new position; Hellicar replied that top of the list was the abolition of the JCB and the CIT. Hellicar had been told by NSWCA chairman Jim Lewis and outgoing CEO Barry Ritchie that the members of the Association wanted the JCB gone. Relations between the Association and the JCB would continue to be difficult.

A major breakthrough for the employers came in 1990 when the Commonwealth Minister John Kerin and NSW Minister Neil Pickard, the two Ministers responsible for the JCB under the joint arrangements, announced that they had appointed Bryan Kelman to undertake a review of the JCB's functions and activities, with Kelman's brief to include reviewing whether its functions should be retained, changed, transferred to some other body or discontinued. Kelman had been the CEO of CSR Ltd in the 1980s at a time when CSR had become a major coal producer and so was well acquainted with the industry. The review, however, did not encompass the Coal Industry Tribunal; that was a still bridge too far for the Commonwealth Government. Kelman was assisted by Denis Casey from the NSW Department of Minerals and Energy and Noel Webb from the Commonwealth Department of Primary Industries and Energy. He canvassed opinions widely across the industry and reported in February 1991.

The NSW Coal Association (NSWCA) submission to Kelman urged him to recommend that the JCB should be abolished, and that some of its functions should to be transferred to other bodies. The CFMEU (then the United Mineworkers Federation) argued for no reduction in the JCB's powers and activities, but in fact an expansion, recommending that the JCB take over the administration of the Coal Mines Regulation Act from the Department. The Department of Minerals and Energy (DME) was most concerned about duplication between its activities and those of the JCB, an issue also raised by individual companies which had to manage the additional regulatory and compliance burdens. The NSWCA submission drew on input from a number of its member companies which had had less than happy experiences with the JCB during the 1980s or earlier. One of these companies was Kembla Coal and Coke (KCC), a part of the CRA group, which operated several collieries between Sydney and Wollongong, including Coal Cliff. CRA had been planning to close Coal Cliff in 1984 and had issued retrenchment notices to 537 employees. The company then became involved in interventions by the Hawke Government and the NSW Government, and the JCB became involved in attempting to broker a solution and ordered KCC to withdraw the retrenchment notices. CRA said that the intervention of the two governments and the JCB prevented it from closing a mine which was not viable.

Costain provided details of its experience with its Ravensworth South development approval process. Costain had been operating Ravensworth under contract to the NSW Electricity Commission since 1971, supplying coal to the Commission's power stations. In 1986, the Commission called tenders for a company to develop and operate a new mine, Ravensworth South, under contract. Costain won the tender and obtained the necessary development approvals from the NSW Government, but late in the process the JCB indicated that it may not be prepared to issue its approval under Order 27 since the coal required by the Commission could be available from export mines that otherwise might have been forced to close. Those export mines had been eligible to bid for the tender and some had done so. The JCB's intervention occurred after it had earlier been involved in discussions with Costain, with every indication that the company's plan was acceptable, and only one month before the company was due to begin the mining process with the removal of topsoil. Costain believed that the JCB's attitude was unreasonable and took the decision to commence operations without the JCB's approval.

Another example in the NSW Coal Association submission was the Ulan mine located near Mudgee in the NSW central west. In early 1990, the JCB had production limits in force on 17 of the 68 coal mines in NSW, one of which was Ulan. In 1982, as part of a strategy to persuade the Federation to lift bans on the start-up of the Ulan and Drayton open cut mines, the JCB had imposed limits on all open cut mines producing for export. Ulan was an underground operation and was seeking approval to develop an open cut mine. The Ulan underground had been subject to production limits over a number of years, with applications to increase production varied by the JCB or subject to certain restrictions. Approval had been given in 1976 for an increase from 250 to 500 tonnes per day, and raised to 3100 in 1980 and to 4500 tonnes per day in 1981. A trial box cut operation was approved in 1979, but when Ulan sought approval to commence a full open cut operation in 1982 it was approved on the basis that the mine could work only one shift per day and annual production was limited to 1 million tonnes and that the underground continued to operate at its then current production rate. An application in 1983 to introduce second and third shifts in the open cut was refused, although the JCB did approve a second shift in 1984, but restricted production to 20,000 tonnes per day. A third shift was approved in 1984, but with a limit on total production of 32,500 tonnes per day. In 1985 Ulan sought approval to increase production from the open cut to 32,500 tonnes per day; the JCB approved unlimited production from the underground, but limited open cut production to 30,000 tonnes per day. Ulan had not only invested in expensive equipment for its open cut operation, including a dragline, but had paid to have a new rail line built to link its mine to the Hunter Valley system, so that it could export through Newcastle. The JCB's controls on the company certainly did nothing to enhance its viability at a time when the market was extremely competitive.

These examples, from the perspective of the producers, were evidence of what they saw as bureaucratic and rather arbitrary controls imposed by the JCB in its attempts to balance supply and demand and protect employment in the underground sector. But this was also a time when the NSW producers were attempting to gain export contracts in competition with Queensland producers (who had no such constraints) and with producers from other countries.

Kelman noted that both the DME and the NSWCA had not maintained adequate communications with the JCB, particularly since the election of the Coalition Government in March 1988, a situation which he attributed in part to the Coalition Government's policy of abolishing the JCB. Kelman also said that the failure of the NSWCA to nominate a new employer Member of the JCB to replace Matt Blackham made many of the differences between the JCB and the NSWCA even more difficult. "Conflict between the main stakeholders has been stifling the operations of the Board and does not contribute to the efficiency of the NSW coal industry," Kelman concluded.

One of the most important recommendations Kelman made was to remove the JCB from its role in controlling the development of the industry through requiring companies to seek its approval for the opening and closing of mines or for major mine expansions. By controlling the pace and nature of development of the industry in NSW, the JCB had been attempting to balance supply and demand. This was an interesting concept in theory, but in practice the JCB's efforts were a constraint on the NSW industry at a time when Queensland industry had no such constraints, and, of course, overseas competitors were also free to develop as they wished. The JCB had also been requiring companies to submit detailed information on markets for their coal, contract conditions and prices. Kelman concluded that "Because of the international nature of the coal market and the extent of competition, it is not possible for the Joint Coal Board to balance supply and demand by controlling the opening of new mines, or mine expansions, in NSW and this situation would not be helped by making this a national responsibility."[369]

The NSW coal companies had been hoping that Kelman would recommend the abolition of the JCB, but realistically that was always unlikely given the fierce opposition that would have generated from the United Mineworkers' Federation, and the fact that Labor was in power in Canberra, with the Labor politicians representing seats in

Newcastle, the lower Hunter Valley, the Illawarra and Lithgow strong supporters of the JCB. The NSWCA, after conferring with its member companies, indicated to the two Governments that it would support a restructured JCB. Amendments were made to the Coal Industry Acts in 1992, removing some of the old powers and restricting the JCB's activities to the following key areas – the NSW coal industry workers' compensation scheme, occupational health and rehabilitation services for the industry, the collection and dissemination of coal industry statistics, the promotion of coal through training programs funded by the UN or the Commonwealth, training schemes and dust monitoring. The Act specified that the last three areas - statistics, training and dust – were in effect temporary functions in that they would continue to be carried out by the Board until such time as the Commonwealth and State Minsters directed otherwise, but these functions continued as a core part of the JCB's programs in the 1990s.

The NSWCA nominated David Sawyer, previously a senior executive with CSR, as its representative on the board of the new JCB. Jack Wilcox retired and Ian Farrar was appointed as the new CEO. Barry Swan continued in his role as the Federation representative. The JCB continued to operate until 2002 when a number of its key functions were transferred to a new company (Coal Services Pty Limited) owned by the NSW Minerals Council and the CFMEU.

The end of the JCB and QCB era

With the election of the Federal Coalition Government in 1996, the issue of the JCB's life arose once more. The Coalition in NSW had been in government since March 1988 under Nick Greiner and John Fahey as Premiers, but lost to Labor in April 1995. The NSW Minerals Council's policy position was to have the JCB abolished and some of its functions allocated to other organisations; the obstacle was the new Labor Government led by Bob Carr.

In 1997 the Productivity Commission (PC) began a detailed review of the Australian coal industry, with all major stakeholders except the

CFMEU making submissions and many appearing in public hearings. The PC issued its report in July 1998, with recommendations on a range of issues, including a recommendation to abolish the JCB: "The Joint Coal Board should be abolished and its functions taken over by the NSW Department of Mineral Resources, WorkCover and other public and private providers as appropriate. If it were decided to retain the workers' compensation role of the JCB, Coal Mines Insurance should be corporatized and required to compete against other insurance options."[370]

The NSWMC welcomed the PC's recommendation as it closely matched its own policy position. In late 1996 NSWMC had written to the NSW Minister responsible for the JCB, Attorney General Jeff Shaw, with a copy of the letter going to the Federal Minister for Resources, Senator Warwick Parer. The timing of this letter reflected the election of the Coalition Government in Canberra and a belief that this could generate some action at state and federal levels to re-consider the future of the JCB and undertake further restructuring. Recognising that the NSW Government was unlikely to do away with the JCB, the letter questioned the need for ongoing Federal involvement in the JCB and argued for reform of the structure of its board (to change it to directors who would be independent of the industry parties) and for a review of its functions.

In 1997, Senator Parer announced the Government's intention to withdraw from the joint arrangements and Canberra then formally advised the NSW Government in February 1998 of that decision. Parer had never been a fan of the JCB and his successor in the Resources portfolio, Nick Minchin, was also not wedded to such a unique organisation. The JCB's days were now numbered but there was still a difficult process to work through for the industry and the two governments. The NSWMC welcomed the Federal Government's decision, but now had to deal with a state government that was determined to continue to have an organisation responsible for the major activities which the JCB carried out, and in particular workers' compensation.

NSWMC initially believed that the NSW Government would opt for the new organisation to be established in the form of a state-owned corporation. However Jeff Shaw wrote to NSWMC in April 2000 advising that the NSW Cabinet had approved new arrangements for the JCB which would involve a new company established under NSW legislation, to be jointly owned by the NSWMC and the CFMEU. The company would run the workers' compensation scheme for the NSW coal industry, and the Mines Rescue Service, which was then a separate entity responsible to the Minister for Mineral Resources. To the NSWMC's surprise, the company was also proposed to include the operations of the coal industry superannuation scheme, Coalsuper. The takeover of the Coalsuper function appeared to NSWMC to be contrary to Federal superannuation law; that part of the proposal was quickly dropped once the Minister became aware of the likely legal issues. Shaw's proposal for a company jointly owned by the CFMEU and the NSWMC was a bolt from the blue for the NSWMC, but it had a precedent with Coalsuper Pty Limited, the trustee company responsible for the NSW coal industry superannuation fund. The shareholders of Coalsuper Pty Limited were the CFMEU and the NSWMC. This arrangement which would bring two organisations, not normally known for their close relations, together to manage a number of functions serving the coal industry.

NSWMC rejected the NSW Government's proposal on a number of grounds, but continued to negotiate on the detailed arrangements for a new company. Not all member companies were necessarily convinced of the wisdom or practicality of entering an arrangement where the producers would be effectively partners with the CFMEU in the new company, with one of the major producers sceptical that the arrangement would work. However, the reality was that if the producers had rejected the Government's model outright, the Government would more than likely have legislated for a statutory corporation, the board of which may not have been to the producers' liking. Better to have influence in the running of the new company than to be potentially excluded or having a much reduced say. With the Minister having past links to the CFMEU[371], and with a key adviser working in his office who was on secondment from the CFMEU, the NSWMC was hardly negotiating from a position of strength. However, it continued to push for key changes to the proposal, including the transfer of remaining regulatory functions relating to training and dust to the Department of Mineral Resources, and a board structured along the lines of public companies rather than being made up of nominees of the union and the NSWMC.

During 2000 it became apparent to the NSWMC that a major reason the Government had chosen to set up an industry owned corporation was that it wanted to remove any potential financial liability on the Government in relation to the workers' compensation scheme. During the 1990s, the WorkCover scheme had been a major problem for the Government, with significant reductions in benefits made in 1997 to reduce its deficit. However, the benefits for coal industry workers, which were well above those enjoyed by workers under the WorkCover scheme, were maintained. The last thing the Government wanted was to inherit potential liabilities in Coal Mines Insurance (CMI) which at that time was under significant pressure because of increasing claims. When the Government legislated the changes to the WorkCover scheme in 1997, it was able to claim that the CMI scheme was fully funded and that this justified leaving benefits for coal industry workers as they were. As an added safeguard the Government later acted to also ensure that CMI was not licensed under the WorkCover system like other insurers. As a corporation set up under NSW legislation, CMI was also not regulated by the Federal insurance regulator, APRA. The coal producers would have to bear all the liabilities under the scheme. NSWMC also continued to push for the Coal Mines Insurance Scheme to be opened up to competition. However, the Government was determined to maintain the scheme's monopoly, a position strongly supported by the CFMEU.

The Federal Government passed its legislation to withdraw from the JCB in 2000, although the legislation was not to take effect until a date proclaimed by the Minister. The NSW Government wanted to get the new company up and running as soon as possible, but with legislation unable to take effect until the Federal Minister had set the date for the Federal withdrawal, the NSW Government was forced to delay its timetable. The NSW Minerals Council still had major concerns about the details of the legislation and asked the Federal Minister, Nick

Minchin, to delay proclamation until these had been resolved. The delay caused a commotion within the CFMEU, as Ron Land, then Northern District secretary, had planned his resignation from that position to allow him to take up the position of the inaugural chairman of the company in 2001. Land was forced to go back to his members to resume his district secretary position until the new company commenced operating. The new company, Coal Services Pty Limited, commenced operating in January 2002. It was essentially responsible for the same programs and activities as the JCB, but with a new board of directors of seven comprising two directors representing the CFMEU, two representing the NSWMC, two jointly nominated by the CFMEU and NSWMC, and the CEO. Ron Land became the chairman as specified in the Act, reflecting the CFMEU's influence in the development of the legislation.

The end of the Joint Coal Board era was marked by a number of speeches in both the NSW and Federal Parliaments by MPs whose electorates covered the major coalfields. There was genuine regret by some of these MPS that the days of the all-powerful JCB had now gone, although arguably that had already occurred by the early 1990s. The member for Maitland, John Price, said that: "The establishment of the Joint Coal Board still stands as the most comprehensive and positive move by two levels of government to reform the coal industry in this State—a significant outcome, given the crucial role of coal in Australian life and in the economy of this nation."[372] Price went on to say that the JCB could "claim credit for the elimination in New South Wales coalfields of the illness known as black lung or pneumoconiosis. In 1946 some 16 per cent of miners suffered from the disease. However, that disease has virtually been eliminated. The Joint Coal Board clearly played a significant role in improving the health and welfare of coalminers and their local communities."

Many of the references in these speeches harked back to the early days of the JCB and its activities in getting the industry back on a steady footing. And while the coalfield MPs did acknowledge that some of the new company's activities, such as workers' compensation and occupational health programs, continued on from the early efforts of the JCB, the speeches served to further highlight how far the industry had come in more recent years, and how less relevant the old JCB had become.

The QCB is abolished

The Queensland Coal Board, set up in 1948, was finally abolished in 1997. While the existence of the QCB had been accepted for many years by the Queensland producers, the Queensland Coal Association (QCA) recommended its abolition in its submission to the Industry Commission inquiry into Mining and Minerals Processing in Australia in 1991, saying that: "Clearly, the expansion and change of the Queensland Coal Industry since 1948 has left the Queensland Coal Board without its intended role. Most of its powers are dormant ... The Association believes the Board should be disbanded and its Act repealed. However, this should not happen while the Joint Coal Board remains in existence. To its credit, the QCB has taken a much less interventionist role than its NSW counterpart, and the Association would be loath to contribute to a situation where the position vacated by the Queensland Board was to any degree filled by the JCB."[373] The Queensland producers were evidently fearful that the abolition of the QCB could lead to an expanded role for the JCB to take in the Queensland coal industry. In hindsight, this was clearly never going to happen, but the Queensland producers traditionally looked somewhat askance at the NSW industry, with the JCB in a powerful position to regulate company activities.

The QCA then made a second submission to the Industry Commission which showed its concern about the Queensland Government's recent decision to expand the role of the QCB. The long era during which the National Party dominated Queensland politics came to an end in December 1989 with the election of the Labor Government under Wayne Goss as Premier. Joh Bjelke-Petersen had ruled for almost 20 years and was followed by Mike Ahern and Russell Cooper as Premiers, but by 1989 the electorate had tired of the Nationals. The new Government had announced that there would be an expanded role for the QCB, and the QCA was fearful of an increasing regulatory burden. The QCA said that the change in policy towards the QCB was a cause of "confusion and consternation" to its member companies who believed the decision was "contrary to the general tide of Government and community thinking towards leaner public administration and less involvement in private business affairs."[374] Any concerns that the coal producers may have had about the QCB assuming a role in the international marketing of coal proved to be unfounded, with Resources Minister Ken Vaughan in April 1991 stating in Parliament that the QCB would have no such role.[375] The Industry Commission's report recommended the abolition of the JCB and the QCB, and said that it shared the QCA's concern that the Queensland Government's changes to the QCB "may be a first step towards duplicating the mistakes made in New South Wales with respect to the JCB".[376]

In 1990 Minister Vaughan commissioned a review of the QCB which resulted in a restructuring of the board of directors, roles which had previously been part time. Kevin Wolfe, who had been the chairman of the board and full time head of the Department of Mines and Energy, was not reappointed. The new QCB board now comprised three full time directors, with Peter Ellis the new chairman; Ellis had previously held senior roles in the Queensland public service, his role immediately before the QCB appointment being head of the Department of Industrial Development. In 1986 Ellis, then a senior officer in the Premier's Department, had been a member of the Savage Committee on business deregulation which had identified the QCB as one of the organisations recommended for abolition. The 1990 review saw the QCB's role expanded, largely to provide the government with expert advice on a range of issues which the old QCB had clearly not been previously competent to provide. As the QCB's annual report for 1989-90 stated: "This review took into account the Government's need to have access to clear policy options in order to maximise the economic and social benefits of the State's substantial coal resources. The Minister required the Board to be competent to discern the requirements of all sections of the Queensland coal industry and associated industries...

employees as well as employers, potential investors, forecasters and analysts. In the light of the economic and coal significance of the industry and the nature and intensity of international competition in the marketplace, the Government required independent, timely and well-considered advice which draws on the professional, technical and industry-related experience of Board members and employees."[377]

The QCB was also subject to a review in 1992 by the Public Sector Management Commission (PSMC). Tony McGrady, then Resources Minister, said that the review had found that the upgraded QCB was "effectively servicing the Government and the industry" although it also recommended that the 1948 legislation under which it was established needed to be reviewed "to ensure its relevancy to Government, union and industry needs."[378] The PSMC found that the QCB played an important policy role, that was "an effective means of providing independent policy advice to the Minister on how to best manage the public resource of coal, that its statistics met the industry's and government's needs, and that the Board's liaison with stakeholders was important on a number of critical issues.[379]

The beefed up role for the QCB and the favourable PSMC review, however, were not sufficient to save the organisation. Peter Ellis recommended to Minister McGrady that the QCB should be abolished, so doing himself out of a job.[380] However, it was not until after the National Party won government in February 1996 with Rob Borbidge as Premier that legislation was actually passed to repeal the Coal Industry (Control) Act of 1948. The legislation abolished the QCB, transferred the QCB's staff and functions in areas including health and coal statistics to the Department of Mines and Energy, and also transferred the Mines Rescue Brigade to the control of a private company funded and managed by the industry.

The abolition of the QCB did not generate the nostalgia and regrets that were expressed when the Joint Coal Board was abolished in NSW. Leading the debate for the Opposition, Shadow Minister Tony McGrady said that the Opposition did not oppose the Bill, and spent most of his time discussing the transfer of the Mines Rescue functions to an industry company.[381] McGrady's major concern about the abolition of the QCB related to the promotion of Queensland coal internationally, with this function to be transferred to the Department of Mines and Energy. McGrady said he believed that there had to be an organisation that promoted Queensland coal, but that "with all due respect to the department ... we need a special type of person to promote the commodity. That does not necessarily have to be a person who has worked for the department for the past 15 years. We need somebody with drive and enthusiasm, and who understands the product being sold. I appeal to the Minister not to ignore that aspect of the work of the former Coal Board."[382] The QCB's time had come to an end and it seems the organisation's passing was largely unlamented.

The end of the Coal Industry Tribunal

A new era for industrial relations in Australia was heralded in April 1993 when Prime Minister Paul Keating gave a landmark speech to a meeting of the Australian Institute of Company Directors, the speech committing to a move away from centralised decisions of the AIRC, with the focus to change towards the individual enterprise. The AIRC was to be the big loser from this change and Keating said that the Commission's role and powers were "not well suited to the system we are trying to create".[383] Enterprise agreements would now substitute for awards in the new system, with awards providing only the basic structure for the new agreements and acting as a safety net for workers who were not able to enter into workplace agreements with the employers.

The AIRC had had an opportunity in 1991 in the National Wage Case to introduce enterprise bargaining, a change which was supported by both the main employer and union parties to the case. However, the AIRC believed that there was no firm evidence that enterprise bargaining would lead to improved productivity, and that the employers and unions parties lacked "the maturity to avoid the chance of a wages breakout".[384] Keating, then Treasurer, called the Commission a "conservative" body that was hindering reform of the wages system. The ACTU did not accept the AIRC decision and this led the Commission to develop an enterprise bargaining principle later that year.

The Coal Industry Tribunal was still firmly entrenched in the early 1990s as the industrial relations arbiter and award maker for the coal industry, although its future had been in doubt for some years. The employer associations had pushed for the CIT's abolition in the 1980s, a move which had also been recommended by the Hancock review. The CIT survived the rationalisation of tribunals following the Hancock review and also the changes to industrial relations in 1988. In April of that year, in his second reading speech on the Industrial Relations Bill, IR Minister Ralph Willis said that "In view of the poor industrial relations performance of the coal industry, the Government believes that there is a strong case for change in, if not the complete abolition of, the Coal Industry Tribunal. The previous New South Wales Government did not favour any major changes to the Coal Industry Tribunal's operations. However, with the change in government in New South Wales, I have written to my New South Wales counterpart seeking his government's view on the future of the Coal Industry Tribunal and to what changes the State Government would be prepared to agree." The new Greiner Liberal Government in NSW had just come to power with a policy of abolishing the JCB, and so a major obstacle to abolition on this occasion would have been Willis' coal electorate Parliamentary colleagues in Canberra, and the pressure they would have been under from the unions to oppose such a move. The coal electorate MPS in the NSW Parliament would also have been vocal in opposing any change. The NSW Government's position was clear, with Mineral Resources Minister Neil Pickard saying that the process to abolish the CIT had begun and that "... no amount of industrial retaliation from mining unions would sway the Greiner Government from its commitment to abolish the coal court."[385] In its 1990 submission to the Industry Commission inquiry into the minerals industry, the NSW Government said that it had no objections to disbanding the CIT and that it was "... willing to co-operate with the Commonwealth Government in absorption of the Tribunal within the general jurisdiction of the Australian Industrial Relations Commission…"[386]

Federal Cabinet agreed in 1988 that the Ministers for Primary Industries and Energy and Industrial Relations would review the industrial relations framework for the coal industry, but this review does not appear to have produced any outcome. However, in March 1991, in a decision which was kept confidential, Cabinet agreed in principle that the CIT should be absorbed into the AIRC.[387] No action to implement this decision was taken at that time, but the sword was now hanging over the head of the CIT. Keating's announcement about the new direction for industrial relations in April 1993 saw the NSW Coal Association and the Queensland Mining Council make a submission as the Australian Coal Association to the Federal Government arguing that the CIT should to be abolished. The NSWCA and QMC feared that the coal industry would be left behind in its own system outside the industrial relations mainstream if Keating's legislation did not specifically address the continued existence of the CIT under the Coal Industry Acts.

The Australian Coal Association's (ACA) rationale for abolishing the CIT was summed up in its submission to the Department of Industrial Relations in response to DIR's issues paper on the proposed industrial relations legislation: "Reform is urgently needed in the coal industry, which still operates under special industrial relations arrangements established in 1946, when the industry was a strife-ridden domestic producer. Since that time, benefiting from a number of natural advantages, it has grown rapidly to become Australia's largest export commodity and the world's largest coal exporter. With the emergence of a number of competitors with similar advantages and lower cost structures, Australian coal producers have recognised the need to continually improve their international competitiveness. The current industrial relations system is inhibiting that process."[388] The ACA also argued that the very existence of the Local Coal Authorities (subsidiary bodies of the CIT) which were located in the major coal mining districts in NSW had encouraged employers and employees and their unions to take the easy way out, abrogating responsibility for settling disputes on the minesite.[389]

The Miners' Federation's rationale for the continued existence of the CIT had been clear from its submission to the 1990-1991 Industry Commission inquiry into the minerals industry. The Federation, now named the United Mineworkers Federation of Australia (UMFA), argued that the CIT was ideally suited to the coal industry's special characteristics. The union said that the CIT system allowed for rapid listing and hearing of disputes, did not involve the complexities of other jurisdictions, and allowed quick settlement of disputes because of the extensive powers of the CIT, with the single member tribunal 's speedy decision accepted as final. The union also argued that the appeal process allowed for decisions by the local coal authorities and the Queensland Board of Reference to be reviewed (that is, by the CIT). The other major union representing non-production employees in coal mining, the Australian Collieries Staff Association, also supported the retention of the CIT.[390] In June 1994, the policy of the NSW Labor Party was amended to reflect the stance of the CFMEU, with the policy supporting "the retention of the specialist Coal Industry Tribunal because of its demonstrated track record of, among other things, facilitating fast, efficient and effective dispute resolution… "[391] The Labor policy also referred to the CIT having minimised the length and severity of disputes, facilitated industry restructuring and encouraging enterprise bargaining within the framework of minimum award conditions.

The ACA's lobbying was not successful and the 1993 overhaul of the national industrial relations legislation by-passed the coal industry. However, it was not long before the Government moved to finally bring the coal industry into the industrial relations mainstream. As we have seen, the strikes by the Miners' Federation in late 1993 and early 1994, and in particular the strikes called after the Government had announced the Taylor inquiry, were the triggers for the Government to announce that the CIT would be absorbed into the AIRC. In his media conference on 19 April 1994 IR Minister Laurie Brereton made his position quite clear: "Let me say today that I believe and the Government believes that the decision of the Mineworkers' Union to take industrial action each time they're dissatisfied with the outcome of coal price negotiations, is a totally unsatisfactory state of affairs. It's doing great damage to this industry, and from the Government's point of view it is totally unacceptable. I think it's high time that, given the contempt that the union has consistently shown for decisions of the Coal Industry Tribunal, that we should bring the coal industry into the mainstream of industrial relations and bring it under the auspices of the Australian Industrial Relations Commission, and I'll be taking steps to put that into effect."[392]   Asked whether he would have any trouble moving the Tribunal into the AIRC after its long existence, Brereton said that "Well, let me say, it's been there for a long time and I think it's done a very good job in many respects, but it's equally clear, as a result of the course of action of the Mineworkers' Union over recent months, that the union is not prepared to take any notice of it at all and its orders. And it's against that background that I think it's appropriate that it should be brought within the mainstream of industrial relations in Australia, and that is the Industrial Relations Commission."

Brereton had been appointed to the industrial relations portfolio by Keating following Labor's re-election in March 1993. According to Keating's speechwriter Don Watson, the IR portfolio was "the pivotal portfolio for a government that balanced on the twin traditions of fraternal alliance with the unions and a proven will to reform, and knew that it must soon turn the latter on the former. Some on the left and in the unions wanted Bob McMullan, but Keating reckoned McMullan was an equivocating character and the last man he needed in the job, or in the cabinet for that matter. He favoured his old mate, Laurie Brereton – 'Dangerman' – who the left and the unions reckoned was a man without honour and the nearest thing to their natural enemy as anyone could find in the entire labour movement."[393]

Following his announcement in April 1994, Brereton would be subject to strong pressure from the unions and coal electorate colleagues to retain the Tribunal. But Brereton proved to be a strong minister and would deliver change; another minister in that role may not have been strong enough to stand up the coal electorate MPs and to the CFMEU. The pressure from Brereton's fellow MPs came almost immediately, with 13 Federal MPs reported to be seeking urgent talks with Brereton on the CIT. The group of Federal coal electorate MPs, largely from the Newcastle, Hunter Valley and Illawarra areas, and called the Federal Labor Coal Consultative Group, wanted Brereton to reverse the decision or delay it until the results of the Taylor inquiry had been delivered.[394] The Group's convener, Illawarra MP Colin Hollis, said he was surprised and dismayed by the decision and by Brereton's attitude, and said the group was "seeking an urgent meeting with the minister to canvass the very serious and long-term implications of the proposal." Another member, Newcastle MP Bob Brown, said that the decision was a "pre-emptive strike" against the Tribunal. In June the annual conference of the NSW Labor Party voted unanimously to support the retention of the CIT, a decision which the CFMEU hailed as a "massive rebuttal" of its new arch enemy, Laurie Brereton.[395] John Maitland, the CFMEU's general president, had also reminded Brereton and Keating that in March the previous year that "Labor had pledged to retain the Tribunal while the conservatives promised the coal owners they would abolish it for them….", and warning that it had not "helped put Labor into power for Laurie Brereton to impose John Howard's policies on us."[396]

By July the pressure on Brereton had forced him to compromise, and his commitment earlier in the year to bring the coal industry into the IR mainstream was looking a little less certain. Brereton issued a draft exposure Bill which contained a two-stage process for the transfer of the CIT into the AIRC, disappointing the producers who had lobbied hard to abolish the CIT. NSW Coal Association chairman and Peabody Coal CEO, Bob Humphris, welcomed the fact that the CIT would be placed under the AIRC umbrella "as a first step in implementing (the Government's) policy of fully absorbing the CIT into the mainstream of Australian industrial relations." However, Humphris said that the producers were disappointed and concerned that there was no fixed time frame for the full absorption of the tribunal" and that "This indefinite delay means more time will be lost before we can bring about the cultural and structural changes which are essential to the long-term survival of the industry."[397]

In October, Brereton advised that he was going to move amendments to the Bill he introduced into Parliament only the week before, thereby heading off a threatened national strike by coal mine workers. The CFMEU's members had held mass meetings in Queensland and NSW and had voted by almost 20 to 1 to support the leadership, giving it "a powerful mandate for industrial action, including an indefinite national strike, to oppose moves by ...Brereton to effectively abolish the Coal Industry Tribunal."[398] Brereton had met with the Labor Caucus economics and industrial relations committee the day before, with Labor MPs reported to have said that at the meeting it was obvious that changes would be made to the Bill. Brereton was under pressure from the employers not to compromise with the unions, from his Parliamentary colleagues, and from the Opposition which was enjoying watching him attacked from all sides. John Howard, then Opposition Spokesman on industrial relations, said that if Brereton capitulated and watered down his "already weak bill", it would represent his "greatest and final humiliation" and that "he might as well give the game away".[399]

Just three days later, after negotiations with the CFMEU, Brereton was close to a settlement with the unions, having gained the support of John Maitland to drop the initial proposal for the two-stage process on integration into the CIT. The CFMEU was now prepared to accept the full absorption of the CIT into the Commission, but wanted to be assured on a number of issues: that mine workers would not be disadvantaged by the move; that CIT chairman David Duncan would retain responsibility for coal industry matters as a member of the Commission; and that legal precedents in terms of past decisions of the CIT would apply to the industry in the future.[400] The issue of past decisions of the CIT and its subsidiary authorities was a major one for the CFMEU, as many of these decisions had been based over many years on custom and practice in the coal industry. They had therefore been critical elements in entrenching the old, established ways of operating, and the unions had traditionally fought to retain these customs and practices. The vast majority of coal producers wanted customs and practices to disappear as far as the industrial relations system was concerned, or if some were to be retained, wanted them to be purely a matter for negotiation at the enterprise level. The CIT's recognition of customs and practices in many of its decisions was also one of the major reasons the producers wanted to see the end of the CIT. It seemed that the CFMEU had had a major win on this issue.

One of the less well known facts about the CIT saga of 1994 was that BHP Coal, the dominant producer in Queensland, had been quietly lobbying Brereton to retain the CIT. The BHP stance was a reflection of its agreement with the coal industry unions in 1992 which provided security of employment to its mine workers in the Bowen Basin, with a range of concessions by the unions allowing the company to better manage its operations. As one former BHP executive explained to the author, BHP and the unions had a very "cosy relationship" at that time, with the unions able to achieve special conditions which were not readily available in the mainstream IR system. And BHP Coal's corporate view was that coal was different in terms of industrial relations, and it did not want to upset the unions. In his contacts with Brereton in 1994 on the development of the legislation, Humphris relayed his concerns about BHP and was comforted by Brereton's assurances that he would not be swayed by BHP's lobbying.

As foreshadowed in October, Brereton introduced changes to the Bill in November, saying that after further discussions with the industry parties and the NSW and Queensland governments he had decided that a two-stage process of integration of the CIT was unnecessary. The November changes omitted the transitional stage and provided for the full integration of coal industrial relations into the AIRC, but there were some critical amendments that required the Commission "to have regard" to past decisions of the CIT and "to give due weight to relevant decisions of the local coal authorities and boards of reference established under the Coal Industry Act." The awards made by the CIT would also continue to be in force.[401]

The legislation amending the Coal Industry Act was passed in December 1994 and took effect from 1 July 1995. The AIRC, now responsible for the coal industry, was required "to have regard to any decisions of the Coal Industry Tribunal that are relevant to the matters before the Commission." This appeared to be a slightly weaker provision than had been agreed by Brereton and Maitland, but it still entrenched a link with past decisions of the CIT. Brereton said that it "hailed the start of a new era for industrial relations in the coal industry", with the Commission establishing a new coal industry panel similar to other panels in the AIRC, and while regard would be had to previous decisions of the CIT, they would be subject to appeal and to the AIRC's general wage-fixing principles. In a classic piece of hyperbole, Brereton said that "Today is an historic day for the coal industry, that such major reform can be achieved without any industrial disruption pays tribute to the constructive approach of both the union and employers. This legislation puts an end to what had been an anachronism, the last of the specialist industrial tribunals. The continued existence of the CIT stood in the way of progress in what is a key Australian industry."[402]

Despite its earlier fiery warnings and threats for industrial action, by December the CFMEU had rationalised its acceptance of the move of the CIT into the AIRC. In the editorial of its monthly Common Cause member magazine, the CFMEU said that the "vexed question concerning the future of the ...Tribunal has been resolved in what can only be described as a sensible manner." [403] With the restructuring of the national industrial relations system, the CFMEU now said that "it was somewhat inevitable that workers in the coal industry would be drawn into the mainstream system.... For us it was always a matter of ensuring that any such move would not penalise employees... Many of the conditions which applied on minesites were "based on customs and practices that have been consolidated in a string of important decisions endorsed by the Tribunal. It was these that we determined to protect in the integration of the Tribunal while the employers sought the opposite."[404] The union's peak policy making body, its central council, had endorsed a resolution to go to the membership recommending incorporation of the Tribunal into the AIRC, but on the condition that certain arrangements were in place by the following July. These included the establishment of a special Coal Industry Panel in the AIRC to be headed by David Duncan, and the appointment of two commissioners to be based in Sydney and Brisbane. The union also said it wanted the right to nominate the Sydney based commissioner. David Duncan was appointed as a Deputy President of the Commission in June 1995 but was not appointed to the coal panel; the CFMEU did not get to nominate a commissioner.

After almost 50 years, the CIT was gone, and the coal industry had survived the change. But with a change of government in 1996, the Howard Liberal/ National Coalition would make industrial relations reform one of its first priorities, and major changes would be in store for Australian business generally and for the coal industry.

The coal and mineral associations unite

In 1990 the Miners' Federation merged with the Federated Mining Mechanics Association to become the United Mineworkers' Federation of Australia. Two years later the UMFA merged with the Building Workers Industrial Union, the Federated Engine Drivers and Firemens' Association, the Australian Timber and Allied Industries Union and other smaller unions to become the Construction Forestry Mining and Energy Union or CFMEU. The Federation was now the Mining and Energy division of the CFMEU. With major mining companies like BHP and CRA having operations in a number of different sectors of the industry, and with membership at the national level in the Australian Mining Industry Council and in the state associations, change was also on the agenda for the state employer associations. The coal associations in NSW and Queensland had been pursuing policies to move the coal industry into the mainstream in terms of industrial relations and regulation. Consistent with this policy it made eminent sense for the separate coal and non-coal employer associations to merge. The first change occurred in 1991 with the merger of the Queensland Coal Association and the Queensland Chamber of Mines. Michael Pinnock, who had been the CEO of the Queensland Chamber of Mines, was appointed as the CEO of the new Queensland Mining Council (QMC).

In NSW, change was a little slower in coming. The NSW Coal

Association and the NSW Chamber of Mines, Metals and Extractive Industries were under pressure to merge in 1994. The membership of the two bodies was largely different, although there were several key companies operating in both the coal and non-coal sectors. In addition to its major coal interests, BHP operated the Minerals Deposits mineral sands mines, CRA had metalliferous mines in NSW, and Exxon Coal and Minerals had interests in both sectors. These companies and other companies saw cost savings in merging the two bodies, and other benefits, not the least of which would be a clear signal to governments and the coal and minerals industries and their employees that coal was not a special industry to be cocooned away from mainstream industries.

The first attempt to merge in NSW in 1993 met with opposition from the Coal Association's members who saw the Chamber as more of a consultants' club than an organisation representing the minerals companies. With the majority of its members not seeing significant benefits for coal producers, and with the membership concerned to push ahead with its efforts to reform the coal industry, the Association declined to proceed with a merger. However, the major companies made another push in 1994 which was successful in convincing the two bodies to merge. The merger became effective in April 1995, and Jane Robertson was appointed the CEO of the NSW Minerals Council. The merger went relatively smoothly but there were major concerns from some of the non-coal members which would take some time to overcome. The CFMEU traditionally had only a small number of members employed by minerals companies in NSW apart from in Broken Hill, and minerals companies tended to look with a mixture of disapproval and fear at the coal industry and its history of bitter disputes and inflexible working arrangements, as well as at the militant stance of the CFMEU and at the role of the CIT, with its decisions effectively unable to be subject to appeal. The metalliferous mining companies were also well aware of the success of the CFMEU in gaining attractive wages for its coal industry members, wages that were on average significantly higher than paid in the non-coal sectors of the minerals industry. Some minerals companies were fearful that a merger would somehow see them drawn into the coal industrial relations web, and potentially open up opportunities for the CFMEU to gain a greater foothold in their industry.

In 1995, not long after the merger, Campbell Anderson, the CEO of North Broken Hill Peko Limited which operated the Northparkes mine in NSW, contacted NSW Minerals Council CEO Jane Robertson about a media article in which the Council was prominently quoted on an industrial relations issue. Anderson was concerned about the Council being involved in and speaking publicly on industrial relations issues and wanted to ensure that coal issues were somehow "quarantined" under the operations of the Council. The Council's committee structure did effectively keep coal issues separate, with those issues handled by its Coal Committee, but there were times when the Council was required to speak out publicly on sensitive issues like industrial relations. The concerns amongst the Council's non-coal members about the CFMEU and coal industrial relations eased over time, although many non-coal mining companies remained determined to resist any push by the CFMEU to increase its limited membership in their operations.

The 1990s also saw the state associations withdraw from their long time roles as registered industrial organisations under relevant legislation. Following the absorption of the CIT into the AIRC, QMC's industrial relations arm, the Coal Operators' Industrial Organisation, was dissolved and the Council ceased to have a formal role in industrial relations, although its staff continued to be available to assist companies on a fee for service basis. The Council was "no longer out there doing things on behalf of the industry but on behalf of individual companies."[405] During the 1990s, the NSW Minerals Council, while not being a registered body, continued to work on behalf of companies and to appear before the Tribunal when requested, and also continued to perform a coordinating function for the industrial relations managers within the member coal companies. In both states, the coal producers – or at least the majority of them – were looking to move towards a greater enterprise focus in industrial relations, something that had been denied until the coal industry's industrial relations system began to evolve from the late 1980s and early 1990s. The abolition of the CIT and the passage of the Howard

Government's new industrial relations legislation gave the employers the opportunity to take a more mainstream approach to employee relations. Some companies pushed ahead with determination, while others were slower to adapt, and major struggles lay ahead.

The Australian Coal Association, the federation of the coal members of the two State councils, continued to operate during the 1990s, with the CEOs of each council alternating yearly to provide the secretariat services. The Minerals Council of Australia more than once during the decade had the urge to absorb the ACA into its own operations, but backed away when it realised the strong support from coal companies for the ACA to continue as an independent body. The ACA was finally taken over by the Minerals Council of Australia in 2013. When MCA CEO Mitch Hooke was appointed in 2001 he said that the member company chiefs who appointed him (Rio Tinto's Barry Cusack, WMC's Hugh Morgan, BHP Coal's Bob Kirkby and Sons of Gwalia's Peter Lalor) told him that "the time was right for rationalising and amalgamating the plethora of industry organisations."[406] Over time the MCA oversaw the abolition of the Gold Council of Australia, the Uranium Council and the Australian Coal Association and the integration of their functions.

The industrial battles of the 1990s

While the coal industry in Australia had traditionally led the field with its abysmal industrial relations record, it would be the 1990s that would see fundamental changes in the way in which coal companies approached industrial relations, leading to some of the most bitter and protracted disputes in the industry's history. However these disputes, and changes to the industrial relations legislation after the Coalition government was elected in 1996, would change the face of the industry for ever. The major disputes in this period included the Gordonstone dispute in Queensland and the Vickery and Hunter Valley No.1 disputes in NSW. But before looking at these disputes, it is important to understand the Howard Government's 1996 Workplace Relations Act and the impact it had on the industry.

We have already seen how the Labor Government led by Paul Keating brought Australian industrial relations into a new era in 1993. The Keating Government's Industrial Relations Reform Act moved industrial relations away from industry-wide negotiations to focus on the enterprise. Enterprise bargaining became the catchcry and many companies throughout Australia took advantage of the new Act to develop their own enterprise agreements. The abolition of the Coal Industry Tribunal in 1995 was a milestone in the coal industry, without which, the coal companies argue, the industry may well have been trapped in a time warp, with industrial relations governed by a tribunal separate from the mainstream and subject to decisions which drew so heavily on the way in which things had been done in the past. But by the mid-1990s, while the coal industry had moved forward a little in terms of its industrial relations and work practices, the coal producers were looking for ways to achieve more fundamental changes to work practices so as to improve productivity and lower costs. Change at the minesite level had been slow, and what work practice changes there had been had involved a significant pay-off to mine workers.

In a working paper published by the National Institute of Labour Studies (NILS) in 1997, based on data available up to 1995 or 1996, the authors were able to draw some telling comparisons between the coal industry and other Australian industries. The authors concluded that "Overall, the evidence suggests that the Australian coal mining industry is distinctive in terms of human resource performance indicators. No other industry in Australia has the same combination of high employee earnings, high levels of paid overtime but zero levels of unpaid overtime, high rates of absence, low rates of labour turnover, low levels of job satisfaction, high rates of industrial disputation, and high work injury frequency."[407] The NILS study noted that the average number of working days lost due to industrial disputes over the 5 year period 1992 to 1996 was 4865 per thousand employees, or almost 5 days per year for every employee in the industry. The all industry average for Australia was only 107 days lost per thousand employees. Coal was 45 times the all industry average.[408] This period, of course, included periods of major industrial conflict, including 1994 when there was a national coal strike and the coal miners marched on the national Parliament House, and 1996 when there were major disputes. However, the industrial record was extremely poor at a time when many industries and companies in Australia were showing a new maturity and cooperation between managers and employees was improving.

Coal companies saw many barriers to moving their operations onto modern, flexible working arrangements that were critical for improving the ability of companies to compete in the international market. These barriers included the LIFO – last in first out - system, the seniority system, restrictions on multi skilling of workers, and lack of access to casual or part time employment. On some minesites a combination of old work practices and the inability of senior management to assert their right to make important decisions meant that the union and the workers on site were effectively running the show, or at least had as much power as the managers.

In its submission to the 1997 Productivity Commission inquiry into the coal industry, the NSW Minerals Council included a summary of the views of one of its member companies in relation to some of the obstacles to change in the industry. The company stressed the important role contractors could play on a minesite, but noted that "At many sites, the requirement to use contractors requires extensive consultation and agreement. If there is a stoppage of work by CFMEU members, then the Union expectation is that any contractors on the site will also cease work, even if they are members of other Unions or non-union employees, and irrespective of the cost implications." Overtime was another problem identified by this company: "The CFMEU, through its Divisions, imposes district overtime limits which are zealously maintained. At many sites, the Union has 'Roster Keepers' who manage the allocation and sharing of overtime. This can lead to difficulties in achieving sufficient people to work overtime to suit business needs...The CFMEU's intransigence on rosters and shifts and roster changes means that many sites are unable to implement alternate rosters and shifts that will better serve their business needs, and may provide additional opportunities for the employees."

The company also lamented its inability to employ casual and part time staff: "The changing needs of the industry would be better served by the ability to employ casual and/or part time employees, as many other industries do." And in a final barb at the CFMEU, the company said: " The view commonly held by the CFMEU officials and strongly supported through their formal and informal communications is that all employers are callous and uncaring, and that anything the employer seeks is to be resisted and fought. This adversarial win-lose philosophy is deeply ingrained and provides explicit and implicit barriers to the development of constructive and supportive relationships at many mine sites…This adversarial approach continues in the delivery and commitment to agreements once they are reached, and the strong resistance and reluctance to negotiate forward thinking arrangements for the long-term benefit of the industry and future employees."

With the election of the Howard Government in March 1996, and the appointment of Peter Reith as Industrial Relations Minister, the new Workplace Relations Act was passed and came into effect in January 1997. The Act was a major break from the past, with restrictions on the powers of the AIRC, a process to limit awards to a narrow range of issues, the introduction of individual agreements (Australian Workplace Agreements or AWAs), provision for both union and non-union collective agreements, and the abolition of the Industrial Relations Court, with the functions of that Court returned to the Federal Court. The old "closed shop" arrangements where unions had exclusive coverage of certain industries and the provisions in award giving preference to union members were also abolished.

The Act now also inserted the secondary boycott provisions back into the Trade Practices Act. In his second reading speech on the Bill, Reith said that: "Secondary boycotts are not an acceptable form of industrial action. Accordingly, meaningful secondary boycott provisions will be restored to the Trade Practices Act 1974, and the relevant provisions of the Industrial Relations Act 1988 will be repealed. The current provisions effectively allow unions to hold sectors of the economy to ransom for three days at a time. This is inimical to an internationally competitive economy. The Commission will be able to conciliate where a federally registered union is involved in a boycott, but there will be no requirement for a specific period of conciliation before legal action may be initiated by firms which are affected by boycotts."[409]

The changes to the secondary boycott provisions were a significant change for the coal industry. The practice of national or district strikes had been a common feature of the industry, and many companies were often wary of the prospect of being caught up in a strike which did not directly concern them. Union threats to bring a whole district, state or the national industry out on strike put fear into the hearts of many companies. But with the change to the secondary boycott provisions, unions were now liable to be taken to the Federal Court for strike action which extended beyond the enterprise, and the Federal Court was regarded as a tougher watchdog on this issue than the IRC. After the Workplace Relations Act took effect, the incidence of national and district strikes declined, and companies showed more willingness to take unions to the Federal Court on a range of issues. Major Industry-wide strikes did occur in 1997 and 1998, but once it became more evident that companies were prepared to take the union to the Federal Court, the practice virtually ceased.

In the past, companies had been reluctant to take legal action against the unions or employees engaged in national or district strikes, partly out of fear of incurring the wrath of the unions. However, the early 1990s saw what appeared to be a change of attitude on the part of some companies. In 1993 for example, CRA, Peabody and MIM took the CFMEU to the Federal Court in relation to a five day national strike held to protest against cuts to coal prices. The CIT had ordered the mine workers back to work; the companies were then successful in having the Federal Court also order the strikers to return to work, and they also indicated an intention to seek damages from the union for the economic losses they had incurred. John Maitland, the union's national president said that the "transnational coal companies have decided to go for the big stick" in taking this action under provisions in the Act which the Federal Government had said it would repeal, and that it was a provocative and heavy-handed action "that is seeking to deny our right to strike."[410]

The Industrial Relations Reform Act of 1993, which was passed in December 1993, created a new court (the Industrial Relations Court of Australia) and transferred the secondary boycott provisions of the Trade Practices Act (sections 45D and E) from that Act to the new Reform Act. Employers saw this as a significant weakening of the secondary boycott provisions, but would welcome changes in 1996 by the Coalition Government which restored those provisions to the Trade Practices Act.

Gordonstone – a fight to the death

Gordonstone was a large underground mine in the Bowen Basin where all of its production workers were members of the CFMEU's Mining and Energy Division. An industrial agreement between the union and the company was certified by the AIRC in October 1996 and it appears that industrial relations on site up to that time had been reasonable. However, ARCO's parent company in the USA announced in April 1997 that its coal assets in the USA and Australia were no longer part of the company's core business, and that it wished to sell these assets. BP had come and gone from the Australian coal industry, and now another major oil company had decided to do the same. The scene was now set for a major dispute between Gordonstone management and the union and its members which would prove to be one of the longest and most bitter disputes in the industry, continuing for the next four years and involving complex legal arguments before the AIRC, the High Court, the Federal Court and the Queensland Supreme Court.

Following the decision from ARCO Australia's parent company, the company decided to scale down its operations at Gordonstone; it also wanted changes to the certified agreement on a range of issues including overall employment numbers, rosters, overtime, bonus payments and the status of temporary employees.[411]   Negotiations between the company and the workforce had begun in early 1997, with the company wanting to negotiate individually, but the workers insisting on negotiating through the union. The certified agreement provided that if a dispute was not resolved, the AIRC's decision would be accepted by both sides, with the status quo to apply while the process was underway. The AIRC rejected ARCO's challenge to the certified agreement. But ARCO refused to be bound by the agreement and by July 1997 the Federal Court had also become involved. ARCO announced that it was planning to retrench 150 employees and refused to be bound by the certified agreement which provided for workers to be retrenched based on the LIFO practice. Then in August ARCO announced that it was planning to retrench the whole Gordonstone workforce.

What was seen as an extreme position from ARCO perhaps should not have come as a surprise, as at the same time the company was also involved in a bitter dispute at its Curragh mine. ARCO was in the process of reducing employment at Curragh by around half and was pushing for major changes including 12 hour shifts and the ability to use contractors without any union imposed restrictions. ARCO was also seeking to overturn the last-on-first-off system with an appeal to the Federal Court, following a rebuff from the AIRC earlier in the year which had ruled that the system was not contrary to the Act. In an attempt to prevent ARCO from proceeding with the Gordonstone retrenchments, the CFMEU launched action in the Federal Court in September. ARCO's position was that the mine was not a viable operation without major changes to work practices and employment numbers, and that it should be closed and put on a care and maintenance basis. The CFMEU said that the company simply wished to downsize the workforce and employ only 190 non-union workers.[412]   On 29 September the Federal Court rejected the union's application, clearing the way for ARCO to retrench the Gordonstone workers. On 1 October 1997 ARCO proceeded with the drastic step of sacking all the mine's production and engineering employees, closing the mine and putting it on a care and maintenance basis. ARCO paid out the retrenched workers 4 weeks' pay in lieu of notice plus 3 weeks' pay for each year of service, and said that they would have to vacate their company housing by the end of the month.

ARCO planned to re-open Gordonstone with a significantly reduced number of new employees who would be employed on merit under a new set of working conditions and wages. ARCO said that its decisions relating to the mine were driven by the need to improve the viability of the mine and by its right to manage the operations. One of the company executives said that in future it would be guided in its relations with employees by six pillars which included a direct relationship with employees, control of the business, hiring and promoting the best, flexibility and adaptability, and continuous operation.[413] In February 1998, following ongoing hearings before the AIRC, Commissioner Hingley slammed the Gordonstone management, saying that its "decision to terminate to avoid obligations was provocative and unnecessary . . . The company would also not have been compelled to accept every volunteer - it could have asserted a final say . . . I am of the view that the terminations had to do with the capacity and conduct of Senior Management not that of employees, and more specifically, an intent on the part of Senior Management to avoid its statutory obligations . . . "[414]

However ARCO kept the mine closed and finally made the decision in late 1998 to sell its 80% equity to Rio Tinto who proceeded to re-open the mine in 1999, which it renamed Kestrel, recruiting new workers and operating under a certified agreement which did not involve the union. The workers who had been sacked in 1997 then took their case to the AIRC which found that they should have been given preference when Rio Tinto recruited employees for the re-opened mine. But Rio Tinto appealed and a Full Bench of the AIRC then overturned the initial decision. The case continued on to the Federal Court which ruled in June 1999 that it had no power to overturn the AIRC decision. The Court said that while the AIRC Full Bench had erred in quashing the original order to reinstate the workers, the Workplace Relations Act prevented it from overturning that decision.

For Rio Tinto the Federal Court decision was a major win, allowing it to proceed to operate with the employees it said it had recruited on merit. Kestrel's general manager, Ross Hannigan, said that "What we did was to buy a mine that was closed and then restart that mine and employ on merit, and on doing so we've had enormous opposition from the CFMEU who have tried to stop the mine restarting. Now this decision is just saying that we can continue to employ on merit and that the mine will continue to produce which I think is a good outcome."[415] Tony Maher, the union's Mining and Energy Division national president, said that the union was likely to challenge the Federal Court decision in the High Court, but acknowledged that that would be a lengthy and uncertain process. The union had already written to the AIRC asking it to accept the Federal Court's finding that it had erred and should reinstate the sacked workers.[416]

The CFMEU's application to the AIRC was heard in July, but was unsuccessful. The remaining problem for Kestrel was the ongoing picketing activity at the minesite which then brought the Queensland Supreme Court into the dispute. Kestrel was successful in obtaining an injunction from the Court which caused the pickets to be lifted in July 1999. An appeal to the High Court was also part of the saga. In March 1999, the Federal Court had ruled against the company (it held that the 1996 certified agreement was valid), but in favour of the company in terms of the procedure to resolve issues before the AIRC. In March 2001, the High Court handed down its decision, with the Court ruling that the AIRC had wider powers to arbitrate in disputes than the Federal Court had ruled. However, as the employees had been retrenched over two years earlier, and ARCO had departed the scene, the High Court ruling had no practical effect on the Gordonstone operations.

Rio completed the purchase of ARCO's 80% equity in the mine in February 1999 and longwall operations recommenced in May 1999. By the year ending June 2000, Rio had wound production back up to 3 million tonnes and was planning to expand production to around 5.5 million tonnes. ARCO's substantial suite of coal mines in the USA was sold to Arch Coal in mid-1998. In 1999, ARCO exited from the Blair Athol mine, with Rio Tinto picking up most of ARCO's 31.4% share, and from the Clermont project, with Rio Tinto also buying ARCO's 19.5% share. In 2000, ARCO's exit was completed when Wesfarmers purchased ARCO's stake in the Curragh mine for around $200 million. ARCO was subsequently taken over by British Petroleum.

Rio Tinto's Vickery and Hunter Valley No.1 disputes

Rio Tinto was formed by the merger of CRA and RTZ in 1995. By the mid-1990s Rio Tinto had achieved significant changes to industrial arrangements at its operations in Australia, notably in CRA's iron ore mines in the Pilbara and its bauxite and aluminium smelting operations. However, CRA's coal mines were still to see the major reforms that the company saw as essential to achieving world class productivity, but that was about to change as a result of the protracted disputes at the company's Vickery and Hunter Valley No.1 coal mines. The Vickery mine, a small open cut employing around 50 workers, was located in the Gunnedah area in NSW. The dispute pre-dated the 1996 Workplace Relations Act, but indicated a much tougher stance on industrial issues by CRA and Rio Tinto than had been common in the industry up to that time. Novacoal, a CRA subsidiary, operated the Vickery mine which produced mainly steaming coal for the export market. In 1995, Novacoal attempted to introduce a new enterprise agreement which provided for 12 hour shifts. Workers at the mine began a strike in August and the mine was closed and put on a care and maintenance basis.

The Vickery mine was not the first to attempt to introduce 12 hour shifts into the coal industry. The Ensham mine in Queensland, which commenced production in 1993, introduced these shift arrangements that year in a new enterprise agreement. The Ensham EA was introduced without the support of the CFMEU and led to an extraordinary situation in 1994 when 30 CFMEU members at the mine obtained a Federal Court injunction to prevent the union from taking disciplinary action against them. The union had demanded that its members employed by Ensham explain why they were working under arrangements which were contrary to the union's rules.

By November 1995, Novacoal had negotiated an agreement with the union for a 42 hour week at Vickery, with 12 hour shifts and an average salary of $67,000, but this was rejected by the striking workers. A couple of months later, the ACTU came out strongly in support of the striking workers, with secretary Bill Kelty warning "that unions would tell CRA's coal customers to expect no resolution of a strike at the Vickery mine near Gunnedah until the company agreed to negotiate."[417] The ACTU's position was that CRA's agenda at Vickery, which centred on the 12 hour shifts and a refusal to agree that new jobs at the mine would be given to unemployed miners, was part of an overall strategy by the CRA group to convert the workforce to a non-union basis.

CRA had been involved in major disputes at its Hamersley iron ore mine in Western Australia and at its Weipa bauxite mine in Queensland, inducing or encouraging employees over to non-union agreements. In October 1995, with the great majority of Weipa employees having been transferred to individual contracts, the workers who had held out and had remained on the award commenced a strike. The award workers were being paid significantly less than those on individual contracts. Sympathy strikes followed, and at one stage in November 25,000 workers were on strike in NSW and Queensland. The strike was settled when CRA agreed to pay the award-based workers the same rates as their co-workers, but following the disputes at Hamersley and Weipa, CRA was certainly out of favour with unions and Labor Governments at both state and Federal levels.

The CFMEU supported the striking Vickery workers throughout the strike with payments of $500 or more per week, and no doubt this was one factor that helped to extend the strike. The dispute was finally settled at the end of July 1996, and employees returned to work on 5 August, almost a full year after the dispute began. The company and the union had agreed on the basis for a resumption of work, and the AIRC had issued orders stipulating that all strike action had to cease. The striking employees voted 27 to 3 to accept the agreement, the substance of which they had rejected back in November. The agreement provided for a three month trial of 8 hour shifts, to be followed by a three month trial of 12 hour shifts. The striking workers went back to work and employment numbers grew to 55 workers by June 1997. But only eight months later Rio Tinto announced that it was closing the mine due to low coal prices; Rio Tinto finally sold Vickery to Whitehaven Coal in 2009. The Vickery dispute was a bitter one and the mine's closure was a sad outcome for the workers who lost their jobs. However, for Rio Tinto, the prospect of new and longer shift arrangements would have been a step in the direction of much more flexible working arrangements for other mines in the group.

The Hunter Valley No. 1 mine (we'll call it HV1) between Singleton and Muswellbrook in the Upper Hunter Valley was developed in the late 1970s by Coal & Allied as the industry responded to the early days of the boom in the demand for steaming coal from Japan and other countries. Coal & Allied, the company with its roots in the industry extending back to the Brown family operations in the 1840s, became part of the CRA group in 1993. CRA secured a major shareholding in C&A in 1991, also winning two seats on the board and appointing John Ralph and Don Carruthers to those board seats, but then had to wait until 1993 to take full control of the company. Rio Tinto saw HV1 as riddled with restrictive work practices, although HV1 was not alone in that regard. As the relatively new owner, Rio Tinto was determined to make major changes, and to wrest control of the mine from what it saw as the iron grip of the CFMEU and the workers on site. After the CRA takeover of Coal & Allied, a new senior management team was appointed. Terry O'Reilly became C&A's CEO, and a new general manager, Allan Davies, was appointed to HV1 in 1996. Kim Tronson, a Rio Tinto executive, was appointed to take over from O'Reilly in 1997. In 1996 HV1 employed around 585 people, and produced 5 million tonnes of coal per year. It was a big operation, but in 1996 made a profit of only $1 per tonne, and when coal prices fell by around $3 a tonne in 1997, it was heading for a loss.[418]

Davies said that he found a mass of restrictions on management's ability to manage the operation, restrictions which would have been unacceptable in many other industries in the mid-1990s. Davies said that there were major demarcation barriers, with staff (employees who were not involved in the production or maintenance operations) not permitted to maintain equipment such as trucks, loaders, shovels, coal preparation plant or the rail loading point. The union co-managed the operation and the union delegates were the main communicators with the workforce and the main source of information for the production employees. Discipline was "ad hoc or non-existent." Shifts were a fixed 8.5 hours, with hot seat changes on all equipment supposed to occur; but Davies said that these changes did not take place. As per the industry award, all overtime was paid at double time, so a worker working 41.25 hours a week was paid for 50 hours. Seniority was also rigidly enforced, with allocations of overtime and rosters favouring the employees with the longest service. When there were retrenchments, as per the industry practice, the last-on-first-off rule applied. Davies summarised the position as "management did not manage and the union ran the operation" and quoted a comment that he made to journalists in 1997 that "if we want to hire, fire, promote or reward and allocate overtime we first need to ask permission of the union."

The scene was set in 1997 for a major confrontation between C&A management and the CFMEU and its HV1 members. Davies said that 1997 was the year when costs were on track to outstrip revenue, sending the mine into a loss and that "we had to do something to ensure the mine remained viable. That 'something' was to improve productivity and that meant eliminating unproductive work practices." Over the next two years, HV1 saw a reduction in employment to less than half of the number employed at the start of 1997, with "the best people (being) retained rather than retrenching based on seniority." Big cuts were also made to the numbers of large trucks and other machinery on site, with the remaining machinery worked more efficiently. However, the path to achieving these changes was long and hard. Other coal producers in NSW were watching every move, some quietly cheering C&A on, hoping that they would be successful, and thereby making further change in the industry much easier.

The very clear message from Leigh Clifford, then CEO of Rio Tinto's global energy division, was that "dramatic change was required, that incrementalism was not enough and that change which was good by Hunter Valley standards would be inadequate." Davies said that this message was maintained constantly from 1997 to 1999. Clifford, who went on to become CEO of the Rio Tinto group worldwide in 2000, had spent time as a senior executive in the NSW coal industry, and understood many of its problems. With this support from his superiors, Davies now had to push ahead. For the CFMEU, the HV1 battle was also one it had to win; at stake was its power base in the industry built up over decades of struggles. The CFMEU had lost the battle to keep the Coal Industry Tribunal in 1994, and the Howard Government's industrial relations reforms in 1996 posed another major threat. And as outlined earlier, Rio Tinto had made huge gains in winning workers over to individual contracts in its iron ore, bauxite and aluminium operations.

In early 1997, the tough new approach from Rio Tinto which emerged with the Vickery mine was now evident at HV1. Rio Tinto had also been offering individual contracts to workers at its coal mines in Queensland, and management at its Mount Thorley mine in the Hunter Valley had taken the step of ceasing to make automatic payroll deductions for the payment of union fees.[419] The company would also soon be offering individual contracts to its HV1 workers.

In April, Coal & Allied took action in the AIRC, seeking a permanent order to prevent strikes at its HV1 mine. Coal & Allied told the AIRC that it had no alternative to deal with "repeated, deliberate and very damaging industrial action taken in breach of established disputes settlement procedures".[420] Coal & Allied's legal representative, Graeme Watson, told the Commission that the HV1 mine had a poor reputation for industrial relations, with 14 days and some weekend work lost in 1996, accounting for 6% of productive time. John Maitland, CFMEU president, accused Rio Tinto of pursuing a return to a "master-servant" relationship with its employees; he also blamed the mine's financial position on poor management and lack of cooperation with the union.

At the end of May, in a further sign of its determination to turn its coal business around, Rio Tinto announced a major restructuring of the business in NSW, with its Novacoal management merged into Coal & Allied, leading to a reduction of over 70 management positions in NSW, and the transfer of all of Coal & Allied's non-marketing functions to the Hunter Valley.[421] Coal & Allied was set a target of achieving a major increase in profitability over the following two years. With after tax profits of only around $4 million in 1996, profits over the next two years were expected to reach between $75 and $100 million, yielding a return on shareholders' funds of around 15%.[422] This was a huge task for the Coal & Allied management, but Rio Tinto had made it clear what was expected.

The dispute then dragged on through 1997 and 1998, with major hearings before the AIRC, ongoing strike action by the CFMEU and its members and picketing of the minesite by striking members. Strike action by mine workers throughout the northern district in support of the HVI workers also occurred. NSW Premier Bob Carr also became involved. Before a meeting with Leigh Clifford in October, Carr accused the company of being on a "rampage" in its battle with the CFMEU. Carr said that he did not want to have any company in NSW whose objective was to engage in industrial warfare. His meeting with Clifford was aimed at getting Rio Tinto to agree to resolve the HV1 dispute through arbitration, with Carr saying that it was the only way to settle the differences between the parties. But Clifford was not prepared to go along with the Premier, believing that working through the IRC would not achieve the fundamental changes to work practices that Rio Tinto saw as essential to raise profitability to an acceptable level.[423]

By October 1998, Rio Tinto was proceeding with major retrenchments at HV1; over 60 employees took voluntary retrenchments, other employees were moved to different roles on the minesite and over 100 employees were retrenched on the basis of no suitable work being available.[424] The CFMEU took legal action against the company for unfair dismissals, and the case was only finalised in May 2002, when 190 former employees of Coal & Allied from the HV1 and Mt Thorley mines accepted a $25 million settlement offer from the company. This was described as the country's biggest unfair dismissal settlement, but it did not involve any of the employees being reinstated.

At HV1 a new collective agreement was finalised in October 2000 covering 90% of the much reduced workforce at the mine; the other 10% of employees were on AWAs. Management, including supervisors, were mainly employed on individual contracts, a major change from the position in 1997 when all non-production employees other than the senior management were employed under one of the various coal industry awards.[425] Rio Tinto had achieved its objective of a massive overhaul of the work practices at HV1, and had asserted its right to manage the mine. The union was still an important player, but with vastly reduced power. Rio Tinto also had to withstand a concerted campaign by the CFMEU in Australia and overseas; its reputation with the Labor movement and Labor governments was severely damaged, but it emerged as a strong and more profitable producer.

The passage of the Workplace Relations Act was an important element in the HV1 story, although Davies said that the fact that HV1 was in dire straits when the Act was introduced was a coincidence. However, he said that "had the legislation not changed, our ability to make change quickly, more lasting and with lower costs would have been a lot more difficult."

Coal mines restructure in both States

Prior to the 1990s, efforts by individual companies to make major changes to working arrangements had generally failed, although some changes such as bonus payments had been adopted rapidly throughout the industry. And in the 1950s the producers and the JCB eventually wore down the resistance by the Miners' Federation and its members to the use of machines to extract pillars. New technology such as continuous miners and later longwall systems also spread through the underground sector once early problems had been overcome. Working arrangements evolved over the years in response to the modernisation of the industry and the pressures of the market. However, it was not until the 1990s that the fairly strict observance of custom and practice broke down and the seniority rule disappeared from the award (although not from enterprise agreements).

The industrial battles in the 1990s involving the CFMEU, Rio Tinto and ARCO were the precursor to a major restructuring of the Australian coal industry. The combined effect of the 1996 Workplace Relations Act, disputes such as at Hunter Valley No. 1, Mount Thorley, Vickery,

Gordonstone and Curragh and Federal Court decisions had made the task of restructuring the industry and overturning decades of restrictive work practices much easier for coal companies. And arguably the absorption of the CIT into the mainstream tribunal was also a factor. The restructuring in Queensland saw employment in coal mining fall from a peak of just under 11,000 in 1996 to under 8,000 by the year 2000, a drop of 27%. In NSW, the cutbacks were even more severe, with employment falling by around 4600 between 1996 and 2000, or by almost 33%. Coal production, however, increased over this period. In NSW production jumped by 17.5% and in Queensland by 32.6%. Production per employee per year soared; in Queensland it rose by 66% and in NSW by 59.5%. Production per employee per hour rose by similar percentages, indicating that there had not been significant changes across the industry in the average shift lengths or average hours of overtime. Productivity, of course, can be measured in a variety of ways and changes in, for example, production per employee, can be a result of many factors. However, for the industry to record such major increases over such a brief period was strong evidence that in the mid-1990s the coal industry was over-manned and plagued by restrictive work practices.

In its July 1998 report on the Australian coal industry, the Productivity Commission noted that: "While many factors influence the competitiveness of black coal mining in Australia, the impact of work arrangements is considerable. Black coal mines operate with a multitude of work arrangements which restrict productivity. Almost all of these arrangements raise unit labour costs and reduce competitiveness unnecessarily. Some arrangements (such as the exclusive use of seniority to allocate overtime, annual leave, promotion and training, and restrictions on the use of contractors) are discriminatory, as well as detrimental to mine performance...Work arrangements in Australian black coal mines appear to be less conducive to improved productivity than those in Australian metalliferous mines and non-unionised United States black coal mines, but similar to those in United States unionised black coal mines...The principal means of enhancing the competitiveness of Australian black coal mining workplaces is to change those work arrangements which unnecessarily restrict productivity. In this way, the productivity of labour and the utilisation of capital will be raised and, in turn, unit production costs will fall. This process should be facilitated by not including such arrangements as allowable matters in black coal industry awards. The Australian Industrial Relations Commission has begun to move in this direction."[426] When the Productivity Commission report was published, the Hunter Valley No.1 and Gordonstone disputes were still underway, and the AIRC had a long way to go in terms of rationalising major industry awards. Much was to change in the following couple of years.

The coal industry's restrictive work practices, and its poor productivity performance compared to leading mining operations overseas, were major challenges for the industry in the 1990s. However, compounding the problems for the industry were the relatively high levels of remuneration. For example in 1993-94, in black coal mining in Australia the average labour cost per employee was $81,263, including wages, superannuation, payroll tax, workers compensation, and fringe benefits tax. The Australian metalliferous mining industry's average was around 17% lower at $67,492, and the average for all industries at $30,022, was only around 37% of the coal industry average.[427]

The National Institute of Labour Studies at Flinders University compared the provisions of enterprise agreements in the coal industry and the metalliferous mining industry in 1996. The differences between the coal and metalliferous agreements in relation to many provisions were stark. The coal industry now had a range of enterprise agreements, but they were generally still tied to the award and were far more restrictive than in the metalliferous sector. Over 90% of the coal industry agreements remained at least partially dependant on the parent award (ie the Production and Engineering Award or the Staff Award). In the metalliferous sector, over 80% of agreements totally replaced the award. Almost three quarters of coal agreements placed major restrictions on the use of contractors, compared with only just over one in ten metal mining agreements. Close to half coal agreements contained a "no forced retrenchments" provision, whereas none of the metal mining agreements contained such a clause. No coal industry agreements provided for employment of part time production or engineering staff, whereas in metal mining over 40% of agreements provided for part time employment. And the study found that it had not uncovered any coal agreement which had done away with the last on first off redundancy provision, while only 14% of metals agreements stipulated that redundancies would be by seniority.[428]

In the Hunter valley, there was also the 5 panel roster system where mines operated with 5 panels (or crews) compared with 4 in Queensland. In order to reduce costs and make the most efficient use of machinery and equipment, many companies in NSW 1997 and 1998 were looking to move to 24 hours per day, 7 days a week operations. The problem was the prohibitive cost of such a move. The CFMEU in NSW was opposed to 12 hour shifts and companies wishing to work on a continuous basis were required by decisions of the CIT and custom and practice to use 5 complete crews, meaning that mines in the Hunter Valley had to employ 20% more employees to operate 24 hours a day, 7 days a week. Producers in Queensland were able to operate with only 4 crews, and hence enjoyed lower costs. The 5 panel roster involved payment for the first 24 hours at the normal hourly rate, with all hours above that paid at double time, or at more than double time for weekends.[429]

BHP acknowledges the role of management

Rio Tinto clearly pursued a tough and at times an arguably brutal approach to reform at its mining operations in the 1990s. However, the other big miner, BHP, took a quite different approach to industrial relations and enterprise bargaining. This was evident in April 1992 with what was hailed as a landmark agreement for BHP's 7 Bowen Basin coal mines. The agreement was struck with 5 unions and gave employees security of employment for three years and the option of yearly contracts including job security after the end of the three years. The CFMEU's Andrew Vickers, who chaired the unions' bargaining committee, said that the agreement provide for continuing implementation of change in the workplace, with employees, unions and the company adopting a process of consultation. Bob Flew, BHP Coal's general manager, said the agreement had far-reaching ramifications for company operations, employees' job security, and profitability", adding that "Unions and employers in central Queensland realise that both groups need the flexibility to change in the face of increased international competition in the marketplace from countries like the US and South Africa."[430] From BHP's point of view the deal was important because it provided for individual minesite agreements, with individual mine bonus schemes replacing the group bonus scheme, and reviews of rosters, working hours and the use of contractors.

However, five years later, in hearings conducted by the Productivity Commission for its 1997-1998 coal industry inquiry, it was clear that BHP Coal saw that it still had a long way to go to achieve changes in its operations. Mike Cupitt, human resources manager for the company's central Queensland region, referred to the critical role for the company's senior managers in achieving what it saw as the necessary changes in a range of working conditions and in the company's ability to manage its operations. Cupitt was candid in acknowledging that management at some of its mines had not been successful in achieving change, and that the fault could be attributed to management. It is also fair to conclude that some of Cupitt's comments on his own colleagues in BHP Coal may well have applied to some of the other coal producers.[431]

Cupitt began his presentation to the Productivity Commission by saying: "First up, what we really want to make clear is that we don't see it as productive to get on to some sort of blame trip, blame government, blame unions or our performance or lack of performance. At the end of the day it's our view that the people who are paid to manage our operations ought to be doing just that. Our view is that our performance to date, I think in your words, Commissioner, has been modest. There has been some change occurring over a number of years, some significant, some smallish by way of comparison with probably what we need to do today, but bottom line our view is that our managers are accountable to manage the operations not government, not the unions as such."

Cupitt said that the coal industry award was "in fact, not a bad working document as such, but has been in place now for 7 years. It would be my view at least, that that document has not been taken full advantage of..." There were certainly some provisions in the award which still restricted management's ability to manage, and Cupitt cited the reduction in hands provision as one example, this provision specifying that retrenchments had to be on the basis of seniority, with the employees with the least service retrenched first.

However, in relation to the ability of management to recruit new employees on the basis of "merit" (selecting the best person for the job based on their qualifications, experience etc) Cupitt admitted that BHP was not constrained by the award: "There's nothing in the award on that subject. What we've got we've created for ourselves, probably with a little bit of help from the unions, but ... again we've created it for ourselves, the current situation." Cupitt said that some, but not all, of the BHP Coal operations had already addressed the issue of merit based recruitment.

Following the passage of the 1996 Workplace Relations Act, awards were required to go through a simplification process, restricting the allowable areas covered in each award to 20 matters. The major coal industry award, the Production and Engineering Award, was reviewed and the new Production and Engineering Award 1997 came into operation, with many of the old and detailed provisions of the award deleted. The process of award simplification continued over the next few years, but the onus was now largely on management to do the hard work to achieve the changes companies believed were necessary.

Shell Coal's IR experience was mixed

As a new mine in the Hunter Valley in the 1990s, Shell's Dartbrook took a professional approach to recruitment and required new employees to complete a 7 week training program before entering the workplace. Dartbrook's mine manager said in 1998 that this approach was "bearing fruit, not only in terms of knowledge and skills but in the development of a unique safety and team oriented culture. This has led to a desire for training and qualifications, which has to be managed. The mine has been in operation for over three years and the energy and enthusiasm is still evident in the entire workforce."[432]

Dartbrook was one of the companies in the 1990s taking advantage of the ability to negotiate enterprise agreements (EAs) with its workforce, with the negotiation actually with the CFMEU. Enterprise agreements gave companies the opportunity to have more flexible working arrangements, with the first Dartbrook EA containing what the company said were "some very innovative and ground breaking changes which … carried through to the second EA which was settled for a three year term…The basis of the first EA was fixed salaries, all inclusive of allowances, overtime and production bonuses. In return employees were required to work rotating shifts of 8.5 hours duration five days per week and commit to work allocated overtime. The only changes to the second EA (are) a step change to level 4 and 6 of the Industry work model, greater use of contractors and the overtime component reduced to allow a separate payment for weekend overtime when worked."[433] Another innovation at Dartbrook was provision for all employees to salary sacrifice to obtain a fully maintained company car, a valuable benefit in a rural area where the great majority of employees had to come to work by car. By 1998, around 60% of employees had taken up this offer. Perhaps more interesting was that the company said that production bonus payments had lost their appeal and that "Dartbrook people are pressing good performances because of personal pride in doing a good job and in their mine."

The company did, however, recognise that, while industrial relations were "very strong and robust", labour costs were high, and it hoped that 'time and technology" would help to make labour costs more realistic. There was also a hope that relationships with employees would become closer over time, without the obstacles imposed by the union: "Although Dartbrook has maintained excellent management / labour relationships over the three years of operation, I believe that it could be enhanced by talking direct with our people without District Union intervention on critical issues. We have built up a strong relationship with our people based on two way trust which could explore mutually agreed new opportunities outside the traditional union structure."

Shell Coal's corporate submission to the Productivity Commission black coal inquiry in 1997 did not give such a positive view of industrial relations in the industry, and called for a number of changes to the coal award. At the time of making the submission (November1997), Shell Coal's Moranbah North Coal mine was under construction. One of the major changes sought by Shell Coal was an end to the practice of recruitment from retrenchment lists (lists of union members who had been retrenched and were looking to be re-employed in the coal industry). The company said: "It is difficult to quantify the impact on enterprise productivity attributable to the recruitment and retrenchment practices of the industry. There is no doubt that the new *Workplace Relations Act* will modify the questionable practice of employing exclusively from union retrenchment lists thereby permitting the employer to select the most appropriate personnel for employment. It is a matter of record that Shell Coal Australia was prepared to withhold capital investment from the Moranbah project until the right to employ the best available personnel was unfettered."

Shell Coal quoted Drake Brockman's observation in his historic judgement in June 1939 that "The last-on-first-off rule is being rigidly enforced by the Federation and apparently quite regardless of such matters as misbehaviour or inefficiency. A miner with several years' seniority is so firmly entrenched in his job that nothing short of closing down of the mine, death or a compensatable illness can prise him out of it." Shell said that observation was apparently as relevant in 1997 as in 1939, and that in calendar year 1997 there had been six major industry disputes involving clause 24 in the award or recruitment and retention issues. Shell said that removal of the clause from the award was urgently needed "on the basis that it is discriminatory and unacceptable when compared with general community standards."

Shell was also critical of some of the EAs in the industry which were very similar in content to the P&E Award, noting that some minesite negotiations had been dominated by a desire to preserve the sanctity of the award and thereby maintain established work practices. The most recent example of this intransigence, it said, was in the circumstances relating to the company's own Moranbah North Coal (MNC) negotiations with the CFMEU between August 1995 and late 1996. Shell did not go into detail about these negotiations but said that "after extensive discussions it became obvious that any departure from the P&E Award was going to be opposed by the CFMEU." This led to the CFMEU applying to the AIRC for a "roping in award" (a decision which would have forced the employees to be covered by the industry award), with MNC in turn applying for a new award for the mine. Shell Coal said that "There was no regard for, or attempt to comply with the (Workplace Relations Act) when it came to the making of a new award. The principal argument appeared to revolve around the proposition that the P&E Award constituted an industry safety net, and it was not open to an employer, or employees to move away from the this award without the express consent of the administrative arm of the CFMEU."

Camberwell was innovative, but still frustrated

In the early 1990s the new Camberwell mine in the Hunter Valley was another interesting example of the changes which were starting to be implemented in the industry, although not without major obstacles and frustrations. Camberwell took an innovative approach to the human resources side of its business and entered into an arrangement which gave the company its own award, one of the few at that time in the industry. The award was negotiated with the CFMEU which had sole coverage of the production and engineering employees on the mine site. The award gave the company some advantages, but its frustration with the industrial relations system was evident in its submission to the Productivity Commission in 1997: "In today's industrial climate which promotes the use of direct bargaining with employees, having a separate award away from the industry's influences should have been a positive influence. This has not always proved to be the case because both in the negotiation and re-negotiation of our award we have been prevented from introducing measures to improve our business competitiveness by the current bureaucratic structure of the union. In order for us to achieve any changes to the current industry practices, issues that we negotiate with our employees are subject to further scrutiny and ratification by the District Branch of the Union. In most instances this stifles productive negotiation with our employees as they know that many issues are opposed to District or National Policy. They are reluctant to consider issues that may incur repercussion from the officials of the union."[434]

Camberwell's frustration was in the context of changes in the international coal market in which, as we saw earlier, first the Taiwan Power Company, Taipower, and then the Japanese power utilities took a more aggressive approach to coal purchasing, with major consequences for companies struggling to be competitive, increasing pressures on mine management to achieve more efficient operations.

Old habits die hard

While the 1990s were a watershed for coal industry industrial relations, with changes to legislation, major AIRC decisions, the removal from awards of seniority provisions and the spread of enterprise agreements and collective agreements, some of the old practices still continued into the new era through enterprise agreements.

In 2015, BHP drew attention to several of its EAs in the coal industry, admitting that there were "numerous examples in BHP Billiton's Queensland and NSW coal business of operational issues outside the scope of employment being brought into EAs." The company added that employers faced "the threat of protected industrial action for the inclusion of clauses permitted under legislation that impede management's ability to operate, and increase the potential for disputes." Three examples within its own coal business included the Mt Arthur North mine in the Hunter Valley, the Appin mine south of Sydney, and the Port Kembla Coal Terminal which was managed by BHP. BHP said that the most recent EA at Mt Arthur North which was signed in 2011 still restricted retrenchments according to the 'last-in-first-out' policy. This policy it said was "inconsistent with an employer's right to decide who they employ, and impacts an employer's ability to ensure the best possible people (e.g. from a merit, skills, culture or diversity perspective) are applied to the task at hand."[435]

The Appin mine's EA, also signed in 2011, specified that "BHP Billiton will not replace employees who resign or retire with contractors and sets a minimum threshold for wage conditions for any contractors that are used." BHP noted that this provision "limits employers from making operational decisions on the appropriate mix of employment, and inhibits competitiveness by creating a floor on labour rates which may be in excess of the market rate for employment." At the Port Kembla Coal Terminal, the EA signed in 2012 "requires employee representatives to be informed of the name and commencement date of new employees", and while the company acknowledged that this was a seemingly minor requirement, it created an "administrative burden for employers, and potential privacy concerns for new employees covered by the agreement."[436]

BHP says goodbye to the Hunter Valley and Newcastle

BHP's connection to the Hunter Valley and the Newcastle region dates back to the establishment of Australia's first integrated steelworks in 1915 and its entry into the coal mining industry in the 1920s through the development of its Elrington and John Darling collieries. BHP moved on to become a major coal miner in the 1930s with its purchase of the Burwood and Lambton collieries in 1932, the takeover of Australian Iron and Steel (AI&S) in 1935 and its development of new mines in the years following World War Two.

The 1980s for BHP began with some optimism for its investments in the Hunter Valley. In 1981 BHP received approval to develop the Saxonvale open cut mine south west of Singleton; the development of Saxonvale was planned to take the mine to an annual output of just under 5 million tonnes per year by 1986-87 and was to involve an investment of around \$360 million.[437] BHP was awarded the lease for Saxonvale in competition with other companies including Peko Wallsend. According to an article in the Sydney Morning Herald the lease was awarded on the basis that Saxonvale coal would supply BHP's Newcastle steelworks.[438] That same article said that BHP later "discovered" that the quality of coal from the Saxonvale lease was not high enough for steelmaking and sought permission to export the output of the mine. "The Government may or may not have believed that the discovery had only Just been made, but it saw nothing wrong with letting BHP enter the export coal trade, provided the company paid the same super royalties and provided the same front-end finance as other exporters."

In the early 1980s BHP was also proposing to construct an aluminum smelter in the Hunter Valley at Lochinvar near Maitland. US company Amax, BHP's partner in the project, pulled out and BHP tried to find other companies willing to invest. With no other investors, with the NSW Government wanting a higher price for electricity and with the world aluminium market in decline, BHP decided not to proceed with the smelter. But the big blow to the region came in September 1982 when the company announced a major restructuring of its steelworks. BHP's steel business was under severe pressure from imports and its productivity did not rate highly compared with steelworks overseas. John Risby, the Newcastle steelworks general manager, was forced to deny a story in the media that BHP was planning to close the whole steelworks, but the key part of the announcement was the closure of one of the 3 blast furnaces, and a loss of over 1700 jobs. This followed other changes which had seen a similar number of job losses over the previous 1-2 years.[439] The long term future of the Newcastle steelworks however was now uncertain.

The development of the Saxonvale mine did proceed, but by 1987 BHP was looking to off-load the mine and thought it had a ready buyer in Peko Wallsend. The purchase by Peko fell through and Elders Resources bought the mine in 1988, only for it and other coal assets to be on-sold to Oakbridge in 1989. BHP's investment in Saxonvale was significant, and it was believed to have recouped only a fraction of this when it sold the mine to Elders. BHP's other mines in the Newcastle area included John Darling Colliery, where the company had invested in a new longwall which was installed in 1981-82, but that mine was closed only five years later. BHP's Macquarie coal group at that time included the Lambton, Teralba and West Wallsend collieries; these collieries were sold to Bond Corporation's Pacific Coal in 1989. The Macquarie group collieries produced around 2.2 million tonnes per year and would not stay long in the Bond group, with FAI Insurance buying them in 1990. Lambton was closed in 1991. Teralba and West Wallsend were later bought by Glencore.

Jim Lewis, who ran BHP's colliery operations in the late 1980s, said that the sale of the Macquarie coal mines was part of the company's program to concentrate on its core business which, in terms of coal, meant the Illawarra coking coal mines. The Illawarra coal was high quality coking coal and, according to Lewis, gave the Port Kembla steelworks a competitive edge.[440] Lewis said that BHP's steelworks in Newcastle at that time was using only coal from the northern district for around 30% of its needs and of the three BHP mines, only Lambton was supplying the steelworks; all the coal from Teralba and West Wallsend was being exported. The sale to Bond Corporation included a long term contract for the supply of coal to the Newcastle steelworks and this was obviously one of the benefits of the sale for the new owners.

BHP's sale of the collieries in 1989 saw its effective end as an underground coal miner in the northern district. Its Newcastle steelworks would continue to be a customer for some northern coal during the 1990s, but in 1997 the end of the steelworks was announced by John Prescott, BHP's CEO; the company would be closing the steelworks in 1999. This decision was announced as part of a broader plan revealed by the company to restructure its business, although the closure of the steelworks, with the loss of around 2500 jobs plus hundreds of jobs of contractors employed onsite, was by far the major element. Prescott said that while it was regretted that the steelworks would be closed, it was a decision that had to be taken if BHP's steel division was to be a viable business, and he said that it had been evident since the 1960s that the future of large scale steel production would be in the manufacture of flat steel products for motor vehicles, ships and other uses. Newcastle produced the wrong sort of products, including some that were being produced from plants using recycled steel. Prescott also said that coal from the Hunter Valley was not suitable for steelmaking.[441] BHP would concentrate on Port Kembla which would now be its only integrated steelworks in Australia. Steel-making operations would also be continuing at Whyalla in SA, with Whyalla focussing on structural steel and rail lines. The decision to close Newcastle was a "boardroom betrayal of working-class Australia" according to NSW Premier Bob Carr, who added that "We knew of the probability of these cuts - the news is not a surprise so much as sad confirmation."[442] There was now no going back and the steelworks did close in 1999. BHP still owned the Mt Owen mine, its last operating coal mine in the Hunter Valley, but this was sold to the hungry Glencore in 1998. Now BHP's only NSW mines were in the Illawarra and included Cordeaux, Elouera, Tower, West Cliff and Appin.

BHP exited the steel industry in Australia in 2000 and 2002 when it floated two new companies out of its old steel business. The long products components of BHP's steel business (including the Whyalla steelworks in SA and other assets) were moved into OneSteel Limited which listed on the ASX in October 2000. After the merger between BHP and Billiton in 2001, BHP Steel Limited was spun out as a separate company which listed on the ASX in July 2002; the company changed its name to BlueScope Steel in 2003. BlueScope's assets included the Port Kembla steelworks, the Lysaght roofing products division and the Illawarra coal mines.

As fate would have it, as a result of its merger with Billiton in 2001, BHP came back in the coal business in the Hunter Valley and the Lake Macquarie area. Billiton brought some major coal assets to the merger, including the Ingwe mines in South Africa, the Mt Arthur North open cut mine near Muswellbrook, and several mines in the Lake Macquarie area which Billiton had acquired when it took over Coal Operations Australia in the 1990s. Billiton had purchased the Bayswater Colliery next to the Mt Arthur North deposit in 1996 and then won government approval to develop Mt Arthur North in 1998, with Bayswater and Mt Arthur subsequently managed as a joint operation.

The final divestment of its Illawarra coal assets by BHP Billiton came when South32 was spun off as a listed company in May 2015, with South32 taking ownership of the Illawarra coking coal mines and other mines in Australia, Southern Africa and South America. BHP Billiton retained the Queensland coking coal mines and the Mt Arthur steaming coal mine in NSW. Mt Arthur now comprises open cut and underground mining operations which produced over 18 million tonnes of saleable coal in 2018-19. BHP transferred its Navajo operations in the USA to the Navajo Nation in December 2013 and ceased as the mine operator in December 2016. Its San Juan energy coal mine, also in New Mexico, was sold in 2016 to Westmoreland Coal. BHP Billiton retains its one third interest in the Cerrejon operations in Colombia, its share of coal produced amounting to around 10 million tonnes. In early 2020, the future of Mt Arthur North is uncertain. BHP is looking to find a buyer for the mine, but progress appears to have stalled with the disruption to global economic activity caused by the COVID-19 virus.

State Governments give up de facto rail taxes

Rail freight rates charged by the Queensland and NSW Governments became a major area for disputes between the coal producers and the governments and their rail operators in the 1980s, with the producers attempting to reduce rates and eliminate the de facto tax element which was providing handsome revenues to the state treasuries. The State Governments came through those battles with their freight rate structures reasonably intact, although the pressure for reform was mounting.

In its 1991 report on the Australian minerals industry, the Industry Commission noted that many of the participants claimed that "State Governments in NSW and Queensland have used their position as monopoly suppliers of rail services, together with their power over licences to mine coal, to charge the export coal industry freight rates far in excess of what would represent efficient costs of supply the necessary services. Such excess charges are effectively a tax (or *de facto* royalty imposed) on the industry, as well as representing a significant source of State government general revenue in some states. The major impact is on coal railed to port in Queensland and NSW, but excess charges also apply to other bulk minerals despatched by rail in Australia."[443]

In Queensland in 1991, rail freight rates were a matter for direct negotiation between the state Treasury and individual coal companies, with the rate including an implicit royalty component. This meant that the actual rates negotiated "were usually based on each mine's ability to pay" which, for many mines," resulted in freight rates being significantly in excess of the cost of haulage". [444] As we saw in chapter 2, there were some rail freight rate concessions introduced in 1984 by the Queensland Government and, as the QMC acknowledged in its submission to the Productivity Commission, these concessions "provided a crucial measure of relief for the industry when the eighties 'coal boom' failed to materialise."[445] QMC said that without those concessions, many of the mines commissioned in the late 1970s and in the early 1980s would not have been viable. No major new mines were developed and commissioned in Queensland from the early 1980s until 1989, at which time lower freight rates were offered to new projects. A further phase of reform of the system commenced in 1994 with the corporatisation of Queensland Rail (QR), and the introduction of commercial pricing principles to the setting of rates. A review of royalty and freight rates for the coal industry had been completed in 1993, and led to a commitment by the Government to phase out the de facto royalty component of rates, with rate agreements for new mines or mine expansions from 1994 based on operating costs of the system and an allowance for a return on QR's capital. For mines which were already operating, commercial rates would apply from the year 2000 or earlier depending on the expiry date of their agreements, or if the company and QR agreed on a new rate structure. While QR also committed to achieving best practice in its rail operations by 2000 (which would see a reduction in operating costs), there was also a sting in the tail for companies, as the Government at the same time lifted the ad valorem royalty rate for export coal to 7%, and for coal for domestic use from 5 cents per tonne in 1994 to 7% by 2000.

With the coal industry a major source of revenue for the State Government, Queensland was clearly not intent on being over generous to the coal producers. According to QR, in 1995-96 the old rail contracts negotiated through Treasury were still generating a de facto royalty of $3.05 per tonne of coal for the Government, or $235 million in total. The Productivity Commission estimated that for the 1996-97 year, the royalty component of freight rates plus the royalty paid directly to the Government totalled $402 million, or almost 5% of the Government's own revenue.[446] The coal industry itself believed that the total royalty revenue for the Government was even higher than this, with a study by Coopers & Lybrand and R G Read & Associates finding that for the 1993 year, the abnormal profit earned by QR through over-charging was in excess of $300 million; including the $270 million in royalties received by QR and acknowledged in its accounts, the total de facto royalties for that year were around $670 million. The Coopers & Lybrand/ RG Read numbers were based on a number of assumptions, including valuing rail assets at written down replacement cost, excluding the value of assets provided by coal companies, and assigning a cost of capital to reflect what was believed appropriate for QR and its risk profile. While the Government and QR may have argued with some of these assumptions, the scale of the Government's royalty revenue from the coal industry was certainly significant, and in excess of the publicly acknowledged levels.

The Queensland producers were still unhappy with various aspects of the new regime in that State, with lack of third party access to the rail system one their major concerns. The producers were looking to competition policy to "breathe new life into rail reform", with third party access one key way of forcing down freight rates. They were seeking other companies to offer rail haulage services in competition with QR, with even the threat of competition expected to see QR behave more commercially or risk losing its only commercial business. Competition could not come soon enough, although QMC recognised that it would depend on how committed the state Government was in developing an effective regime to govern access and related issues. Third party access would not come for several more years in Queensland, with Asciano winning its first Queensland coal contracts with Rio Tinto and Xstrata in 2008, followed by its first contract with Anglo in 2009 and a contract with the Isaac Plains mine, then owned by Aquila Resources and Vale. QR itself was able to expand into NSW much earlier, when it won its first contract in 2004 to rail BHP's Mt Arthur mine to the port of Newcastle, with haulage commencing in 2005.

By the late 1980s, it was obvious that the state run rail system in NSW needed to significantly lift its efficiency. A review of the State Rail Authority in 1989 by consultants Booz Allen & Hamilton had found that although the SRA operated an efficient system for coal freight, its "equipment and facilities were generally outdated, poorly designed and worn out."[447] The consultants also concluded that the SRA was "grossly over-staffed and inefficient with poor productivity due to outmoded work practices, excessive overheads and obsolete plant." In relation to the SRA's overall freight operations, Booz Allen found that the labour productivity was less than 20% of the level achieved in North American rail freight operators, and this was despite what it said was its favourable mix of freight traffic.[448] The SRA was also in a dire financial position, as made clear in a report in 1988 by PA Consulting which concluded that "by any normal business standards, absent its support from the State Government, (the SRA) would be considered bankrupt." PA also concluded that the only potential commercial activities of the SRA were coal, minerals and grain.[449] Pressure on the SRA's freight revenues from the demands of the coal industry threatened to exacerbate that already unacceptable financial situation and helps to explain why the process to reform coal rail freight rates and open up the rail system to competition would be a difficult and drawn out process.

In the early 1990s, the NSW Government had not yet publicly admitted that its rates contained any de facto tax element, although Hunter Valley coal producers were in no doubt that this was the case. The major impetus for reform came in 1995 when the Commonwealth, State and Territory Governments agreed to a landmark agenda of reforms to competition policy whereby government owned business enterprises would be subject to competition. However there was a major catch – the carriage of coal was excluded from these reforms for 5 years until November 2000. This provision to exclude coal was necessary to protect the arrangements already put in place by the Queensland Government.

The exclusion of coal from the National Competition Policy reforms fortunately did not see the NSW Government excluding competition, and to its credit it announced in 1995 that it would open up its rail freight network to competition from other freight providers from July 1996. The 1995 year also saw the SRA split into four businesses, one of which was FreightCorp, the new above rail freight haulage provider. The track and associated assets were now owned by Rail Access Corporation (RAC). The new regime governing access to the rail system and issues such as pricing and rates of return on rail assets commenced the following year. The first competitor to FreightCorp to emerge was the National Rail Corporation, in which the NSW Government was a one third shareholder, along with the Commonwealth and Victorian Governments. NRC won a contract to haul coal to the Bayswater and Liddell power stations which were owned by the NSW Government's Macquarie Generation. NRC later became part of Pacific National when it and FreightCorp were privatised.

NSW used the 5 year coal moratorium period to achieve a phase out of the monopoly rent (excess profit or de facto royalty) component of its coal rail freight rates, with a 25% reduction commencing in July 1997, and the full phase out by July 2000. According to the NSW Treasury, the monopoly rent generated by Hunter Valley freight rates totalled $50.8 million in 1996-97, and $40.1 million in 1997-98, with zero the target for the 2000-01 year.[450] This monopoly rent was collected through the access charges coal producers had to pay RAC for using the track. The figure of $50.8 was a significant amount at that time, particularly for coal producers facing severe competition in export markets, and with coal prices generally depressed. However, Treasury's figures significantly understated the real excess profit component of freight rates. Since August 1996, the regime had allowed RAC to earn a maximum rate of 14% (nominal after tax) on the value of its assets, although the regime did not disclose on what parameters this figure was based. RAC as the owner of the assets converted the 14% figure to a pre-tax figure of 21.88%. It was the Hunter Valley coal producers located in the central part of the region (between Bloomfield, 38 kilometres from the port and Dartbrook, 153 kilometres from the port) which were paying this maximum rate as allowed under the regime. The cost of actually hauling the coal to the port, a service provided by FreightCorp, was an additional charge. This group of mines however accounted for over 80% of all the coal railed to the port of Newcastle. FreightCorp itself recognised that the 14% after tax return was excessive and was "substantially above a return commensurate with the inherent risks within RAC's business …"[451] The NSW Minerals Council was also fighting the use of such an excessive rate and pointed to the much lower regulated rates of return which applied in the gas and electricity industries – 7.5% and 9.5% real respectively.

IPART's 1999 report on the RAC's access regime April 1999 recommended that the maximum rate of return under the regime should be set at 8% real pre-tax (where the real rate is the nominal rate less the rate of inflation). IPART also noted that allowing RAC to set a nominal rate of return based on depreciated replacement cost, was "highly unusual" as RAC would then be compensated twice for inflation. Under the IPART recommended changes, the rate of return which RAC could charge would fall from 21.88% pre-tax to around 10 to 11% pre-tax (assuming inflation of 2 to 3% per year).

The $50.8 million monopoly rent which Treasury acknowledged, based as it was on an excessive rate of return allowable under the regime, clearly understated the true level of excess profit being earned by RAC and passed on to the Treasury. There were other significant issues of concern to the coal producers regarding the access regime which would take the next few years to resolve. For example with RAC required to move to charging based on operating costs which were efficient, there was considerable scope for further reductions in access charges. RAC's own strategy was to achieve a 30% reduction in operating costs between 1996-97 and 2000-01, resulting in a reduction of over $100 million in total annual costs.[452]

The battle to achieve rail freight rates which were considered reasonable by the NSW coal producers had been waged over many years, commencing in the late 1970s. However by the mid to late 1990s, real progress was now being achieved, but there would be several more years ahead before monopoly rents were eliminated and a much more efficient rail system emerged. The time was just around the corner when further major changes would occur, including the privatisation of FreightCorp (which became Pacific National and later became part of Asciano), the takeover of the ownership of the Hunter Valley rail infrastructure by the Australian Rail Track Corporation and the move into NSW by Queensland Rail as a major competitor to Pacific National. Queensland Rail would also be the subject of a major battle between the coal producers and the Queensland Government in the 2000s when the Government moved to privatise QR.

Coal research – a success story for the industry

One of the success stories of the modern Australia coal industry has been its research programs. A history of the industry would not be complete without a brief overview of the various research programs and projects which have significantly benefited the industry in terms of safety, environmental performance and mining productivity.

Coal research by the Australian coal producers goes back many years, with the establishment of coal laboratories in 1945 by the NSW producers a major step forward by the industry. The south coast and western district producers registered a new company, Coal Research Pty Ltd, in December 1944, and built a laboratory at Bellambi north of Wollongong for the company's operations. The major coal producers were members of Coal Research, with the exception of Australian Iron & Steel, which had its own research facilities, and the State Coal Mine in Lithgow. The Bellambi facility's initial focus was work which would enable the producers to comply with the requirements of the coal mining regulations, in particular in relation to mine gas and other hazards. The main work of Coal Research Pty Ltd in 1946 involved surveys of dust in the various southern and western mines to ascertain the highest concentrations of dust associated with the various phases of mining operations and the causes of the dust emissions. Back at the laboratory, the dust samples were examined under microscopes to estimate the number of particles per cubic metre of air.[453] The Newcastle producers, through the Northern Colliery Proprietors' Association, built their laboratory at East Greta Junction near Maitland, primarily to carry out analysis of coal dust and air quality in underground mines; the coal regulations were again the impetus for this laboratory.

The problem of dust in coal mines in 1946 also saw the Commonwealth Government ask its research organisation, the Council of Scientific and Industrial Research (to become the CSIRO in 1949), to investigate the issue. The CSIR established an advisory committee with members including the coal producers, the Miners' Federation and State and Federal departments. The CSIR brought a British expert, Professor T. David Jones from the University of Wales, to Australia in early 1947 to review the industry. Jones made a range of recommendations relating to mine mechanisation and dust measurement and control. He also recommended that an organisation be set up for pre-entry medical examinations for new coal mine workers and for periodic on-going examinations, functions which became one of the Joint Coal Board's responsibilities in NSW. The CSIR then moved in 1948 to establish its Coal Survey Section; this Section was renamed the Coal Properties Section in 1949 and then the Coal Research Section later the same year; and in 1960 it became the Division of Coal Research. In 1956 BHP opened its new Central Research Laboratories facility in the Newcastle suburb of Shortland, with coal research one of its key areas.

The coal producers in NSW and Queensland formed Australian Coal Association Research Ltd in 1955; the priorities of the new organisation were research into coal preparation and coal utilisation. In 1964, in an agreement involving the Commonwealth, State Governments and major coal consumers, Australian Coal Industry Research Laboratories Ltd (ACIRL Ltd) was set up as a non-profit company. ACIRL Ltd took over the functions and the assets of ACA Research. ACA Research Ltd and ACIRL Ltd were chaired by Sir Edward Warren, who continued in this role until 1976. ACIRL had facilities in Ryde in Sydney, Bellambi, Maitland, Ipswich and Rockhampton.

During the 1970s, in response to the increasing problems of the emissions of fly ash from the smoke stacks of power stations, the CSIRO made a major advance in the understanding of the technology relating to electrostatic precipitation, developing a technique for reliably estimating the design parameters for a precipitator. The CSIRO based their technique on a small scale coal combustion furnace, and its technology was soon adopted by the electricity industry, and by 1980 Japanese organisations had built two combustion facilities along the lines of the facility developed by the CSIRO.[454]

Coal research in Australia saw a major change in 1977 when the Federal Government budget introduced a levy of 5 cents per tonne on black coal production. An excise on coal production to fund the industry's long service leave scheme was already in place and so it was administratively simple to amend the legislation to increase the excise and to create a new pool of funds for coal research. Energy related research in Australia saw another major change in 1978 when the Government established the National Energy Research Development and Demonstration Council (NERDDC) to oversee a new National Energy Research and Demonstration Program (NERDDP). Seven advisory committees were also created, two of which were directly related to the coal industry (Coal Mine Site Technology and Technology of Coal Utilisation), with members including the major coal producers, the CSIRO, ACIRL, BHP's steel division and the Victorian and NSW Electricity Commissions. These committees were responsible for advising the Council and the Minister on projects to be funded, with the coal research fund one of the sources of finance for these projects, together with funds provided by the Government.

The NERDD Program and its technical committees continued their work during the 1980s, with the coal industry a major participant. However by the late 1980s, with growing industry dissatisfaction with the

Program, there was a strong mood for change amongst the coal industry's senior management. Some companies would have been happy to see no scheme, believing that individual companies should be responsible for their own research as was the case with most other industries. There was also a feeling that continued government oversight of the scheme was paternalistic; the NERDDP was seen as another special coal industry arrangement, along with the Coal Industry Tribunal, the Joint Coal Board and coal export controls, which was no longer justified.

Negotiations began in 1989 and, led by its then chairman Ian McCauley, the Australian Coal Association put three options to John Kerin, the Commonwealth Minister responsible for NERDDP and the coal industry. The Government could do away with the scheme (which was acknowledged as most unlikely), reduce the 5 cents per tonne levy, or allow the coal industry to take responsibility for its own scheme. As a result of the efforts of the Association, the Government finally accepted that it should change the scheme and it signed a Memorandum of Understanding with the Association which allowed a new industry managed scheme to commence in January 1992. ACARP, the Australian Coal Association Research Program, was born. ACARP was funded by a "voluntary" levy of 5 cents per tonne, the same as the coal excise; the levy was voluntary in the sense that there was no longer any legal requirement to pay as was the case with the excise. Ian McCauley said that Kerin "was pretty sympathetic to the view that the coal industry should be managing its own research fund, but one of the conditions he put on the ACA was that 100 per cent of our members had to be willing to come into the scheme, replacing the compulsory levy with a voluntary one. As I remember we had a little difficulty getting 100 per cent agreement, but eventually we were successful."[455]

The transfer of the scheme to the industry was publicly announced in January 1992 along with a range of other reforms and policies by new Energy Minister Simon Crean and Industrial Relations Minister Peter Cook.[456] The policy statement was a watershed for coal industry policy, and included the reform of the JCB, the removal of the coal export levy on BHP's operations, and revised funding arrangements for coal long service leave. However, BHP did not see the removal of the coal export levy without any strings attached. As a trade-off for the abolition of the levy, BHP committed to spending an additional $65 million over five years on coal industry research and development in collaboration with the Commonwealth and scientific institutions, and $50 million over 3-5 years on accelerated mine site rehabilitation and environmental research. As we saw earlier, in late 1991 Cabinet had also approved that the integration of the Coal Industry Tribunal into the AIRC would proceed, but this did not see its way into the Statement; another two years would elapse before Prime Minister Keating and IR Minister Laurie Brereton would finally bite the bullet to do away with the CIT.

ACARP was initially set up as a three year trial program, running until June 1996, to provide strategic leadership for the industry's R&D, to act as a catalyst to stimulate interest in R&D and to foster a collective and integrated approach. ACARP was administered through Australian Coal Research Limited, and its first Executive Director, Ross Graham, was appointed in October 1991. The first round of 45 projects was approved for funding in 1992. Graham recounts that a major factor in getting ACARP supported by the coal companies in the early years (when coal prices were low and profits meagre) was the focus on minesite research which allowed individual mines to see the benefit of projects for their operations and their bottom lines. The early 90s was also a period when the coal industries in countries including the UK and Germany which had traditionally been leading centres of coal research were declining rapidly; coal research was no longer a priority and key sources of technological support, particularly for the underground sector were disappearing. The ACARP program assisted to fill this void.

After the initial three year period, Graham worked with his board to see ACARP extended for a further two years. One of the potential obstacles at the time in securing the extension was the attitude of CRA, at least at the corporate level. CRA had its own major research program and questioned the value of the ACARP program. However CRA's minesite executives understood the practical value of ACARP, and the company moved to swing its support in behind its extension. Graham moved on to set up a similar industry research fund for the Australian meat industry and in 1998 the coal program was reviewed and a new Executive Director, Ross McKinnon, was appointed. McKinnon was successful in getting the industry's agreement twice to extend the program on a five-yearly basis, and in introducing the concept of landmark projects, large projects involving a major budget allocation and which would involve research over several years.

As the program developed and as the results of the industry managed scheme began to be more widely appreciated by the companies, ACARP received increasing industry support, with the program designed to avoid one of the major problems with the previous NERDDP. While NERDDP projects were valuable, companies sometimes saw the results as too focussed on the researcher and difficult to implement or transfer to different mine sites. ACARP was also seen as successful because of the willingness of key technical staff employed by the coal companies, including geologists, metallurgists and mining engineers, to share information for the benefit of the whole industry. In the 2012 publication reviewing the first 20 years of ACARP, Bruce Robertson, a Shell Coal executive, who chaired the ACA's Research Committee which had oversight of ACARP, said that there was "no other example like ACARP in terms of industry-funded research, and the critical issue of representation is envied by everyone that I speak to in the technical and research world. It's unique and valuable."[457]

ACARP has funded a wide range of projects, some of which are featured in the 20 years of ACARP publication. A fine-coal separator developed by Kevin Galvin at the University of Newcastle and commercialised by Ludowici was initially funded by ACARP, with Hunter Valley producer Bloomfield Collieries undertaking critical trials of the new technology. The separator is able separate the coking coal and thermal coal from coal fines; it can also separate coal at much finer size than spirals and can produce lower ash levels in fine coal. These major benefits have revolutionised coal preparation in Australia and have meant that millions of dollars' worth of coal has not been discarded, but has been able to be sold.

Underground coal mining and the operation of underground machinery and equipment have historically been associated with high rates of injury. ACARP funding was provided for what has been described as the world's most comprehensive analysis of underground mining equipment-related injuries. The research was undertaken by ergonomist and researcher Robin Burgess-Limerick between 2004 and 2010 and it provided manufacturers with the information they needed to improve equipment design and enabled mining companies to understand the health and safety risks inherent in buying that equipment. The need for the research was identified by Dave Mellows, then the Group Safety Manager for Xstrata Coal. Mellows said that he and associates realised that the undergrounds sector "lagged 10 to 15 years behind the open cuts in the development of machinery ergonomics and that the injuries were concentrated around roadway development activities and employee transport."

As the 20 years of ACARP publication noted, the "Australian coal mining industry has seen a dramatic improvement in underground coal mine safety" over the past 20 years, and "the effective management of geotechnical risk has been driven by consistent, world-class research..." Jim Galvin, Emeritus Professor of Mining Engineering at UNSW, said that Australia's mining industry had developed into the safest and most technologically advanced mining industry in the world, with "ACARP's strategic and sustained commitment to basic and applied research (contributing) enormously to this success."[458] Galvin noted that "Nowhere is ACARP's contribution more evident than in ground control where eliminating fatalities and injuries is approaching fruition... These days science provides a clear understanding of what is happening underground and a suite of tools to manage those conditions."

These examples are just a small component of ACARP's funding in areas directly related to mining and mine safety. Funding has extended across a wide range of other areas including greenhouse gas emissions from coal mining, minimising the effects of blasting on local communities, management of mine subsidence, water quality, river diversions, rehabilitation of land after mining and gas drainage.

ACARP is continuing to operate and in 2019 the program was funding 258 research projects for which it had made a total commitment of $67 million. In addition, new funding totalling almost $21 million was approved in 2019 for 89 new projects; including the value of in-kind support by the researchers and the host mines, the total value of funding for those 89 projects was around $37 million.[459]

Of course coal related research prior to the ACARP program also led to many valuable advances, with Australia in some cases leading the world. An excellent example is the development of technology to analyse the ash content of coal. In the early 1980s a major improvement was achieved with the development of on-line analysis of coal though the work of the Australian Atomic Energy Commission (AAEC), the CSIRO, the Julius Kruttschnitt Mineral Research Centre at the University of Queensland, The Australian Mineral Development Laboratories (AMDEL), and Mineral Control Instrumentation Pty. Ltd. By 1987 there were 33 of the new Coal-scan Ash Monitors in use in four countries measuring the ash content of the coal on a conveyor or the content of a sample. Later on, the technology was refined to enable on-line analysis of coal slurries in washeries, leading to a 10% increase in the coal content of the washed product.[460]

4

Mine safety – disasters, inquiries and progress

In the post war era, the Australian coal industry has generally seen marked improvements in its safety record, particularly during and since the 1990s. But along the way there have been many tragic fatalities including those involving multiple fatalities such as at Box Flat in the Ipswich district in 1972, at Appin colliery south of Sydney in 1979, the three events at the Moura/ Kianga collieries in the Bowen Basin in 1975, 1986 and 1994, and at the Gretley colliery near Wallsend in 1996. The tragedies in 1994 and 1996 proved to be a watershed for the industry, with major inquiries followed by new legislation and regulation.

For much of its history the industry's main indicators of safety were the numbers of fatalities and the numbers of serious accidents and reportable accidents. In the 1980s, the lost time injury frequency rate (LTIFR), a ratio of the number of injuries involving time off work to the number of hours worked, became a major indicator. For many years the regulators have also published data on the numbers of fatalities in relation to the industry's coal production or employment, expressed as a ratio of fatalities per million tonnes or fatalities per thousand workers employed. These measures have also been common to coal mining in other countries. Regulators in NSW and Queensland also published detailed information on the serious accidents, broken down by the type of accident.

Changes in the ways in which accident and injury data were recorded and reported over the years, and questions about the reliability of the data, make long term comparisons difficult, and that is still the case today to some extent. But in more recent years, a clearer picture has emerged of the trends in the industry and much more detailed metrics are now available covering not only the LTIFR, but a range of other metrics including, for example, the serious bodily injury rate and the total recordable injury rate.

The coal industry certainly has a grim history in terms of its safety performance, but it is clear from a number of measures that safety has dramatically improved since the 1990s. While fatalities still occur in coal mining, as they do in metalliferous mining, quarrying and in many other industries, and one fatality is accepted as one too many, the record is clear. Coal mining, while still potentially dangerous above and below ground, and while still far from perfect, has made commendable strides. Many mines can now claim periods in which their safety record has been first class, while others have some way to go to lift their performance.

The LTIFR in NSW coal mining fell from 168 in 1990, to 34 in 2000, and to around 4.5 in 2012-13. The NSW underground sector, which has always had the poorest record, saw its LTIFR fall from 79 to 22 and 7.15 over this same period. LTIFR data are now published for rolling 5-year averages; the NSW industry's average in 2008-09 was 13.2 and has fallen to 4.5 in 2017-18. The rolling 5-year average for the total recordable injury frequency rate (TRIFR) has fallen from 28.4 in 2011-12 to 15.1 in 2017-18.[461]

In NSW the serious bodily injury rate (SBIFR) was 2.5 in 1997-98, but dropped to 1.9 by 2002-03 and to 0.4 by 2012-13; it rose again to 0.77 in 2014-15 and to 0.86 in 2015-16. The total recordable injury rate, for which data are available only for a relatively short time, fell from 45.3 in 2007-08 to 19.5 in 2012-13; the rate has since declined to 14.7 in 2015-16. The 5-year rolling average SBIFR was 1.2 in 2008-09, falling to a low of 0.6 in 2018, but since then it has steadily risen to 1.0 in 2017-18.

In Queensland coal mining the LTIFR has also dropped, but not as dramatically as the state has traditionally had a lower rate. Queensland coal's rate has dropped from 63 in 1989-90, to 12 in 1999-2000, to 4 in 2012-13, and to 3.2 in 2014-15 and 2.7 in 2015-16, rising again to

3.2 in 2016-17.[462] Of course the LTIFR is by no means an adequate measure of safety performance. It has been criticised on a number of grounds over the years, including, for example, by the NSW Review of Mine Safety in 1997. The Queensland annual mine safety performance report still contains data on lost time injuries, but does not highlight the LTIFR, focussing instead on serious injuries and high potential injuries. Queensland's disabling injury frequency rate (the number of disabling injuries per million hours worked) has not shown improvement over the last ten years. For open cut mines, the rate was 3.9 in 2004-05, 9.0 in 2014-15 and 5.6 in 2015-16. For underground mines, the rate was 22.3 in 2004-05, falling to 8.9 in 2012-13, but rising to 27.41 in 2014-15, and falling again to 16.9 in 2015-16.

In NSW the record in terms of fatalities during the period between 1950 and 1980 was not something to be proud of. The numbers of fatalities in each decade were as follows: 1951 to 1960: 145; 1961 to 1970: 113; 1971 to 1980: 116. In the 1950s, the year with the lowest number of fatalities was 1959, with 9 fatalities. The total for the 1970s includes the 14 fatalities at Appin in 1979. The 1990s, a period which included the 4 fatalities at Gretley in 1996, saw a big improvement on previous years, with deaths down to 39, or close to 4 per year on average. The first decade of the 2000s (the years 2000-01 to 2009-2010) saw further improvement with 11 fatalities; and in the years since 2009-10, there has been 1 fatality in 2010-11, none in 2011-12 or 2012-13, 4 in 2014-15, and none in 2014-15 and 2015-16.

The 1950s (1951-1960) in Queensland saw 29 fatalities, the 1960s 24, the 1970s 61 and the 1980s 28. The 61 deaths in the 1970s included two disasters, Box Flat and Kianga, which accounted for 33 fatalities, and the 1980s figure included the 12 killed in 1986 at Moura. In the 1990s there were 21 fatalities, including 11 killed at Moura in 1994. The decade from 2000-01 to 2009-10 saw a reduction to 7 fatalities, and the years 2010-11 to 2014-15 another 7 fatalities, but with 4 in 2014-15 alone. There were no fatalities in the coal sector in 2015-16 or in the non-coal sector; this year was a milestone, being the first year since records began in 1877 when no fatalities occurred in the Queensland mining industry.

However, there was one fatality in both 2016-17 and 2017-18, but the last two years have seen a rise, with 3 fatalities in 2018-19, and a further 3 fatalities in 2019-20.

The trends in key safety measures in both NSW and Queensland are not clear. While the LTIFR for both States has continued to trend down, the other key metrics which are now regarded as more important and better indicators of safety trends have seen some reversal in recent years.

Box Flat 1972

The Box Flat colliery at Swanbank near Ipswich was opened in 1969 to supply the nearby Swanbank power station. In July 1972 a fire developed in a section of the mine and attempts were made to extinguish it. These attempts were unsuccessful and explosions in the mine caused the deaths of 17 men. The official inquiry into the disaster found that the causes were a spontaneous heating in a pile of coal in an area of the mine which was exacerbated by a stoppage of the ventilation fan for a period of 11 hours, during which the reduced air flow allowed the heat generated to reach a point where the coal actually ignited. "This heating then developed into a large fire which was assisted by the increased air flow when the mine's fan was (re)started. Efforts to extinguish the fire were unsuccessful. An explosive mixture of gases generated by the fire, and possibly accompanied by water gas, was ignited. Coal dust was active in the explosions that propagated throughout the mine."[463]

The inquiry found that the requirements of the Act and Regulations in relation to stone dusting were not complied with, but acknowledged that whether full compliance would have made any difference in the circumstances was not known. It recommended that, with non-compliance also common in the Ipswich district, urgent steps needed to be taken to rectify the situation. The inquiry was also critical of the standard of training in the industry, with none of the supposedly experienced and qualified men who formed the Mines Rescue Team on the night of the explosion apparently aware of the potential for an explosion.

The inquiry's first recommendation was that a Safety in Mines Organisation be established in conjunction with the NSW coal mining industry, that body's role being to provide practical demonstrations of safety matters, so giving mine personnel first-hand experience in how to deal with major underground fires and hazards. It made a range of other recommendations including some relating to training in the industry on fire fighting, a review of the provisions in the Act on coal dust, the requirement for gas detection equipment to be available at all mines, and ventilation and sealing of mines. It would take another 16 years before the mines safety organisation that was recommended was officially opened.

Appin 1979

Appin colliery, south of Sydney, was part of the group of mines owned by AI&S, a subsidiary of BHP. It was recognised as a gassy mine. On 24 July 1979 an explosion occurred during planned changes to the ventilation system at the mine resulting in 14 deaths; 10 of the men who died were in the crib room at the time of the explosion. The NSW Government appointed Judge Goran to conduct an inquiry into the tragedy, with Goran acting as the Court of Coal Mines Regulation; the judge reported in May 1980. Goran found that an explosive mixture of methane and air had built up in one heading in the mine and that the gases were ignited by an electric spark in the starter box for a fan, the box not being flame proof due to work being undertaken by an electrician. A subsequent coronial inquiry found that there was a possibility that the ignition source may have been a Deputy's safety lamp, a possibility that Goran had specifically ruled out.

Goran's report highlighted practices at the mine involving methane gas, noting that "Deputies at Appin appear to allow substantial quantities of methane gas to collect in standing places, upon the basis that there is no danger if there is no apparent source of ignition and no great problem if they do not have stop mining. The management regards methane gas problems of this kind as inevitable under the conditions which exist at Appin. This in itself is a dangerous attitude, leads to complacency and usually is in breach of the Act. The attitude must be changed. It has been permitted to continue by Inspectorial tolerance. Hand in hand with this principle is the fact that the management can easily be misled by its own deputies, if it chooses to rely upon their General Rule 4 reports. These are vague in the extreme and give no real indication as to actual methane gas conditions or ventilation."[464] Goran also noted that deputies were issued with methanometers (devices which measure methane concentrations in the air) only after the explosion; prior to it they had relied on their safety lamps which could not measure methane concentrations of less than 1.25%, and so could not determine if the mine was complying with the Act (which specified that the concentration limit in intake airways was 0.25%).[465]

The Department of Mines' senior inspector suggested to the inquiry that the Act should be changed to allow a slightly higher level of methane than the 0.25%. Goran, however, was adamant: "As a result of the Appin experience I believe that the strictest control of methane gas percentages in intake airways should be maintained. If there is any tolerance, it should be limited to a low departure from the statutory provision and only given on written application to the Chief Inspector for exemption."[466]

Moura and Kianga 1975, 1986 and 1994

The Moura and Kianga mines are located towards the southern end of the Bowen Basin, south west of Gladstone. The Kianga mine was initially developed by Thiess Bros from the late 1950s, with Les Thiess winning the first major export contract for Queensland coking coal to be supplied to the Japanese steel mills. The tragedy at the underground Kianga No. 1 mine took place on 20 September 1975 when 13 men were killed in an explosion that occurred when the men were engaged in attempting to seal a heating in a section of the mine. The men's bodies were never recovered. The force of the explosion was such that a motor vehicle parked at the top of the mine was damaged beyond repair. Spontaneous combustion caused the explosion. In 1975 the operator of Kianga was the Thiess Peabody Mitsui consortium.

The explosion at the Moura No. 4 mine, operated by BHP, occurred on 16 July 1986 and killed 12 men. The evidence presented to the inquiry failed to demonstrate that spontaneous combustion had occurred, but it was believed that the most likely source of ignition was a flame safety lamp or frictional ignition.[467] The inquiry into Moura No. 4 recommended that the legislation should be amended to specifically prohibit the use of flame safety lamps underground, with an exemption possible for a brief period and only with the written permission of the Chief Inspector. The inquiry also recommended that officials should be required to carry equipment underground at all times to detect methane and other gases. Other recommendations covered training, emergency procedures, and the need for continuous monitoring of underground air to measure methane and carbon monoxide concentrations.

The explosion at the Moura No.2 mine, operated by BHP, occurred on 7 August 1994; it killed 11 men and was caused by spontaneous combustion. The bodies of those killed have never been recovered. The Warden's inquiry into the Moura No.4 disaster reported in January 1996. Perhaps the most damming words in the report related to the inability of the industry to learn from previous disasters: "The previous three Inquiries into major explosions in Queensland coal mines have consistently made recommendations aimed at addressing perceived deficiencies in the coal industry's arrangements for training, or the state of knowledge of industry personnel. There has also consistently been the conduct of seminars and symposia as a response to those disasters, accompanied by the production of publications about the hazards of underground coal mining revisited in the course of those Inquiries... These measures have, however, clearly not been effective in the longer term with the industry displaying, as it does, a capacity to lose sight of the lessons of the past and to not maintain an adequate knowledge base among key personnel."[168]

The inquiry found that management had been guilty of neglect: "It is the opinion of the Inquiry that events at Moura surrounding assumptions as to the state of knowledge of the night shift on 7 August, and the safety of those at the mine, represent a passage of management neglect and non-decision which must never be repeated in the coal mining industry…Mineworkers place their trust in management and have the right to expect management to take responsible decisions in respect to their safety. They also have the right to expect management to keep them informed on any matter likely to affect their safety and welfare…It is regrettable that the air of caution, arising out of uncertainty, which was exhibited at the mine in order to bring forward the sealing of 512 Panel did not extend to the general safety and welfare of the workforce and, in particular, to informing and keeping persons out of the mine for a time subsequent to that sealing."[469]

However, despite these rather damning words, the Inquiry did not find the Moura No.2 management guilty of any negligence. As Hopkins notes in his book on the Moura No. 2 tragedy, the chairman of the inquiry made it clear that he was not interested in pursuing any potential issues relating to negligence: "I really do not want to hear that word (negligence) again in these proceedings. It is a distraction which we do not need and it is not a consideration in these proceedings."[470]

Releasing the Inquiry report, the Warden, Frank Windridge, was also scathing about the Queensland Department of Minerals and Energy and its lack of funding for the mines inspectorate: "The department was prepared to accept a level of death and injury in the industry so long as budget targets were met… Governments have no moral right to walk away when a disaster happens and decline to accept any responsibility. They are, by association and legislation, clearly involved. Put bluntly, they must either regulate the industry properly or they hand the regulatory duties over to some other authority."[471] The Warden's report did not recommend any criminal charges against the company or any individuals, but made a wide range of recommendations relating to improved mine safety and was influential in subsequent changes to the Queensland mine safety legislative and regulatory system.

Over the course of just over 20 years, explosions at Box Flat, Kianga No.1, Moura No. 4 and Moura No. 2 mines caused 53 deaths, with spontaneous combustion the identified cause in three of the explosions. Spontaneous combustion is a process well known in coal mining, as coal has a tendency to heat up when exposed to the air. This heating process is more common with certain types of coal, but is exacerbated when there is a continuous supply of air and the heat is not dissipated. When the heating occurs in the presence of a sufficient quantity of methane or coal dust explosions can occur.

The mine safety organisation that had been recommended by the Box Flat inquiry was a major step forward, but it commenced to be built only in the 1980s, and was under construction when the Moura No.4 disaster occurred. SIMTARS, the Safety in Mines Testing Station, at Redbank near Ipswich was officially opened in September 1988.

Gretley 1996

The Gretley colliery, located near Wallsend west of Newcastle, was operated in 1996 by Oakbridge, then a subsidiary of the American mining company Cyprus. On 14 November 1996 four miners were drowned when their coal cutting machine broke through into old flooded workings of an adjacent mine, the Young Wallsend Colliery, which had been abandoned in 1912. The NSW Government ordered an inquiry into the tragedy under the Coal Mines Regulation Act which was conducted by Acting District Court Judge James Staunton.

At the time of the tragedy, the company was operating in that section relying on maps that had been obtained from the Department of Mineral Resources. Tragically, those maps contained errors, and the miners, rather than working 100 metres or more from the old Young Wallsend workings, were actually hard up against them. At the time of the disaster, the company had been planning to drill towards the Young Wallsend workings to check whether there were any problems, with this work planned for the following weekend. Tragically, the workings were breached just days before this drilling which would almost certainly have found evidence of the flooded Young Wallsend workings.

Justice Staunton's report released in July 1998 found that Department was at fault for the original errors; Staunton also found that the company's managers and surveyor should have checked that the maps were accurate. Staunton was scathing about the lack of a documented Department policy in relation to prosecutions and its attitude towards safety issues. The Department's chief inspector, Bruce McKensey, had told the inquiry that his preferred response to fatalities was not to prosecute but to carry out a "system safety accident investigation" with a view to making recommendations to avoid a recurrence. However McKensey insisted that the Department did in fact have a prosecution policy, albeit an informal one, which was well understood by the inspectors; he also said that he saw prosecution as the last card in the pack, with the expectation or fear of prosecution necessary for the inspectorate to be effective.[472]

Staunton also referred to the independent review of mine safety in NSW conducted by Susan Johnston in 1996-97 which was also critical of the lack of a prosecution policy. The review reported that "… a range of stakeholders expressed concern at what they saw as confused Inspectorate approaches to enforcement of safety breaches. Several company stakeholders from varying levels expressed the view that the Inspectorate was too 'laid back' in this regard. This view was generally shared by union representatives." Johnston's report went on to say that "The Inspectorate seem reluctant to take action. There have been serious incidents where they did very little." And her report quoted a mine manager: "As a Mine Manager I expect the people who come to the mine to be bastards, I want them to be unreasonable if needs be - I don't want them not to care."[473] Johnston's report also noted that: "Several Inspectors expressed concern that Departmental policy on enforcement and prosecution was either vague or contradictory, and that this lack of clarity hampered the Inspectors in their being able to send clear signals to industry."[474]

Staunton concluded in his report that: "No mining company (or senior official) has been prosecuted under either the Coal Mines Regulation Act 1982 or the Occupational Health & Safety Act 1983 since April 1990. The Court suspects that before 1990 the position was little different. Since 1990, however, there have been more than 33 deaths, and a number of serious incidents. Many of the fatalities involved gross negligence, and breaches of the law. The Department's inaction is in part the consequence of its not having a documented prosecution policy. The Court believes that such a policy is now being drafted. An attitudinal change is also required. Prosecution has a place in securing mine safety. The statutes create offences. Mining companies and senior officials must be made aware, by timely prosecution, that they are accountable under the law for their actions."[475]

The judge made a wide range of recommendations relating to education and training, risk assessment, the Department's approval process for underground mining, and the need for a written prosecutions policy. He also recommended the establishment of an autonomous unit in the Department to investigate fatalities and serious injuries and dangerous events, with the unit to report directly to the head of the Department. Staunton also effectively recommended that the company be prosecuted, saying that the papers should be referred to the Crown Solicitor to determine whether offences had been committed under the OH&S Act (which requires employers to ensure the health and safety of their employees).[476] No specific recommendation was made in relation to the prosecution of individuals as opposed to the company.

The CFMEU took an extremely hard line from the start against the company and the Department. Northern District secretary Ron Land told the ABC on 12 November 1996 that: "Well, we regard the Department of Mineral Resources as equally culpable as the company in those deaths, and quite simply, had—it is our belief that had the Department of Mineral Resources done their job properly, those men would not have died. We believe the Department's got a duty of care to the industry's workers, and anyone who was conversant with his Honour's findings could not help but come to the opinion that the Department was culpable in discharging its duties in the industry, and especially at Gretley Colliery."[477]

The CFMEU launched a strong campaign after the tragedy to have the company, its managers and the Department prosecuted. It was ultimately successful, with the NSW Government commencing a case in the NSW Industrial Relations Commission in April 2000, charging the company and 8 employees with offences under the OH&S Act. One of the 8 was no longer with the company at the time of the tragedy, but as mine manager when key actions were taken earlier, was seen to bear some of the responsibility. Despite the inquiry finding "serious shortcomings" in the Department's policies and procedures, the Government decided not to prosecute the Department, a decision which Attorney General Jeff Shaw said was based on legal advice that such a prosecution would not be successful. The prosecution of the managers was based on the provisions of the OH&S Act and the related Coal Mines Regulation Act. The legal action took several years to reach a verdict, with a change of ownership of Oakbridge and its subsidiary, Newcastle Wallsend Coal Company, occurring in 2000, when Xstrata Coal (then called Glencore Coal) became the new owner of Gretley. In August 2003 Xstrata offered to enter a guilty plea if the charges against its employees were dropped, but WorkCover, the agency prosecuting the case, and its lawyers did not respond to this offer. Finally, in August 2004, Justice Patricia Staunton recorded convictions against the company and its officers for breaching the OH&S Act by failing to ensure the health and safety of employees.

The CFMEU welcomed the convictions, its general president Tony Maher saying that they were long overdue: "It is the first time in the 200-year history of the NSW coal industry that anyone has been convicted for the loss of life...." Maher said that the convictions were "a great relief for the families of the four miners, who have had to endure a protracted 7-year ordeal in their search for justice. Companies and management will be held to account for the health and safety of their employees, and our union will continue to vigorously pursue those who endanger workers' lives." [478] This judgement was handed down at a time when the CFMEU was demanding changes to legislation to provide for prison terms for managers responsible for workplace fatalities.

But this was not the end of the story. Xstrata Coal appealed on behalf of its employees, with CEO Peter Coates saying that he believed that there was "an absolute obligation on the company to ensure that these people are given the very best opportunity to avoid being convicted of criminal charges associated with the incident." [479] In March 2005 the appeal court found the company and three officers guilty of various offences, fining the company almost $1.5 million, and imposing fines on the two mine managers and mine surveyor. The company again appealed and the case was finalised in December 2006 when the Full Bench of the Industrial Court dismissed most of the appeals, but set aside the charges against the former mine manager and against the mine surveyor. The Full Bench also expressed concern that WorkCover had never fully investigated or prosecuted the Department which had provided maps that were incorrect.

The Moura No.2 and Gretley tragedies and the inquiries that followed proved to be a profound wake- up call for the industry and, together with the review of mine safety in 1997, led to widespread changes in the industry's safety legislation and regulation.

NSW Mine Safety Review

Just prior to the Gretley disaster, the NSW Government announced a review of mine safety to be undertaken by ACIL Economics and Policy Pty Ltd. The ACIL review team was led by Susan Johnston, with expert input from Professor Frank Roxborough (University of NSW School of Mining Engineering) and Emeritus Professor David Rowlands. The review's terms of reference were broad, covering the coal and metalliferous sectors, and the review commenced after a period when, in addition to the Moura No.2 and Gretley tragedies, there had been other fatalities, serious injuries and "potentially catastrophic near misses at NSW mines" and when the industry's lost time injury frequency rate had been improving.[480]

Johnston and her team spoke to a range of people across the industry, including company executives, mine managers, middle managers, union leaders and mine workers. They also reviewed the Inspectorates for both the coal and metalliferous sectors and the existing mine safety legislation.

The review report was provided to the Government and industry parties in 1997 and found that the LTIFR, the industry's key indicator, was seriously flawed for several reasons: it could be a poor reflector of actual safety numbers and of the extent of serious injuries on a minesite; it could be a poor indicator of how mines were managing major risks; and it was viewed with deep scepticism by many people in the industry, including mine workers. The review also found that while a strong focus on reducing the LTIFR by a mine could lead to a significant reduction in this measure, such a reduction was not necessarily matched by the actual safety performance of a mine.

Bob Martin, the NSW Minister for Mineral Resources, tabled the review report in Parliament on 9 April 1997. Martin told Parliament that the review had been approved by Cabinet after a growing number of deaths and injuries in the state's mines, and that feedback from the industry had suggested a dangerous culture had crept into some workplaces, offices and boardrooms. He said that: "In too many operations lip-service was being paid to workers' safety", adding that "Just days after the review was commissioned, as if to underscore the bloodstained history of the mining business, four workers were killed when water flooded the underground workings at the Gretley mine near Newcastle." Martin went on to say that "The report paints a grim picture of mine safety in New South Wales. It says that deep-seated problems have developed not overnight but over generations.... A deeply disturbing aspect of the report was the manipulation of safety performance indicators and figures, or what is referred to as cooking the books. Some mines artificially managed safety figures to give outsiders and inspectors the impression that the mines were safer than they actually were. In at least one mine some injured miners with broken limbs were chauffeured to work to sign the time book, only to be told to lie down in the change room for their shift or to be ferried back home. Others were coerced not to report injuries. Generous food and gift hampers and other incentives were offered to workers and mine managers as a means of hiding true accident and injury rates. One manager achieved his lost time safety target with an attractive hamper scheme to workers. However, the reality of actual injuries at this mine was unchanged and at an unacceptably high level. Let there be no doubt that this practice will be stopped. It will be one of the first issues to be targeted. The Government will ensure that the coalmines insurance operators eradicate this practice."

Martin had effectively accused the industry of being dishonest, highlighting some of the findings of the review, and implying that the rorts were widespread and indicative of the whole industry. The newspapers picked up his speech and media release in their reporting the following day. The Sydney Morning Herald's coverage began with the following statements: "Injured NSW miners were chauffeured to work so they could sign time books, then left to spend their shifts in changing rooms or returned home so employers could conceal injury rates, an investigation has found....The investigation's damning report, Review of Mine Safety in NSW, details for the first time alleged bribery and coercion to stop workers reporting mine accidents." The Herald went on to repeat some of the allegations Martin had made in Parliament and overall could be read as an indictment of the industry.[481]

The NSW Minerals Council and key member companies were incensed at what they saw as the politicisation of the issue. Many companies felt that they were making strong efforts to improve their safety and saw Martin's handling of the report as unfair. Following a meeting between a number of company CEOs and Council staff, the Council issued a media release accusing Martin of presenting misleading information to the Parliament. The Council said that Martin had sensationalised the ACIL report and had used terms such as coercion and bribery which were not in the report.[482] Coal & Allied CEO Terry O'Reilly, in his capacity as chairman of the Minerals Council, said that Martin's statement to Parliament contained allegations that were serious and misleading and that the ACIL report did not reach the conclusions stated by Martin. O'Reilly called on the Premier to "set the record straight" and retract any inference of coercive practices in the industry. Martin categorically denied that he had misled Parliament.

The Council wrote to the Premier demanding an apology, and Carr responded with some soothing words. But the damage had been done. The review had found serious shortcomings in the industry and major improvements were urgently required, but the mining industry, and particularly the coal sector, had been tarnished and would be made to suffer the consequences.

More often than not, the review said, there was a difference between the targets set by CEOs for safety and what mine managers, middle level managers and the rank and file workers felt was achievable. In relation to incentive schemes, the review actually came to a more measured conclusion than Minister Martin, with Johnston saying subsequently that there was "distortionary media reporting"of this issue, although she diplomatically did not mention that the media reporting appeared to be based on Martin's speech in Parliament and the media release he and the Premier issued.[483] Johnston said that the review did not suggest that incentive schemes were "a deliberate and nefarious attempt on the part of companies to hide true safety statistics" and that there was no reason to believe that these schemes were anything other than an attempt to improve safety performance. In fact the review found that many in the industry, from CEOs down, provided examples of where incentive schemes had led to improved LTIFR numbers, but no improvement in on-site safety performance. So, far from some sort of cover-up, the review found that the industry was being up-front about the problems with incentive schemes and the use of the LTIFR.

A key conclusion of the review was the need to expand the range of safety performance indicators, and the wealth of detail now available in both NSW and Queensland on fatalities, serious injuries and other saftey issues is evidence of how far the companies and the regulators have come in recent years. The review also had a fair bit to say about the Inspectorates, with the Coal Inspectorate found to be "over-worked, confused about its role, prone to sending messages to stakeholders, poorly organised, affected by poor internal relationships and under-resourced."

At the time of the Gretley disaster, the NSW Parliament was considering amendments to workers compensation legislation. The mining industry employers, through the NSW Minerals Council, were lobbying to bring the benefits for coal mine workers more into line with benefits enjoyed by workers in other industries. But the timing of events proved to be disastrous for the Minerals Council. The memorial service for the miners killed was held in Newcastle on 27 November 1996, with the NSW Governor and Premier making speeches to the service. The Opposition Leader, Peter Collins, was also present. The night before, the Opposition and independents had made changes to the legislation in the Upper House relating to benefits for coal mine workers. Reflecting the anger at what had happened and the union's fierce opposition to any changes to workers compensation benefits, the union called an immediate strike which shut down the industry in NSW for 4 days. The Opposition was forced to back down from its support for changes to the legislation and agree to reverse its vote.

The background to this back down was that the NSW WorkCover scheme was in a poor financial position, and the Government was legislating to reduce workers' compensation benefits for all NSW workers. The exemption for the coal industry, which has a separate insurer, was to ensure that coal mine workers did not suffer any reduction. It was introduced as an amendment to the draft legislation just before the legislation was introduced into Parliament; previous drafts had not contained such an exemption. The NSW Minerals Council's position was to oppose this exemption on the basis that coal industry workers' compensation benefits were already well above other industries, and that such an exemption reinforced the special treatment for the coal industry which the Council had been working for years to change.

There was considerable media focus on the Gretley tragedy and the battle over the coal exemption in the legislation. The general tenor of the debate in the media (and particularly from radio commentators such as Alan Jones) was that coal miners were a special case, that their work was dangerous and that they fully deserved additional workers' compensation benefits.

The industry post Moura No.2 and Gretley

The Australian coal mining industry was never the same after the Moura No.2 and Gretley disasters and the subseqent inquiries and reviews, and the processes which were put in place in both states to move the industry forward to improve safety. As the authors of a paper on how mining disasters changed mining in Australia concluded , after the Moura No.2 disaster "…the whole approach to mining in Queensland was turned on its head. This was a watershed moment in the history of mining in

Australia and the drive towards zero harm gained momentum."[484]

In Queensland the Mines Inspectorate underwent a major review and was restructured and provided with more resources; the Inspectorate has also been subjected to further periodic reviews. New mine safety and health Acts were passed in 1999 for both the coal and non-coal (metalliferous mining and quarrying) sectors, and a new safety and health regulation for each sector was introduced in 2001. The new Acts incorporated a risk based approach to safety under which mines are required to develop safety management plans for principal hazards. While there continues to be a strong element of prescription in the legislation, companies must now take into account the particular conditions in their own mines, including the geology, the gases and the types of machinery and equipment, and develop plans to manage the risks, rather than relying on lists of dos and don'ts in the Act or in regulations.

It is also interesting to note that flame safety lamps were finally banned in Queensland in the years after Moura No.2. Their banning was a key recommendation of the Moura No. 4 inquiry in 1986, but it took research by Simtars to settle the issue. The cause of the Moura No.2 explosion was never conclusively established, although it was suggested that a roof fall in an unventilated part of the mine had driven methane which had collected there into workings where a safety flame lamp carried by a deputy ignited the explosion. Simtars carried out research which concluded that this had been a likely cause of the explosion. The Queensland Government then banned these lamps and they were replaced by electric gas detection equipment.

In NSW the changes have been no less dramatic. Major changes to safety legislation were made in 1998 and in subsequent years. There was also an important mine safety review headed by ex Premier Neville Wran in 2004 which made a wide range of recommendations leading to further research and a number of industry working parties. Clearly the industry has learnt a great deal over the years as a result of these tragedies and extensive research into spontaneous combustion. Major advances in technology have also been critical, with methane drainage now common in gassy underground mines, and the technology used to measure and monitor methane levels is much more accurate. Senior managers of mines are more attuned to the dangers, attitudes which in the past accepted risks are no longer tolerated and regulators have better qualified and resourced inspectorates.

Black lung disease reappears

One of the most disturbing health issues in the industry in recent years has been the "re-emergence" of lung diseases, devastating conditions which include pneumoconiosis and silicosis. Black lung disease was common in the coal industry in the pre-war and early post-war years, but had seemingly been eliminated in Australian coal mining by the 1980s. Lung disease in coal mine workers stems largely from exposure to coal dust and crystalline silica, with the dust generated during both underground and open cut mining operations. It was in September 2015 that this "re-emergence" came to the attention of the industry, with the reporting of the first case of coal workers' pneumoconiosis in a Queensland coal miner in 30 years. The worker in question had been diagnosed with the condition in May 2015.[485]

A Parliamentary committee on the issue was set up in 2016 and concluded that pneumoconiosis in the coal industry had never actually been eradicated in Queensland. The Committee's interim report in March 2017 stated that rather than "re-emerging" in 2015, it was actually "re-identified after more than 30 years of responsible Queensland authorities failing to look for it or properly identify it." The report found that, as at March 2017, 20 Queenslanders had been diagnosed with "this insidious and entirely preventable disease."[486] The Committee reported that the overwhelming weight of evidence pointed to the likelihood that many more miners and former miners would be diagnosed with pneumoconiosis or related conditions as a result of what it said was "a catastrophic failure of the regulatory and health surveillance systems intended to ensure the protection of coal industry workers" and that the industry as a whole would see "many more cases of this totally preventable disease in the very near future." The Parliamentary

Committee's inquiry looked at a range of evidence, including a 2016 review by a specialist group from Monash University and the University of Illinois in 2016 which found that the Coal Mine Health & Safety Scheme (MMHSA) which covered coal mine workers in Queensland was deficient in a number of respects. While the Monash University report was rather diplomatic in its findings, the Parliamentary Committee's final report in May 2017 did not mince words, slamming the Department of Natural Resources and Mines (DNRM), and saying that the evidence to the inquiry had "clearly demonstrated that DNRM did not adequately administer the CMSHA to ensure coal mine workers were not exposed to the serious health hazard of respirable coal mine dust." And in doing so, the Committee said, "DNRM failed to protect the health of coal mine workers with respect to respirable coal mine dust." [487]

The Queensland industry and its regulators have made major changes in the last few years in the ways in which dust in mines is managed and reported, and the levels of dust in mines have dropped significantly. There have also been other major reforms including compulsory chest x-rays and lung function tests for coal workers, medicals for retiring workers and the mandatory reporting of coal mine dust lung diseases. However, with 81 cases of mine dust lung disease among current and former coal mine workers as at June 2019, and a further 27 cases among workers with both coal and non-coal mining experience, it may be some years before the full impact of past industry and regulatory failures is evident.

New manslaughter provisions in Queensland

The surge in fatalities in the mining industry in Queensland in 2018 and 2019 has seen the passage of a new Resources Safety and Health Queensland Act in 2020. The new Act and changes to related legislation involve a number of major changes and initiatives, the most controversial being the introduction of industrial manslaughter offences. The offence of industrial manslaughter will apply when an employer's or senior officer's criminal negligence causes the death of a worker in the resources sector.

According to the Department of Natural Resources, Mines and Energy, the industrial manslaughter provisions "will strengthen the safety culture by bringing into focus the conduct of employers and senior officers." The provisions have applied for some time in non-mining industries and so bring mining into line with other sectors. Serious penalties will now apply, including up to 20 years in prison for an individual and fines of over $13 million for a corporation. A commencement date for these provisions has not yet been proclaimed and so it may be some time before their impact on the mining industry can be assessed.

A further blow to the industry in Queensland occurred on 6 May 2020 when a gas explosion in Anglo Coal's Grosvenor mine (part of the Moranbah Grosvenor complex) seriously injured 5 workers. That incident followed a number of other incidents involving gas at the Grosvenor mine and at other mines in 2019 and 2020 and led to the Minister for Natural Resources establishing a board of inquiry to examine these incidents. That inquiry will provide a final report to the Minister by 30 November 2020.

The industry has made major strides forward, but clearly still has a lot more work to do. Mining, particularly underground, has inherent dangers because of factors such as the potential for catastrophic roof falls and problems with gas. And surface mining continues to see many accidents, including deaths, as a result of machinery accidents and other causes. As in any workplace, there is also always the human factor; mistakes will be made, even with the best systems and technology in place. The coal industry and its regulators have made significant gains since the 1990s, and the challenge for the industry is to continue to learn from each and every accident and near miss.

5

The New Century

In the 1980s and 1990s, we saw the emergence of major oil producers as significant players in the Australian coal industry, with Shell, Exxon, ARCO and BP the most prominent. However, by the year 2000 or soon after these companies had largely exited coal mining in Australia and overseas. They found the industry had produced poor profits and had not lived up to the hype following the oil shocks of the 1970s which were expected to herald a new golden age for coal mining, particularly for steaming coal. Demand for steaming coal had certainly grown, but not as strongly as many were initially forecasting. Competition from other suppliers, including Indonesia and South Africa, had been fierce and coal supply for the export market was often well in excess of demand, and with resistance to change and entrenched work practices, good profits had been the exception. Many other companies also became part of the passing parade of coal miners. US miner Cyprus came and went in just a few years. CSR burst onto the scene in the 1970s and created an outcry in Queensland, taking over the local icon Thiess, but sold its coal assets in the 1980s, retreating back to its core business of building materials and sugar; CSR subsequently sold its sugar interests to the Singapore company Wilmar International in 2010. The Australian coal assets of US miner and construction company Utah, a pioneer in the Bowen Basin along with Thiess, were bought by BHP in 1984 following an earlier change in ownership of the Utah group, when the new owner General Electric decided mining was not to be part of its core business. The BHP takeover was the largest ever case of an Australian company "buying back the farm", and proved to be one of the bargain buys of the century,

helping to underpin BHP's operations and providing a profitable base
to offset some poor decisions in the 1990s and the meagre profits of its
steel division.

Anglo enters the industry

Shell's Australian coal assets were sold to Anglo American in 2000 for
$1.6 billion, bringing one of the oldest major mining companies in the
world into the Australian coal industry. Anglo has operations spread over
South Africa (its original base), Australia, Asia, Europe and the Americas;
it produces a wide range of minerals including platinum, copper, nickel
and diamonds, as well as metallurgical and thermal coal. Anglo was now
a major new player in Australia with major interests and management
rights to the German Creek and Moranbah North underground
metallurgical coal mines in Queensland and three open cut thermal coal
mines (Callide in Queensland; Drayton and Dartbrook in NSW).

Shortly after the purchase from Shell was finalised, Anglo found
serious problems at Moranbah North and was reported to have been
negotiating with Shell for a refund of some of the price it had paid. A
consortium of MIM, Rio Tinto and Glencore offered only $900 million
for the Shell assets, and other producers including Billiton were also
not prepared to pay what was seen as an excessive price.[488] Despite the
apparent high price for Shell's assets, Anglo was able to enter the Australian
coal industry a few years before the booming Chinese economy would
transform global commodity markets. Anglo was able to make strong
profits in a number of years. However, in December 2015, following a
major write down in the value of its assets worldwide, Anglo announced
that it was planning to restructure the company, cutting 85000 jobs out
of its total workforce of around 135,000. The Australian coal assets were
now up for sale. The Dartbrook mine was closed in 2007 and placed on
a care and maintenance basis, with the company saying that geological
problems had made the mine unviable. In late 2015 Anglo agreed to sell
Dartbrook to Australian Pacific Coal which plans to re-open the mine
and operate it again as an underground operation. In June 2016 Anglo

announced the closure of Drayton following a long struggle to gain NSW Planning and Assessment Commission approval for its Drayton South project. The refusal of consent for the new project by the PAC on the basis of conflict with nearby horse studs saw Anglo close the mine in September 2016 with the loss of around 200 jobs. It then announced the sale of its 88% interest in the Drayton mine and the Drayton South project to Malabar Coal Limited.

With the recovery in coal prices in 2016 and 2017 Anglo backed away from its plans to sell out of its Australian coal assets and continues to operate its Queensland mines which include the Capcoal mines, Moranbah North, the recently developed Grosvenor mine, and the Dawson mine complex. The Dawson mines date back to the Thiess Moura/Kianga mines developed in the early 1960s. Anglo also has the Moranbah South project for which it is progressing environmental approvals and the Aquila underground project. Mitsui is a major partner in a number of Anglo's mines and projects.

BHP and Billiton merge

BHP wound back the clock in 2000, when it announced a bid for QCT Resources, with Mitsubishi an equal partner in the bid. QCT Resources had begun life as the Queensland Coal Trust, set up by BHP as part of its financial engineering for the takeover of Utah in 1984 and became the owner of the CQCA shareholding in BHP. QCT Resources expanded in the 1990s beyond that initial holding of BHP shares and became a miner in its own right, owning and operating the South Blackwater mines - the South Blackwater open cut and the Laleham and Kenmare underground mines. At the time of the bid, QCT Resources held a stake of just under 33% in the Blackwater, Goonyella, Norwich Park, Peak Downs, Saraji and Gregory Crinum mines. BHP was also able to induce Santos to sell its 36% stake in QCT Resources, giving it control of the company. The price received by shareholders in QCT Resources from BHP looked a little on the low side in 2001 when BHP and Mitsubishi secured an increase in the price for their Queensland coal of 16%. BHP had tended to talk down the prospects for price increases during the QCT takeover process in 2000, but with the market tightening, BHP Billiton and Mitsubishi had secured a bargain when they purchased QCT.[489]

In 2001 BHP merged its operations with South African based Billiton to form BHP Billiton, the new dual listed entity, with BHP Billiton Limited listed on the ASX and BHP Billiton plc listed on the London Stock Exchange. The newly merged group would become the largest mining company in the world, with assets spanning the globe and including coal mines in Australia, South Africa, South America, Indonesia and the USA. The merger meant that the new company was not only the major coking coal exporter in the world, but now also a major steaming coal producer.

BHP Billiton proclaimed the benefits of the merger in its 2001 annual report, proudly pointing to its leading position in key minerals markets; it said that it was now the world's largest exporter of energy coal and of metallurgical coal, the third largest exporter of iron ore, copper and aluminium, and the world's largest producer of manganese, chrome and ferroalloys.[490] The company added that it also had substantial interests in oil, gas, and LNG. The annual report's title was "A New Day Begins", and page one featured the smiling faces of its CEO, Paul Anderson, and Deputy CEO Brian Gilbertson. Anderson, who was recruited from the USA, played a key role in restructuring BHP after its poor performance in the 1990s, but would return to the USA in 2002, with Gilbertson becoming the new CEO. Gilbertson, previously the CEO of Billiton, would remain BHP Billiton CEO for only a short time before being shown the door and replaced by Chip Goodyear. Marius Kloppers, who came to the merged company after a number of years with Billiton, was appointed CEO in 2007; he was replaced by Andrew Mackenzie in 2013. Mike Henry became CEO of what is now BHP Group Limited in 2019.

Under Chairman Jac Nasser and CEO Mackenzie, the company was split in 2015, and South32 was listed on the ASX. BHP Billiton's assets now comprised four key "pillars" – metallurgical coal, oil and gas, copper and iron ore; potash was listed as a potential 5th pillar. Many of the assets which Billiton brought to the merger in 2001 were now housed in South32, leading some commentators to question the rationale for the 2001 merger.

Billiton had entered the Australian coal industry initially through its takeover of Coal Operations Australia Limited, a company which operated the Chain Valley, Moonee and Wallarah underground mines south of Newcastle and the Mt Arthur mine in the Hunter Valley. Mt Arthur produced 4.5 million tonnes in 2001-02 and had the potential to be expanded to a much bigger operation. Billiton also came to the merger with major coal mines in South Africa, its Ingwe Collieries Limited group of mines becoming part of Billiton in 1998, and then part of BHP Billiton in 2001. Ingwe was re-named BHP Billiton Energy Coal in 2007, and in 2015 it became part of South32.

A major restructuring of BHP Billiton's Queensland coal business took place in 2001 with BHP reducing its share of the CQCA assets and the Gregory joint venture assets to 50%, and Mitsubishi increasing its share to 50%. BHP Billiton said that it was "a further step in the strategic alliance" formed between the two companies at the time of the joint acquisition of QCT.[491] The restructure did not affect the management of the mines, BHP Billiton retaining responsibility, but gave Mitsubishi a bigger share of the business for what now looks to be a modest investment of $1.005 billion. BHP Billiton and Mitsubishi became "jointly responsible" for the marketing of coal from the BHP Mitsubishi Alliance (BMA) mines. BHP Billiton and Mitsui continued their joint venture operations, Riverside and South Walker Creek, which were owned by BHP Mitsui Coal, with BHP Billiton having an 80% interest and Mitsui 20%.

BHPBilliton's urge to merge with Rio Tinto

One of the more controversial moves in recent years occurred in 2007 when BHP Billiton proposed a merger with Rio Tinto. There had been consideration of a merger in 1999 of the two companies' iron ore operations, and discussions about a full merger in 2001. BHP Billiton's then CEO, Brian Gilbertson, is reported to have sought support from his board for a merger in 2003 but was unsuccessful and was soon given a golden handshake by the board. In September 2006 the chairmen of the two companies, Don Argus of BHP Billiton and Paul Skinner of Rio Tinto, met informally to discuss a possible deal, but the Rio Tinto board proved to not be interested.

By 2007 BHP Billiton CEO Marius Kloppers and chairman Don Argus were ready to proceed and made a formal offer to Rio in November: three BHP Billiton shares for every Rio share. The bid valued Rio Tinto shares at around $130, a modest premium of 13% on the pre-bid price. BHP Billiton promoted the benefits of the deal to Rio as having potential cost savings of $3.7 billion per year, a figure which would arouse the interests of most CEOs and directors. Rio Tinto had only months earlier paid an eye-watering $US39 billion to buy aluminium company Alcan, out-bidding other companies including the world's largest iron ore exporter, Brazilian company CVRD (Vale). The Alcan purchase would soon cause major problems for Rio, but Rio's board saw BHP Billiton's offer as under-valuing the company and rejected the offer.

Asian customers for iron ore were clearly concerned about the proposed merger, with some speaking quite openly about those concerns. From the perspective of the buyers, to have two of the three largest exporters of iron ore merged into one company, creating a supplier with 35 to 40% of the market, had the potential to reduce competition and push up prices. And if the merger went ahead, the two major suppliers - BHP Billiton/ Rio Tinto and Brazil's Vale - would control around two thirds of the market.

Regulators in a number of countries also became involved in looking at the possible implications of the deal. In Australia the ACCC gave the go-ahead, but the European Commission (EC) would prove to be a major obstacle. In November 2008 the deal was officially abandoned, following serious issues raised by the EC. But by late 2008 the reasons for BHP Billiton calling off the deal were more numerous than just the EC's concerns. Rio Tinto's purchase of Alcan had required the company to take on massive debts, and at a time when the global economy and the Chinese economy were looking at a serious slowdown, the debt-laden Rio Tinto was being seen increasingly as a much less attractive takeover target. The takeover, had it proceeded, would have also had some significant effects on the Australian coal industry, as Rio Tinto in 2008 was still a major coal producer, operating mines in Queensland and NSW including the Coal & Allied mines in the Hunter Valley, and Blair Athol, Hail Creek, Kestrel and Meandu in Queensland.

Rio Tinto's proposed joint venture with Chinalco implodes

With Rio Tinto under pressure as a result of its purchase of Alcan, the Rio board announced in February 2009 that it was recommending to its shareholders that they approve a deal to bring the Chinese state-owned Aluminium Corporation of China (Chinalco) in as a "strategic partner" through the creation of joint ventures in aluminium, copper and iron ore. Chinalco would be issued with convertible bonds which would allow it to increase its shareholding in Rio. The benefit to Rio Tinto would be an injection of $US19.5 billion, which would dramatically reduce the company's debts. If Chinalco exercised its right to convert the bonds into equity, it would increase its shareholding in Rio Tinto plc to 19%, and to 14.9% in Rio Tinto Limited; for the combined dual listed companies, the average shareholding would be 18%.

Rio Tinto heralded the deal as enabling the company to lead the resources industry into "the next decade and beyond by ensuring the continuity of its strategy with the benefit of Chinalco's relationships, resources and capabilities." It said that its board had "extensively considered a range of strategic options, and has concluded that the opportunity offered by the strategic partnership with Chinalco, together with the value on offer for the investments by Chinalco in certain of Rio Tinto's mineral assets and in the convertible bonds, is superior to other identified options and offers greater medium term certainty and long term value for Rio Tinto's shareholders."[492] Chinalco's investment of $US19.5 billion was to be through a $US7.2 billion purchase of convertible bonds and investment of $US12.3 billion in strategic alliances in iron ore, copper and aluminium. The investments in these three strategic alliance areas would give Chinalco a major stake in key Rio assets, including 30% in Weipa (bauxite), 49% in Boyne (aluminium smelting), 49% in the Gladstone power station, 49.75% in Escondida (copper) and 15% in Hamersley (iron ore).

The deal needed the approval of the Australian Government, with the Foreign Investment Review Board (FIRB) due to give its decision in June 2009. However, just days before the FIRB deadline for a decision, Rio walked away from the deal. Rio's share price was rising, debt markets were opening up and Rio now had more options to improve its balance sheet. Chinalco was reported to have been prepared to be flexible and to negotiate possible changes to the deal, but the Rio board was no longer supportive, instead planning to make a rights issue to raise $A15 billion.[493] The Chinese were not impressed and Rio now had to find a way to restore its credibility in China.

But BHP Billiton was happy the Chinalco deal had fallen through, with Kloppers saying that he was still looking to consummate a deal with Rio to combine the iron ore operations of the two companies. "I believe it has been worth the wait" Kloppers said, referring to the previous unsuccessful attempts to merge the companies or at least merge their iron ore businesses in the Pilbara.[494] "Combining these world-class assets and associated infrastructure which operate side by side - if we combine these, we can get very, very substantial production, development and financial synergies." BHP and Rio were now set on a path to merge their iron ore operations, but would need the approval of regulators in a number of countries.

It emerged in 2010 that BHP had been lobbying behind the scenes in Canberra to sink the deal between Rio and Chinalco. "According to a confidential US embassy cable obtained by the WikiLeaks website and released to The Age, federal Treasurer Wayne Swan's chief of staff told American embassy officials that BHP had out-manoeuvred its rival to orchestrate the collapse of the Chinalco deal. The revelation is embarrassing for BHP Billiton, which has consistently refused to be drawn on suggestions it had engaged in a campaign to persuade federal government ministers not to approve Chinalco's push to double its stake in Rio Tinto."[495] On the same day as the deal collapsed, BHP Billiton and Rio Tinto announced their plan for the iron ore joint venture, indicating that BHP may have had plenty of advance notice that the Rio-Chinalco deal was doomed. The collapse of the deal also saved the FIRB and Treasurer Wayne Swan from having to make a ruling. Australia's foreign investment policy was uncertain at that time, not having been sufficiently debated and considered by the Government when China's well-resourced and asset hungry state-owned corporations were actively looking for investment opportunities in the minerals and resources industries.

However, by mid-2010 the BHP - Rio Tinto iron ore joint venture proposal was looking shaky to say the least, and by August was effectively dead, with Fairfax Media reporting that senior executives involved in seeking regulatory approval said that there was no longer any chance that the Chinese regulators would accept the joint venture in the light of the new price setting system for iron ore. Led by BHP's Marius Kloppers, Asian iron ore buyers had had to accept the move away from annual price negotiations between the exporters and buyers to a quarterly system.[496] The Chinese buyers had obviously convinced their regulators that BHP and Rio would have too much pricing power under the joint venture arrangement. The EU was also reported to have sent hundreds of questions to BHP querying the new system and BHP's claim that they did not determine the prices. The BHP – Rio proposal of course did not involve coal, but it is interesting to speculate on what effect, if any, there might have been if the iron ore joint venture had been given the go ahead. Would the Chinese have moved to discriminate against the coal divisions of BHP and Rio, or would the dominance of the Bowen Basin exporters of quality hard coking coal have been such that the Chinese would have had to accept the reality of the merged iron ore operations?

BHP's major restructuring in 2015 saw the new South32 taking ownership of a range of assets which no longer fitted into BHP's strategy to focus on its four key pillars. Assets transferred to South32 included

South African thermal coal, Illawarra metallurgical coal, alumina, aluminium, manganese, nickel, and silver, lead and zinc assets. BHP also maintained ownership of the Mt Arthur thermal coal mine in the Hunter Valley and its one third interest in the Cerrejon coal operations in Colombia. Having exited from thermal coal in South Africa and the USA, BHP is looking to also sell off Mt Arthur.

The rise of Glencore – and the demise of MIM

Up until the mid-1990s, Glencore's role in the coal industry was essentially as a trader, but that began to change in 1994 when Glencore purchased equity in the small Queensland miner Cook Colliery. This was followed by the purchase of an interest in Cumnock Colliery in the Hunter Valley, with Glencore acquiring a majority interest in Cumnock in 1998. 1998 marked the time when Glencore started to get really serious about Australian coal, with the purchase from BHP of the Mt Owen open cut mine in the Hunter Valley. Mt Owen was a mine producing around 3.5 million tonnes a year of thermal and semi soft coking coal and was expanded to 5 million tonnes after NSW Government approval was received in 2001.

Glencore made several more major purchases in 1999, including the Oceanic group of mines near Newcastle (Westside, West Wallsend and Teralba) which had been owned by FAI, then part of the HIH Insurance group which spectacularly collapsed in 2001. Other purchases in 1999 included the Liddell mine and the Glendell project in the Hunter Valley, and the Togara North project in Queensland, assets which it purchased from Pasminco (Pasminco having acquired these assets when it took over Savage Resources). In 2000 Glencore took advantage of the exit of US company Phelps Dodge from the industry; Phelps Dodge subsidiary Cypus Amax sold its Oakbridge interests to Glencore and Centennial. Glencore's new assets included the Bulga and South Bulga mines in the Hunter Valley and the Baal Bone underground mine near Lithgow. In 2001 Glencore purchased Exxon's equity in the Ulan mines near Mudgee and Xstrata's interest in the United mine in the Hunter

Valley; it also increased its equity in the Oakbridge mines. In late 2001, as part of a further rationalisation of mines in the Hunter Valley, Glencore announced it had secured the Ravensworth and Narama mines from Coal and Allied, mines previously operated by Peabody. Peabody's exit from Australia earlier in 2001 had seen its NSW mines sold to Coal and Allied. By 2001 Glencore was operating mines in NSW with production of around 25 million tonnes (not including the Ravensworth and Narama mines which did not formally become part of the Glencore stable until 2002).Glencore Coal was briefly called Enex Resources in 2001 when it planned an initial public offering, but this IPO was cancelled following the September 11 attacks in the USA and the impact of those attacks on equity markets.

One of the more controversial corporate takeovers in the mining industry occurred in 2003 when Xstrata, partly owned by Glencore, purchased MIM for what was later seen as a bargain price. MIM had a history in the coal industry dating back to 1952 when it acquired an interest in Collinsville. MIM and Dacon Collieries merged their coal operations in the Collinsville area in 1975, with the new company Collinsville Coal Company acquiring MIM's Bowen Consolidated Coal Mines and Dacon Collieries. MIM moved to full ownership of the Collinsville Coal Company in 1977. That year also saw MIM gain Queensland Government approval to buy the company which owned the Newlands steaming coal deposit situated around 80 kilometres south of Collinsville. MIM had begun to get serious about the export market for steaming coal.

In 1978 MIM and Thiess announced that they would develop the Wandoan deposit in the upper Dawson Valley, about 350 kilometres north west of Brisbane. The project was a contender to win the contract to supply a major new power station at Tarong, but that contract was subsequently awarded to CRA subsidiary Pacific Coal and involved the supply of 66 million tonnes of coal over 18 years from its new Meandu mine. The Wandoan project only proceeded to the stage where a box cut was developed and has never been developed into a mining operation. MIM's first foray into coking coal occurred in late 1979 when it took a

40% interest in the Oaky Creek coking coal project in the Bowen Basin, the other partners being Houston Oil and Minerals and Dutch company Hoogovens. Oaky Creek commenced as an open cut mine and produced its first coal in December 1982; underground mining commenced in 1989, with the first longwall installed in 1990. Development work for a second underground at Oaky Creek North commenced in 1995 and coal production began in early 1999. After buying into the project in 1979, MIM soon increased its equity in Oaky Creek to 79% and moved to 100% ownership in 1993.

MIM's NCA project (Newlands, Collinsville, Abbot Point) was given the go-ahead in June 1981. Valued at $550 million, it involved the construction of coal loading facilities at Abbot Point near Bowen, development of the Newlands mine and expansion of the Collinsville mining operations. Exports to Japan and other markets were planned to commence in 1984, with 4 million tonnes per year from Newlands and 1 million tonnes from Collinsville. The first coal from Newlands was exported through Abbot Point in 1984. When the NCA project was announced in 1981, there was still considerable optimism in the coal industry and MIM was looking to a strong decade ahead for its metal mining and smelting and its coal operations. It was one of the leading companies in Australia, with a market capitalisation putting it at number 4 in Australia behind BHP, CRA and CSR. As Murray McMillan later wrote: "MIM was one of the glamour mining houses in the early 1980s, riding the so-called resources boom. It had the great Mount Isa Mine and a growing world demand for its minerals. The world was its oyster and the company was bent on expansion. Its strategy for growth was to supplement its copper, lead, zinc and silver operations by getting into coal, the black gold of the 1980s. It would set up a series of strategic relationships overseas with investments in selected base metals companies. But the coal investments failed to perform and base metal prices began to slide, hitting the returns from its overseas investments."[497]

Coming on-stream in the mid-1980s, the NCA project's timing was not ideal. It was a difficult time for base metal mining and smelting, the core business of MIM, and also for coal producers. At the company's annual meeting in November 1984 chairman Bruce Watson said that MIM's coal projects were "operating well", although he admitted that the financial situation was "less satisfactory", with Oaky Creek and Collinsville making profits, but Newlands operating at a loss. Coal sales had reached 8 million tonnes per year, with exports to a range of markets in Europe and Asia, but the low coal prices were hitting the company hard.[498] The company had borrowed heavily in the early 1980s to finance its expansion, with half of the $1.6 billion in loans for coal projects. The NCA project's funds were raised on a non-recourse basis, with revenue from the project servicing the loans.

In the early 1990s MIM was continuing to struggle, and was fortunate that the non-recourse financing structure was in place for the NCA project, as a conventional debt financing structure may have created even more problems for a company that was not in a strong financial position. MIM survived the traumas of the early 1990s and undertook some major restructuring of its operations, shedding a number of its investments overseas. In 1993-94 it made a loss of $195 million, but by that year it had reduced its gearing to manageable levels and had set the basis for a more sustainable future. By the late 1990s, although the going was still difficult for most mining companies, MIM's future was looking a little brighter. Its assets included its Mt Isa mine, its half share in the Alumbrera copper and gold mine in Argentina, its coal mines and other promising projects including the Ernest Henry copper-gold deposit. But by 2000, MIM was starting to be seen as a potential takeover target.

MIM released its annual results In August 2000, announcing a profit of $112 million, a turnaround from the loss the previous year. Takeover rumours in the mining industry were in the air, fed by a battle between Noranda and Billiton for Rio Algom, a Canadian company which held a 25% interest in Alumbrera. As reported in The Age, Alumbrera was "not central to the bidding war over Rio Algom but has prompted a rethink by the market on MIM's break-up value…"[499] However, MIM was positive about the future, with CEO Nick Stump saying that the company was looking to use "its improved financial outlook as a base to improve shareholder returns and take advantage of growth opportunities." Stump said that the company was looking to expansion in "copper, copper-gold and coking coal, through acquisition, exploration and organic growth of present operations, while maintaining a strong zinc-lead-silver business."[500]

Some other major changes and surprises in the minerals industry also occurred in 2001. BHP and Billiton merged their operations and announced a profit of $4 billion for the year ended June; Shell make an offer to buy Woodside, but was refused approval by the Federal Treasurer, Peter Costello; Pasminco was placed into voluntary administration and then broken up; Rio Tinto announced a profit of $1.6 billion for the six months to June; and Anglo Gold of South Africa and Newmont of the USA fought to gain control of Normandy Mining, then Australia's largest gold miner. These were just some of the major corporate moves in 2001, and followed an interesting year in 2000 when Peabody sold its Australian coal assets to Coal & Allied, with Coal & Allied subsequently on-selling two of the mines (Ravensworth and Narama) to Xstrata.

Gauci fights a losing battle

It was therefore not surprising that MIM would soon become a takeover target. In 2002, Xstrata CEO Mick Davis made an unsolicited approach to MIM which led to discussions between the two companies. The market reacted, with MIM's share price jumping by over 20%. MIM was now clearly in play and there was speculation as to whether other big miners such as Rio Tinto and Anglo would consider making an offer.[501] No other offers emerged and negotiations between the two companies continued for the next few months. In June 2003 MIM shareholders voted to accept the $1.72 per share offer by Xstrata. MIM's CEO, Vince Gauci, had fought against the offer and against his own board of directors, arguing that it seriously under-valued the company. Gauci argued that "Just as you should not sell a house that is in good order and the real estate market is depressed, we should not sell the company now...Commodities markets are depressed, the world has just witnessed a war in Iraq and the major economies of the US, Japan and Europe are still struggling. The market environment is not right for selling our house today, most certainly not at $1.72 per share."[502] Gauci said that the company had worked through its major challenges, selling its smelters in Europe, resolving problems from currency hedging and improving its copper and coal operations. But Gauci had lost the argument and MIM was quickly absorbed into Xstrata which now had a significantly larger stable of coal mines. Xstrata then on-sold 20% of its interests in the NCA and Oaky Creek assets to Itochu and Sumitomo for around $550 million, as well as 25% of four projects in Queensland (Rolleston, Wandoan, Red Rock and Pentland).

Gauci was proved correct. MIM's directors had sold the company just as both coking and thermal coals were about to experience five years of price rises. More spectacularly for MIM, the copper price was about to take off. Dismayed long standing retail shareholders of MIM, who had endured a long period of depressed metal prices, watched as the copper price shot up within months of the Xstrata acquisition, quadrupled within three years in 2006 and quintupled as it peaked in 2011. At the same time, all the value gained by MIM's corporate improvements was reaped by Xstrata which, along with mining companies around the world, generated greatly increased profits during the ensuing commodities boom.

The privately owned Glencore went public in 2011, with an IPO which generated around $US10 billion from the sale of 20% of its shares; the company listed on the London and Hong Kong Stock Exchanges in May. There was already speculation that the new listed company had its partly-owned Xstrata in its sights for a takeover, speculation that would soon prove to be correct. Glencore had been a major shareholder in Xstrata since it was listed in 2001, its holding of around 34% of the shares giving it effective control of Xstrata. In November 2012 shareholders of both companies voted to proceed with the merger. The merger was completed in May 2013, with the shares in the new company, GlencoreXstrata, listed on the London Stock Exchange on 3 May. Ivan Glasenberg, Glencore's CEO, became the CEO of the new company. Mick Davis, the Xstrata CEO, who had originally expected to be the CEO of the merged company, lost the battle with Glasenberg and left. GlencoreXstrata was now the number four mining company in the world (behind BHP, Rio Tinto and Vale) with sales revenue of over $US236 billion and assets of $US169 billion.[503] Just after the merger became effective, Glasenberg touted the benefits of the way in which his new company was structured, with executives and board members holding almost one quarter of the shares, and with almost 11% owned by other employees. Glasenberg contrasted this major stake in the company with the very low percentage of shares owned in their companies by the executives and board members of BHP Billiton, Rio Tinto and Anglo, saying that Glencore's management were "owners and not caretakers of assets" and that his team had a proven track record of creating value resulting from this ownership. Glasenberg himself was GlencoreXstrata's largest shareholder with 8.3% of the company's shares.[504]

Glasenberg accuses other mining CEOs of creating excess supply

While Glencore maintained a positive stance on the long term prospects for coal, in 2013 it took a very hard-nosed approach to the question of expansion at a time when prices were easing and the short term outlook was anything but promising. In February Glasenberg launched a direct broadside at the major mining companies, slamming them for their decisions to expand production over the previous few years, decisions which he said had led to excess production capacity and lower profits. He said that "The big guys really screwed up" and had continued to invest in new assets; he said that the mining industry had to stop doing that and that it had "to learn about demand and supply."[505] Glasenberg noted that the CEOs of all the major companies had recently lost their jobs and now hoped the new generation of CEOs had learnt from the experience of their predecessors who had built new mines but failed to get adequate returns for shareholders. It was, he said, time to stop building.

In early 2013 spot prices for thermal coal were around US$95 per tonne (FOB Newcastle), still reasonable but well down on the peak reached in 2011-12. However, the steaming coal market was becoming over-supplied, with new mines and expansions coming on stream, including a significant increase in production from Indonesia. Also, as noted by the Department of Industry later in the year, "Rather than reducing output in response to declining prices, many high cost producers have increased production in order to reduce their unit cost. Some of these producers, largely in Australia, are locked into fixed take-or-pay contracts for infrastructure services and it has been more cost effective to increase production than to close. This extra production has placed further downward pressure on thermal coal prices."[506] Glencore announced that it had put the big Wandoan project on hold, its position at that time being that it would not develop new capacity unless it was confident that this would not materially add to oversupply in the market and have a negative impact on global coal prices. The Queensland Government granted mining leases for Wandoan in 2017, following earlier environmental approvals by the State and Federal Governments. While Glencore welcomed the granting of the leases, it noted that it would "continue to assess the project's timing against the global coal market" and said that the company was "taking a responsible approach and (would) only bring on new large scale production volumes if we are confident that market conditions support such a development." Wandoan's development will also require a multi-billion dollar commitment for the mine and associated rail infrastructure.[507]

In May 2013 Glencore also withdrew from a major port project in Queensland, the $1 billion 35 million tonnes per year capacity Balaclava project which was to have been developed north of Gladstone. Balaclava would have been the port for the Wandoan coal mine project.[508] Glencore was concerned about the short and medium term outlook for steaming coal and the expected over-capacity in Queensland coal ports, as well as limitations on Balaclava's potential to expand; the project was shelved "because the risk-reward balance of investment greenfields coal infrastructure just doesn't add up."[509] The company's reticence to invest in new capacity, however, did not prevent it from taking a major interest in the Clermont steaming coal mine in Queensland in 2013. Glencore and Sumitomo agreed to buy Rio Tinto's 50.1% interest in Clermont for

$US1.015 million. Clermont was Australia's third largest steaming coal mine, producing around 12 million tonnes a year, blessed with a low strip ratio of 3.2:1 and, as its coal had relatively low ash content, the mine was able to avoid passing 90% of its production through a washery.[510] The purchase was obviously a good business decision for Glencore and Sumitomo; with the mine sitting in the top quartile of mines in terms of profit margin, it was well positioned to ride out short term price pressures.

In 2014, as coal prices fell, companies continued to produce and export at a high rate, with the "take or pay" contracts with rail and port providers still a major burden on the industry. Glencore at one stage that year said that up to one third of the Australian industry was operating at a loss; another company executive said that it was costing more to mine and ship coal than the industry was able to sell it for.[511] In November 2014, with spot prices for thermal coal down to around $US52 per tonne, Glencore announced that it was ceasing production at all its Australian coal mines for three weeks over the Christmas period, with this move estimated to take production of 5 Mt out of the market. In 2015 the company went further, with cutbacks to production totalling 15 Mt, and its coal CEO, Peter Freyberg, saying that while the company's capacity was programmed to operate at 110 Mt for that year, it would not "push incremental tonnes into the markets that don't want or need them – doing so would force prices down further…The market doesn't need all this coal…so we've cut, we've decided not to produce anywhere near that volume this year."[512]

Having listed Glencore successfully in 2011 and merged with Xstrata in 2013, Glasenberg still had ambitions for Glencore to become even larger on the world mining stage. He approached Rio Tinto in August 2014 regarding a merger, but was rebuffed, with Rio's chairman, Jan Du Plessis, saying that his board had rejected the offer as not in the interests of Rio's shareholders.[513] Glencore finally secured a major stake in the Rio Tinto's Hunter Valley Operations (HVO) mines in NSW in 2017. The company had earlier expressed interest in these mines in view of the potential synergies with its own operations, and had competed with

Yancoal when Rio Tinto formally put the mines up for sale. It was outbid by Yancoal, but then did a deal with the winner to secure 49% of HVO. In 2018, Glencore sold the Tahmoor coking coal mine located south of Sydney to SIMEC Mining, a part of Sanjeev Gupta's GFG Alliance which had purchased the Whyalla steelworks. Glencore's next purchase was Rio Tinto's 80% interest in the Hail Creek coking coal mine in the Bowen Basin, an acquisition that significantly strengthened its standing in the coking coal trade.

Life is never dull in Glencore, and the company stunned the industry in February 2019 when it announced that it would cap its global coal production at around 140 to 150 Mt per year. The company's said that "To meet the growing needs of a lower carbon economy" it aimed "to prioritise its capital investment to grow production of commodities essential to the energy and mobility transition and to limit its coal production capacity broadly to current levels."[514] Bowing to pressure, in particular from companies involved in the Climate Action 100+ group, Glencore said it "recognises the importance of disclosing to investors how the company ensures that material capital expenditure and investments are aligned with the Paris Goals" with these investments to include "each material investment in the exploration, acquisition or development of fossil fuel (including thermal and coking coal) production, resources and reserves, as well as in resources, reserves and technologies associated with the transition to a low-carbon economy. And from 2020, the company says that it intends "to report publicly on the extent to which, in the Board's opinion, this was achieved in the prior year and the methodology and core assumptions for this assessment."[515] The industry's reaction to these commitments has been varied, with one producer saying that it was "an astute commercial decision cloaked in climate change PR spin" and adding that he was not overly concerned as over time it could assist to put a floor under coal prices by restraining coal supply.[516]   Other reactions were less positive, with one producer reported as saying that the Glencore decision "screws every other major provider in Australia, from a development point of view", but that from the perspective of coal prices it would be beneficial.[517]

Glencore's is the world's largest thermal coal producer and exporter and, together with BHP, is the dominant force in the Australian coal industry. Glencore is responsible for an estimated 12% of the world trade in thermal coal, and around 35% of the trade in high quality thermal coal.[518] It has certainly come a long way from its days in the 1990s when its role in the coal industry was that of a not-so-humble trader. How the company's commitments in relation to capping production and reducing emissions will impact on its operations in Australia is difficult to assess at this time, but the company's new policies are certainly indicative of the realities confronting coal producers in an era of heightened pressure from governments, communities and importantly financiers in relation to climate change and greenhouse emissions.

Rio Tinto exits coal

While Anglo initially looked to sell out of its coal mines but has now maintained most of them, Rio Tinto in a short space of time went ahead and sold its mines in Australia. Rio's exit from coal began in 2013 with the announcement of the sale of its Clermont and Blair Athol mines, although that year also saw the official opening of its $US2 billion Kestrel mine extension. However, the process gathered steam in 2016 when Rio finalised the sale of its interest in the Bengalla mine in the Hunter Valley to New Hope and in 2017 with the sale of its Coal & Allied interests to Yancoal. The C&A assets included its major shares of Hunter Valley Operations, Warkworth and Mt Thorley, and its 36.5% of Port Waratah Coal Services. This brought an end to Rio's interests in thermal coal mines in Australia, and was followed in 2018 by the announcement of the sale of its coking coal mines in Queensland, with its 82% share in Hail Creek going to Glencore, and its 80% of Kestrel going to ERM Capital and Indonesian miner Adaro Energy. Rio's undeveloped coal projects were also sold in 2018, Winchester South to Whitehaven and Valeria to Glencore.

Rio Tinto, through Kembla Coal and Coke and later CRA and Coal & Allied, had seen a long involvement in the Australian coal industry.

It had been a major producer, although never matching BHP in total production. The total sale proceeds of Rio's Australian coal assets generated around $US8.7 billion for the company, helping it to "reshape our business for long-term success in a low-carbon economy."[519]

Peabody in, out, in again

Peabody, one of the major coal companies in the USA, has had an eventful history in Australia. Peabody became part of the Thiess Peabody Coal Company Pty Ltd in 1962, joining with Thiess to develop the Kianga and Moura mines and becoming part of the first major export development in the Bowen Basin. Mitsui formally became part of the group in 1963, the company then becoming known as Thiess Peabody Mitsui, or TPM.

Peabody was purchased by Kennecott Copper Corporation in 1968, and by 1970 was the largest producer of coal in the USA. In 1971 the US Federal Trade Commission ruled that the purchase of Peabody violated US anti-trust legislation. Kennecott challenged the ruling, but lost, and in 1976 the Federal Trade Commission ordered Kennecott to divest itself of Peabody. Peabody was then sold to a consortium of companies including Newmont Mining with its Australian assets sold to BHP. BHP was now the 58% owner of TPM, which was re-named Thiess Dampier Mitsui, BHP's subsidiary Dampier Mining being the legal vehicle for the purchase of the Peabody assets. So Peabody was no longer part of the Australian coal mining industry and it would be some years before the name reappeared in the Australian industry and this would occur via British company Costain.

Costain won a major contract in 1967 with the Electricity Commission of NSW to supply the new Liddell power station from a mine at Ravensworth near Singleton to be developed and operated by Costain. Costain was also part of a consortium which won the right to develop the Warkworth mine in the Hunter Valley in 1976, the other partners including local oil company HC Sleigh, Mitsubishi Development, Mitsubishi Mining & Cement and T&G. The Warkworth lease, near Singleton, containing soft coking and steaming coal was developed from

1980, with initial production commencing late in 1981. Warkworth had been planned to be in operation by around 1980, but was delayed by the market uncertainties and the need to secure firm export contracts to underwrite the development. Warkworth won a contract to supply 5 million tonnes of steaming coal over 10 years to Japanese power utilities in 1978, but it was not until April 1980 that it had secured a similar sized contract to supply soft coking coal to the JSM. In June 1980 the company also signed a contract with Mitsubishi Mining & Cement to supply 5.5 million tonnes over 12 years, Mitsubishi's membership of the joint venture having clearly opened the way to secure that contract. The contracts now in place enabled the consortium to give the go-ahead for development of the mine.

UK investment company Hanson, known at the time as one of the leading "corporate raiders", bought Peabody in 1990. In October 1992 Costain announced that it was selling its Australian coal assets to Peabody, those assets including the Warkworth, Ravensworth and Narama mines in the Hunter Valley. Costain had accumulated significant debt and the sale allowed it to reduce that debt to a much more comfortable level. Peabody was now back in Australia under British rather than American owners. Bob Humphris, who had come to Australia from the UK in the early 1970s, and rose to head Costain's mining division in 1983, was appointed by Hanson as CEO of the Peabody group in Australia. Peabody Australia grew during the 1990s, developing the Bengalla mine in the Hunter Valley, which it managed on behalf of its joint venture partners Wesfarmers, Kepco, Mitsui and Taipower. Bengalla commenced operating in 1998, and the following year Peabody bought BHP's interest in the Moura mine, teaming up again with Mitsui who had remained part of the project since 1963. Peabody also developed the Narama mine from 1993 to supply Macquarie Generation.

Ownership changed again in 1998 when US investment bank, Lehman Brothers, purchased Peabody. Peabody's Australian operations continued under Lehman Bros, with Humphris continuing as the CEO. Humphris recounts how as CEO he had an agreement with his American boss Irl Engelhardt, that Engelhardt and his CFO would only visit Australia if they had been invited by Humphris. Many Australian CEOs reporting to overseas owners would no doubt have envied such an arrangement which kept such visits to a minimum. But Lehman moved in 2001, putting its Australian assets up for sale, with Coal & Allied, now part of Rio Tinto, successful in the competition to buy Peabody's mines. Humphris bowed out of Peabody in 2001, but went on to take on director roles outside the coal industry. Humphris had played a major role in the Australian coal industry, not only leading Costain and Peabody for many years, but also as a chairman of the NSW Minerals Council and Australian Coal Association.

While 2001 saw Peabody once again exit the Australian industry, its departure would prove to be fairly brief. Peabody Energy was floated on the New York Stock Exchange in 2002, and that same year Peabody Australia was back in the action again, buying the Wilkie Creek mine in the Surat Basin in Queensland from Allied Queensland Coalfields. It then moved on to buy the North Goonyella underground mine and the Burton open cut mine from German miner RAG Australia in 2004, and in 2006 bought Excel Coal Limited. Excel Coal under managing director Tony Haggarty had built up a stable of mines which included the Wambo open cut in the Hunter Valley, the Chain Valley underground near Lake Macquarie and the Metropolitan underground mine north of Wollongong. Excel also owned the Wilpinjong mine north east of Mudgee which was nearing completion, and was developing the North Wambo underground and the Millennium mine in the Bowen Basin. Peabody's stable expanded again to include the Coppabella, Middlemount and Moorvale open cut mines in the Bowen Basin which it purchased in 2011 when it acquired Macarthur Coal for $US5 billion. However, lower coal prices and the string of acquisitions, and in particular the expensive acquisition of Macarthur near the top of the coal market, would soon place major strains on the Peabody group's balance sheet.

The acquisition of Macarthur Coal, following the earlier acquisitions in Australia, was consistent with the company's expansion plans announced in 2010. Peabody's Australian mines, according to CEO Greg Boyce, were "poised to supply increasing demand for coal throughout the Pacific Rim". The company, he said, was "advancing projects that could double its metallurgical and thermal export coal platform by 2014, targeting 35 to 40 million tons of production, assuming a favorable investment climate in Australia."[520] That same announcement also talked about the pipeline of projects and partnerships in other countries in the Asia Pacific region, including China, where it was advancing joint ventures with potential partners including utilities, steel producers and coal producers. In India, the announcement stated, Peabody had been selected by Coal India as a key partner for long term supplies; in Indonesia it was exploring partnerships with local companies; and in Mongolia, its joint venture was "driving exploration of dozens of resource licenses", with the company also continuing its efforts to be part of the Tavan Tolgoi coalfield.[521]

In its annual report for 2011, Peabody said that "enormous needs around the world" pointed to what it expected to be a "long-lived super cycle for coal" and "a period of sustained market expansion to meet the requirements of an emerging global middle class".[522] Boyce was still extremely positive about the future in 2012, when he said that the international coal market outlook was very strong, with China expected to increase its imports of coal by around 100% between 2011 and 2016. Peabody was forecasting that by 2016 China would be importing 300 to 350 million tonnes per year of steaming coal and 100 to 110 million tonnes of metallurgical coal. The company's forecasts for India had its imports increasing to between 200 and 220 million tonnes by 2016. Total world trade over that 5 year period was forecast to increase by between 385 and 515 million tonnes, with more than 85% of that growth to be in the Chinese and Indian markets.[523]

Peabody's acquisitions and ongoing plans for new projects and joint ventures were among the most aggressive and optimistic in the industry. The Peabody forecasts proved to be dramatically wrong and in April 2016 it filed for Chapter 11 bankruptcy in the USA, a process that it hoped would allow it to restructure and trade its way out of its financial difficulties. The company referred to an "unprecedented industry downturn" caused by a range of factors including slowing growth in China, low prices and burgeoning production of shale gas in the USA, with shale gas seriously reducing the consumption of thermal coal by American power stations.[524] Peabody was also suffering under the mountain of debt caused in part by its purchase of Macarthur Coal in 2011. Peabody said that its Australian operations would continue to operate and that they were not included in the Chapter 11 filing. Peabody was not the only major US coal producer to go through the bankruptcy process, with Arch Coal and Alpha Natural Resources also hit by the squeeze on the coal industry. The Sierra Club, a US conservation organisation, rejoiced in the fall to earth of Peabody, saying that its bankruptcy should serve as a wake-up call to anyone promising that coal's glory days would return.[525]

However, Peabody weathered the storm and emerged from Chapter 11 in April 2017, barely twelve months from filing, to list on the NYSE. The company had restructured its debt and reduced the stake of its major shareholder, Elliott Management, to 28%. Peabody was still optimistic on the future for thermal coal, expecting continuing growth in the Asia Pacific. In April 2018, CEO Glenn Kellow said that the company expected demand for imported coal from Southeast Asia and the Pacific to outstrip demand from China, with growing demand from the high efficiency, low emission power stations, with Australia well placed to benefit.[526]

Centennial – from small beginnings to a major producer

The coal industry in Australia has seen a number of companies grow from very modest beginnings to become significant producers, with some taken over by larger companies. Among these are Centennial Coal, Macarthur Coal, Felix Resources and Excel Coal. There have also been investors like ex mine electrician Nathan Tinkler whose star soared for a period of time before crashing to earth.

Centennial Coal began in 1989 as a small producer with one mine, Preston Colliery, in the Gunnedah area producing around 200,000 tonnes per year. Preston had been owned by Coal & Allied and was bought by

Robert (Bob) Cameron with the backing of other investors. Cameron, a mining engineer, had worked for fifteen years for Coal & Allied and so understood the potential of the Preston mine. Centennial then seized the opportunity to expand in 1994 when it purchased three Blue Circle coal mines from Boral Ltd in a public float of the company after Boral had made the decision to exit coal mining. The Charbon, Berrima and Ivanhoe coal mines supplied coal to the Blue Circle cement works and had export contracts, including a 500,000 tonne per year contract with Korean electricity giant, KEPCO. The company went on to purchase major interests in two other mines in the Western district, Springvale (in a joint venture with Korean interests) and Clarence.

The next major phase of Centennial's expansion came when it was the successful bidder for the NSW Government's stable of Powercoal mines in 2002. The Powercoal group of mines, with local and export contracts, included the Newstan, Myuna, Cooranbong, Awaba, Munmorah, Wyee and Angus Place mines which were supplying around 8 million tonnes under long term contracts to the NSW power stations, and a number of projects which could be developed. The jobs of the Powercoal employees were guaranteed for three years under an enterprise agreement certified by the AIRC. Centennial Coal said that it would be spending $200 million over the following two years to develop Mandalong and on other major projects, with Mandalong expected to commence deliveries of coal to power stations by January 2005.

Centennial was essentially a thermal coal company, but in 2005 it made the move into coking coal by acquiring Austral Coal which owned the Tahmoor mine south of Sydney. Tahmoor would prove to be a problematic acquisition for Centennial and was sold to Xstrata in 2007.

One takeover by a foreign company which came as a surprise to the industry was the purchase of Centennial Coal by Thai company Banpu in 2010 for around $2.5 billion. Banpu has interests in coal mining, power generation and alternative energy in Thailand, other Asian countries and the USA. By 2010 Centennial had grown into a significant medium sized coal producer, its mines in the Newcastle/ Lake Macquarie area and in the NSW western district producing around 15 million tonnes per year. Banpu made its first move on Centennial in May 2010, buying 14.9% of its shares; it then made a formal offer for the company in July which received the support of the Centennial board. The offer came just days after the Federal Government had negotiated with BHP, Rio Tinto and Xstrata Coal to introduce the mining resource rent tax to replace the minerals super profits tax, but this did not deter Banpu from making a major investment in the Australian coal industry. The takeover was a friendly one, with Cameron retiring as CEO to become chairman of the company.

NSW Government exits coal mining after 90 years

The role of the NSW Government in coal mining began with the early days of the Colony, but ceased in 1831 when AACo took over the operation of the state mines in Newcastle. The Government's involvement re-commenced in 1912 with the passage of the State Coal Mines Act which gave it the power to develop coal mines. The development of the Lithgow State Coal Mine was given the go-ahead in 1916, although financial constraints and the First World War saw development suspended in 1917. Development of the Lithgow mine proceeded in 1921, with the mine now under the control of the NSW Railways, and by 1923 it was one of the largest in the State, employing around 500 workers. In 1932 the State Coal Mines Act transferred the Lithgow mine from the Railways to a new State Coal Mines Control Board. The Act was amended again in 1948, creating the role of Director of State Coal Mines and re-naming the Board the State Mines Control Authority (SMCA). Jack Baddeley resigned as a Member of Parliament and from his positions as Deputy Premier, Chief Secretary and Minister for Mines, and was appointed the first Director of State Coal Mines in 1949. In the years after World War Two, as the electricity industry grew to supply a growing population and industrial base, the NSW Government adopted a policy of developing its own coal mines to supply its new power stations and the SMCA was the initial vehicle for these new mines.

The SMCA developed the Awaba mine south of Newcastle from 1947, with the mine in production by 1948, supplying coal by rail to the White Bay and Ultimo power stations in Sydney and a power station in Newcastle (these stations then under the control of the NSW Railways) and later to the new Wangi Wangi power station. In 1950 it began development of the Oakdale State mine in the Burragorang Valley south west of Sydney, developed to supply the Bunnerong power station at Matraville in Sydney. Oakdale continued under SMCA control until it was sold to Clutha in 1968 by the Askin Liberal Government which was concerned about the mine's unprofitability. The next major development, announced in 1951, was the Liddell State coal mine between Singleton and Muswellbrook in the Upper Hunter Valley.

The second NSW Government body to become involved in coal mining after the war was the Electricity Commission of NSW which was established in 1950, although its entry into the coal mining business did not come until 1958 when it purchased the Newstan and Newcom collieries from the Joint Coal Board. Newstan was located at Fassifern south of Newcastle, and Newcom at Lidsdale near Lithgow. These two mines produced around 1.5 million tonnes of coal in 1959-60; together with the SCMA mines' production of around 1.1 million tonnes, the Government mines accounted for one sixth of the State's total production. The Electricity Commission and the SMCA would continue as separate coal mining operators until 1973.

The development of new power stations at Vales Point and Munmorah on the NSW central coast in the 1960s saw the parallel development of the Wyee and Munmorah mines by the Electricity Commission. Munmorah opened in 1964, and Wyee saw production from a continuous miner commence in 1965. The early to mid-1960s also saw the development of the Newvale and Newvale No.2 mines to feed the Commission's new power stations near Lake Macquarie. By 1969-70 the Electricity Commission mines were producing 3.7 million tonnes, and the SCMA mines 3.2 million tonnes. The Government mines had increased their share of the State's production to 20% and were now were equal to Clutha in production and close to matching the BHP/AI&S group in size.

The NSW Government took the logical step in 1973 of amalgamating its coal mines, transferring responsibility for the SMCA mines to the Electricity Commission. The Commission now had a significant stable of mines - Liddell State, Newcom, Awaba, Wyee State, Munmorah State, Newstan, Newvale and Newvale No.2. The old Lithgow State Mine had been closed in 1964. Newcom was renamed Angus Place in 1978. The development of the new Eraring power station in the early 1980s also saw the development of the Myuna and Cooranbong mines in the Lake Macquarie area. Eraring also drew coal from Newstan and Awaba and from privately owned mines.

In 1981 the Electricity Commission's powers were amended to allow it to diversify its operations to compete for sales in markets outside its traditional power stations. The changes to the Act allowed the Commission, in addition to being a coal miner, to produce and sell coal to power stations or to produce and sell coal commercially (or as the Act specified "exclusively, principally or partially with a view to profit"). A review of the Commission by Gavan McDonell in 1987 had recommended that the mines should be separated from the Commission, but no action was taken on this recommendation. Following the election of the Greiner Coalition government in 1988, a review of the State's finances by James Curran contained a number of recommendations relating directly to the Electricity Commission and its coal mines. The Curran review said that the costs of coal from several of the Commission's mines (Huntley, Newvale and Liddell) compared unfavourably with the costs which the Commission could incur if it sourced coal from other mines.[527] Curran recommended that the Commission be allowed to buy coal from the lowest cost suppliers, be they its own mines or privately owned mines. Curran also recommended that the Commission's mines should be run independently rather than as a subsidiary of the Commission. Under a new Government, elected with a mandate to shake up the public sector and bring a more business-like approach to its management, the future of the Electricity Commission's mines was now uncertain.

It did not take long for the Greiner Government to decide to privatise some of the Commission's mines. In September 1989, Cabinet approved arrangements to create two new companies, provisionally named Power Coal and Premier Coal, with Premier Coal to be sold to the private sector. Premier Coal would own the mines able to compete on the export market, including Awaba, Liddell, Newstan, Newvale, Newvale No.2 and Huntley. Power Coal would own the other mines (Angus Place, Cooranbong, Munmorah, Myuna and Wyee) and would continue as a subsidiary of the Electricity Commission.[528] In December the Government awarded Jarden Morgan, a Brisbane based stockbroker, a contract to carry out an initial study on the privatisation of the mines. Electricity Commission general manager, Barry Flanagan, said that he hoped the mines would be up for sale by June 1990.[529] The Commission's Huntley mine, located at Avondale near Dapto and which supplied the Tallawarra power station, was closed in 1989 when the Government also announced that the power station would close.

This plan for the creation of the two new companies did not eventuate, but just over 12 months later, Premier Nick Greiner and his Cabinet decided to put all the mines up for sale, even though the Government acknowledged that it would be difficult to sell the mines given the current market conditions. Commission chairman John Conde noted that for the first time its mines had been able to compete successfully with privately owned mines for contracts to supply the power stations. Seven of the Commission's nine mines had been successful in winning contracts; however, two mines – Awaba and Newvale – did not win contracts, and their future looked grim. Despite Conde's optimism, Greiner was no strong supporter of the Government's mines, saying they had traditionally been very inefficient and that "the best way of ensuring they remain efficient is ensuring there's some private equity interest." Greiner believed that the mines should be privatised "because the Government has no business owning coal mines."[530] The search for a buyer or buyers for the mines commenced, and in 1992 one of the mines – Liddell – was sold to Cumnock Coal, the sale no doubt assisted by a three-year contract to supply coal to the Electricity Commission. However, with potential buyers clearly not prepared to pay the price the Government was hoping to achieve, in October 1992 the Government announced that the privatisation of the mines would no longer be pursued. But in a blow to the workforce of Newvale, the Minister for Energy also announced that as Newvale had failed to gain a power station contract, it would cease production on November 30 and would be closed by 31 January 1993. Almost 200 jobs would be lost from the closure of Newvale and from cutbacks at Awaba, and the future of the remaining employees at Awaba was to depend on whether export contracts would be secured for coal from Newstan, allowing Awaba to take over part of Newstan's contract to supply the Eraring power station.[531]

Newvale was given a temporary stay of execution when the Labor Opposition succeeded in having a Parliamentary Select Committee established to inquire into the proposed closure and to look at the winding down of the Awaba mine and the proposed sale of Munmorah and Liddell. Following the announcement of the closure, nine miners at Newvale had staged a sit-in, and remained underground for nine days until they learned about the establishment of the Committee. The Committee moved quickly to begin hearings. Phillip McCarthy, the CEO of Powercoal (the coal mining subsidiary of Pacific Power as the Commission was now called) appeared before the Committee in December 1992, defending the proposed closure of Newvale, and referring the Committee to the low productivity of the mine. McCarthy said that annual production per worker had increased to around 3500 tonnes, but that this was still well below the 4400 tonnes per worker for all Pacific Power mines, and the 4500 tonnes per worker for the NSW underground industry as a whole.[532] McCarthy said that the company's management had worked over the previous year to get the mine's workforce to understand the challenges facing the mine, with its cost of production at over \$38 per tonne. With the average export price for steaming coal in 1991-92 around \$48.50, Newvale would have been uncompetitive on the export market, allowing for costs of transport to Newcastle and port loading and other charges, and was also obviously unable to match the competition to secure any power station contracts.

The Select Committee handed down its report on 23 December 1992, but it was not the Christmas present any of the Newvale workers may have been hoping for. Ray Chappell, chairman of the Committee, said that the Committee supported the decision by the Government to sell or close some of the mines owned by Pacific Power. Chappell acknowledged that the workers had been "making great efforts to improve productivity in recent times" and that "The workers, the United Mine Workers' Union and management (had) progressed a long way in facing the new commercial realities for the State-owned coalmines." However he said that the reality was that there was "too much coal and too little demand."[533]

There was, however, one slight glimmer of hope for Newvale in the Committee's report which recommended that Pacific Power and the union discuss the possibility of leasing the mine to a private operator, an option that had been proposed by some of the Newvale miners. Several months later, one of the Committee members, Andrew Humpherson, said that while the Committee had not been optimistic about the possibility of leasing the mine, it had wanted to see the option investigated. He added, however, that with several months having passed since the Committee handed down its report, it was disappointing that nothing had been concluded.[534] A number of meetings were held between Pacific Power, Powercoal and the union and offers to take over the mine were discussed, leading to a second Select Committee inquiry which reported at the end of April 1993. Offers to lease Newvale had been made by Ron McCullough of Alliance Investments Pty Ltd, but were rejected by the Committee which "concluded that it was not possible, on the financial data put before it, for Newvale to operate under the proposed leasing arrangements without a significant subsidy by the taxpayers."[535] The way was now clear for Pacific Power and its Powercoal subsidiary to close Newvale.

The Electricity Commission had been corporatised in the early 1990s and was renamed Pacific Power in 1992. Its coal mining group was managed by ENC Management Pty Ltd, which later became Powercoal, and included the two groups of Commission mines controlled by Newcom Collieries and Elcom Collieries as well as Huntley Collieries. The Powercoal mines were required to compete for supply contracts to the power stations and by 1997 were supplying only 36% of coal used for power generation in NSW, and at prices well below levels prevailing at the start of the 1990s. The power stations were also looking to further falls in coal prices as competition in the national electricity market intensified.[536]

During the 1990s, in the increasingly competitive domestic coal market, Powercoal made major progress in improving the performance of the group and lowering mine costs. The Newvale No.2 mine which had been renamed Endeavour, was closed in 1999. Awaba was closed in March 2000, but it re-opened in December 2001. By 2001, with the exception of Wyee, the Powercoal mines had lifted productivity significantly. Among its non-longwall mines, Munmorah's production per employee (around 5800 tonnes per employee) was now equal to the State average; Cooranbong at around 5900 tonnes was marginally higher than the average, and Myuna at almost 6800 tonnes was well above average. Powercoal's longwall mines included Wyee (around 4100 tonnes per employee), Angus Place (8000 tonnes) and Newstan (9100 tonnes). The NSW average for all longwall mines was just under 8600 tonnes per employee. Wyee was closed in 2002.

Powercoal was now much better placed to be put up for sale, but first the Carr Labor Government needed to get the United Mineworkers Union (UMW) and its Powercoal members onside, or at least not strongly opposed to privatisation. In August 1991, a mass meeting of mine workers opposed the proposed sale of the mines to private owners and wanted written guarantees from the Government before the miners would consider supporting a sale. The issues of concern were job security, current pay rates, working conditions and leave entitlements, the sale of all the mines as a group, and retrenchment benefits.[537] The UMW understood the challenges facing the mines, not least the need to fund the development of the Mandalong underground mine which would replace the ageing Cooranbong mine. The ongoing maintenance of the mines and the development of Mandalong were estimated to require in excess of $350 million over the following few years. The UMW also understood that unless action was taken soon, there would be further mine closures.[538] Senior UMW official, Peter Murray, said that the NSW Government did not appear to be prepared to provide the funds needed, and that the realistic options for the group were for it to be transferred away from Pacific Power as a separate State-owned company or for it to be purchased by a private sector company which was prepared to provide the necessary capital expenditure. Privatisation in fact was the only option.

Another threat emerged in 2001 when the State Government said that if the mines were not sold they would require Powercoal to cease selling to the export market. Peter Murray said that a Government committee had written to the UMW advising that exporting coal in competition with the private sector was not a core business for a government body.[539] The mines were put up for sale and Centennial Coal was the successful bidder, paying $331 million. The sale was announced on 30 July 2002 and completed the following month.

The era of NSW Government mining had ended in 2002, or so it was thought. In 2010 the NSW Labor Government announced that it wanted to own and develop a mine between Mudgee and Dubbo called Cobbora. The plan was for the mine to sell coal to the Government-owned power generators at prices which reflected the costs of production, rather than at prices which would be determined by a tender process involving a number of bidders. The generators would benefit from the lower coal price, enhancing the price the Government expected to obtain from its planned sale of "gentrader" contracts. In 2009 the Government had authorised a joint venture involving three state-owned electricity generators to apply for an exploration licence for Cobbora, with the joint venture to seek tenders from the private sector to develop and operate the mine and supply coal to the generators over a 15-year period. The tender process saw only one complying tender submitted, with that tender specifying a high ex-mine price of coal, so high in fact that the price would have had a significant impact on the profitability of the generators. A review of the Cobbora project by Frontier Economics described the price in the tender as "exorbitant" and "dramatically higher than those faced by (national electricity market) competitors and the anticipated price when the process was commenced".[540] In August 2010 the Government decided not to accept the tender; instead it would now own and develop the Cobbora mine. However, the Government's plans for Cobbora did not survive when Labor lost the 2011 election, with the new Treasurer Mike Baird announcing in July 2013 that the Government had terminated a contract to develop the mine. Baird also announced that the Government was selling the Eraring power station to Origin Energy, which had already purchased the output of the station under the gentrader contracts entered into by the Labor Government; those contracts generated funds for the Government and were effectively a partial privatisation of the electricity system. Origin Energy had now agreed to the cancellation of a 17 year agreement under which coal would have been supplied from Cobbora, with that cancellation costing the Government $300 million. Baird said that the net cost to the Government of the new arrangements was $75 million recognising the potential return from the sale of Cobbora and the damages it may have had to pay Origin.[541] The Government also cancelled contracts requiring Cobbora to supply coal to Macquarie Generation and Delta Electricity. Baird claimed that the decision to cease development of Cobbora and the new deal with Origin meant that the state had "dodged a fiscal bullet" and that it had avoided a potential liability of $1.75 billion. Centennial Coal now had a commitment with Origin to supply 24.5 million tonnes over eight years.

The Tamberlin inquiry concluded that the Labor Government's decisions on Cobbora were justified based on the facts that it had at the time, and that "Whether each of the decisions will prove, when and if the mine is sold, to have been better than the alternatives with the benefit of hindsight is presently unknown. The time has not yet arrived when such a judgment can be made." Tamberlin also concluded that the major benefit to NSW from the Cobbora project was that coal supply agreements with the generators would have ensured that they had access to coal to allow them to operate, with the coal price unlikely to disturb their merit order in the national electricity market.[542] However, the Cobbora project would have involved significant potential liabilities for the Government and, when the Coalition Government decided not to proceed to develop the mine, the major costs to develop the mine were still to be incurred. As at

June 2011 the Tamberlin Inquiry report showed that the capitalised costs to date had totalled $130 million, including land purchases ($92 million), water licences ($5 million) and mine development ($31 million).[543] The Cobbora project may well be the last coal mine any government in Australia seeks to develop.

China and India become major coal importers and investors

When Chairman Mao died in 1976, China was a poor and backward country, still struggling to recover from the turmoil that he had created with his Great Leap Forward, Cultural Revolution and other disastrous initiatives. But by the end of 1978, Deng Xiaoping, at 74 years of age, had emerged as the most powerful man in China and in 1979 Deng's reforms began to open the Chinese economy to the world. China now commenced an era of economic development and social change arguably unprecedented in world history in terms of its scale and speed.

In 1980 China's GDP was almost US$190 billion, only 26% bigger than Australia's. By the year 2012 China's GDP had soared to over US$8000 billion, compared with Australia's GDP at just over US$1500 billion.[544] In real terms, China's economy grew by over 9% pa in the 1980s, by over 10% pa in the 1990s and by over 10% pa in the first decade on this century. Since 2010, the economy's growth rate has slowed significantly, but has since averaged over 6.5%. In the 1980s and 1990s, while its economy expanded by around 550%, much of China's growth drew on internal resources and, as the economy was growing from a relatively low base, there was not the major impact on the demand for mineral, energy and other resources from other countries that we have seen in more recent years. During the 1990s China's economy grew much faster than its imports, with the ratio of imports of goods and services to GDP falling from 20% in 1994 to only 16% in 1998. Imports then started to grow more rapidly, reaching 23% of GDP in 2002, and then from 2003 imports boomed, the ratio hitting 32% in 2005, before easing back to 30% in 2007. The global financial crisis had an impact in China and the ratio fell to 22% in 2009, but rose to 25% in 2012. But by the early 2000s, China had become one of the world's major economies, and in 2004 was around one sixth the size of the USA. Double digit growth in an economy of that size does make a difference to world commodity markets. And as the major western economies recovered from the "tech wreck" of 2001, when world economic growth slowed to only 2.4%, the impact of Chinese growth began to be felt.

China's coal industry is huge. In 1991 its output was around 1.1 billion tonnes, and by 1996 it was almost 1.4 billion tonnes. But as this growth led to the domestic market being oversupplied the Government took steps to reduce production by ordering the closure of illegal and unsafe small mines, reducing production from other mines, delaying expansions and new developments and promoting exports. These policies saw production levelling out for a few years, but by 2000, output had started to grow rapidly again. In 2012 total production was around 3.95 billion tonnes, or over 8 times the size of the Australian industry; by 2013 year production peaked at 3.95 billion tonnes. Production then declined to around 3.87 billion tonnes in 2014 and to 3.41 billion tonnes in 2016; it has since risen to 3.85 billion tonnes in 2019.[545]

In 2016 China announced plans to reduce the size of the coal industry, with major cutbacks in employment in both the coal mining and steel industries totalling 1.8 million jobs. China said it planned to cut annual coal production by around 500 Mt over 3 – 5 years and not approve any new mines over the following 3 years. This cut was increased to 800 Mt in the next Five Year Plan. The excess production capacity in the coal industry in April 2016 also saw China introducing restrictions on the number of days coal mines were permitted to operate, with the standard 330 days per year cut to only 276. These restrictions on operations were eased towards the end of the year with the 330 days standard reinstated. However, with the domestic industry still struggling, the Government began to approve new coal projects subject to certain conditions. In 2017 new capacity was allowed provided it was replacing older less efficient capacity, but on the basis of 1Mt of new efficient capacity replacing 1.2 Mt of old capacity, a policy which was tightened in 2018 to 2Mt of new replacing 1Mt of old.

The Chinese Government's efforts to reduce the capacity of the industry have focussed on smaller mines. Of the 6163 legal coal mines in 2015, over half (3497) were producing less than 300,000 tonnes per year. More than 2000 small mines have been closed in recent years, leaving around 2000 small mines still operating; consultants Wood Mackenzie believe that most of these mines will need to be closed or consolidated into larger mines if the country is to meet its target of 800 million tonnes of capacity.[546] Nevertheless new capacity is still being approved, with 340 Mt of projects commissioned in 2017 and around 58 Mt in the first half of 2018.[547] A massive industry will continue in China for many years, albeit one that has been streamlined and is far more efficient.

Chinese coal exports rose during the 1990s, reaching 32 million tonnes in 1997, and in 1998 the Government introduced subsidies to stimulate coal exports; exports rose to around 59 million tonnes in 2000. In the early 2000s, most energy industry forecasts had China continuing to be a major exporter, and did not foresee that it would become a much more significant importer. For example in 2002, the Australian Bureau of Agricultural and Resource Economics (ABARE) forecast that Chinese coal exports would continue to be a major competitor to Australia in the thermal coal market, with exports potentially in the range 60 to 100 million tonnes per year by 2010.[548] Rio Tinto had a similar view; in an internal paper in 2001, the company put total Chinese coal exports at 145 million tonnes by 2010.[549] Forecasts of China's total coal demand also missed the mark, with, for example, the US Department of Energy in 2002 forecasting demand in 2010 at around 1630 million tonnes.[550]. China's total coal production in 2010, which was roughly equivalent to its consumption, in fact grew to around 3400 million tonnes.

In the early 2000s, China's State Economic and Trade Commission actually set a target each year for coal exports, with coal considered an important source of foreign exchange earnings. That was to change in the following few years as China's internal demands for coal from its steel mills and power stations outgrew the capacity of its coal industry to supply. China's coal exports in fact declined after peaking in 2003, a year which saw thermal exports at around 81 million tonnes and coking exports at around 13 million tonnes. By 2009, thermal exports were down to 21.5 Mt and metallurgical exports down to 1.5 Mt.

The year when international coal and other commodity markets noticeably changed was 2004, and the changes caught coal producers in Australia and other countries by surprise. China, which had emerged in the 1990s as a modest importer of coal, started to become a major importer. In the 1990s the rapidly growing southern provinces, including Guangdong, were hungry for coal, particularly steaming coal for their power stations. The major coalfields in China in the north west, far from the coast, and with poor rail links to the ports, were unable to supply the south's needs. By 2009 China was importing 92 million tonnes of thermal coal; imports have continued to rise, reaching 241 Mt in 2019.

China's steel industry, which produced almost 1 billion tonnes of steel in 2019, was a relative minnow in the 1980s. Its growth in the intervening years has been nothing short of phenomenal, with production exceeding 100 million tonnes for the first time by 2001 and over 500 million tonnes by 2008, by which time it accounted for 38% of global production; by 2019 its share had grown to 53%. While domestic production provides the bulk of its metallurgical coal needs, the growth in China's steel industry has also seen it import increasing quantities of metallurgical coal, with imports hitting a record 75 million tonnes in 2019.

China's investments in Australian coal

China, with massive trade surpluses to invest overseas, has also become a significant investor in coal and other mining industries in Australia and other countries in recent years, with Yangzhou Coal Mining, Shenhua Energy and China BaoWu Steel Group Corporation the major investors to date. Yangzhou Coal Mining Company is a major Chinese coal mining company which is listed on the Hong Kong stock exchange. Its Australian subsidiary, Yancoal Limited, is listed on the ASX, although the great majority of its shares are held by Yanzhou Coal (around 65%) and Cinda International, a Chinese investment and trading company (around 17%). Yancoal owns some of Yanzhou's Australian coal assets, others being held directly by Yanzhou or held through another company.

The first significant investment by a Chinese company in the Australian coal industry came in 2004 when Yanzhou purchased the Southland mine near Cessnock in NSW and subsequently renamed it Austar. The Austar mine introduced the first longwall top caving technology to Australia, which is suited to mines with thick seams and allows a greater recovery rate of coal than conventional longwall mining. A much bigger investment came in 2009 when Yancoal purchased Felix Resources for $3.3 billion, gaining interests in the Moolarben mine near Mudgee in NSW which produces thermal coal for export, and the Yarrabee mine north east of Blackwater in the Bowen Basin which produces semi-anthracite PCI coal for the steel industry. In 2012 Yancoal merged with Gloucester Coal, a deal which brought the Stratford Duralie, Donaldson and Ashton mines into the group. Yancoal purchased an additional 30% of Ashton in 2011. Ashton is an underground mine located near Singleton in NSW and produces semi soft coking coal for export. The Stratford Duralie mines in the Gloucester Basin north west of Newcastle produce semi soft coking coal and thermal coal for export.

The 2011 year also saw Yanzhou the company purchasing 100% of Premier Coal in WA from Wesfarmers and Syntech Resources owner of the Cameby Downs mine. Premier Coal is managed by Yancoal and operates an open cut mine near Collie supplying local power stations. The Cameby Downs mine in the Surat Basin produces thermal coal for the export market and ships out of the Port of Brisbane. Yancoal also entered a joint venture with Peabody Energy to develop the Middlemount mine in the Bowen Basin which produces PCI and hard coking coal for export. Middlemount commenced full scale mining in 2011 and now produces around 4 million tonnes per year.

In January 2017, Yancoal agreed to buy the Coal & Allied assets from Rio Tinto for $US2.45 billion, including 67.6% of Hunter Valley Operations (HVO), 80% of Mt Thorley, 55.6% of Warkworth and 36.5% of Port Waratah Coal Services. In May 2017 it extended its offer to buy Mitsubishi's 32.4% share of Hunter Valley Operations for $US710 million. Yancoal was successful in out-bidding Glencore for the Coal & Allied assets, but subsequently agreed with Glencore to establish a joint venture for the HVO assets, with Yancoal holding 51% and Glencore 49%. Yancoal completed the purchase of Mitsubishi's almost 29% interest in the Warkworth joint venture, giving it ownership of around 84.5% of the joint venture and increasing its share of coal production from the integrated Mt Thorley Warkworth operations from around 64% to 83%.

In 2008 Shenhua Watermark Coal Pty Ltd paid $300 million dollars to the NSW Government for an exploration licence in the Gunnedah Basin. Shenhua Watermark is part of China Shenhua Energy Company Limited, a large coal producer which also has major interests in electricity generation, rail transport, ports and shipping. Shenhua Watermark has been pushing ahead with the project to seek approval for developing an open cut mine which would produce up to 10 million ROM tonnes per year and employ over 400 people once in production. The company has invested significant money in exploration of its lease areas on the Liverpool Plains and on its environmental studies. In July 2017, the NSW Government announced that it was reducing the area of the exploration licence granted to Shenhua to just under half the original area, refunding around $262 million of the $300 million paid by the company in 2008. Resources Minister Don Harwin said that the area excised from the licence encroached on fertile agricultural land. However, still being so close to the rich farming area, the Watermark project continues to be a controversial one and is meeting ongoing resistance from community and other groups.

BHP Billiton was granted an exploration licence in the same area in 2006, paying $100 million to the NSW Government, and was planning to develop the large Caroona underground mine. However, following environmental objections and changes to the licence conditions by the NSW Government (which prevented any mining under aquifers or the floodplain or open cut mining on the floodplain), the company agreed to hand back its licence and was paid $220 million by the Government. Premier Mike Baird said that the "After careful consideration the NSW

Government has determined that coal mining under these highly fertile black plain soils …poses too great a risk for the future of this food bowl and the underground water sources that support it."[551] BHP's then Minerals Australia president Mike Henry claimed that his company would have developed Caroona responsibly, but accepted the Government's decision and said that it appreciated its willingness to work with BHP "to agree an acceptable financial outcome for the cancellation of our exploration licence…The Caroona Coal Project was studied extensively and developed cautiously for almost 10 years. We carried out extensive planning to ensure there would be no mining under the black soil plains, consistent with the conditions contained in our Exploration Licence… It was also subject to extensive scientific research which showed the proposed project could have been developed in an environmentally sustainable manner."[552] It seems clear that the Caroona project became too difficult in terms of the community reaction and the politics for BHP; the company was reimbursed a reasonable sum and moved on.

China's BaoWu Steel is the second largest steel producer in the world and, through its subsidiary Aquila Resources, has a 50% interest in the Eagle Downs coking coal project in the Bowen Basin, the other 50% owned by South32. That project began with Aurizon holding a 15% interest but was put on ice in 2015, with South32 entering the picture in 2017. Substantial investment in the project was made prior to 2015 and it has the potential to produce 4.5 Mt per year once operational. Another coal project in Queensland, Bluff, is being planned by China Kingho Energy Group, a company involved in coal mining, processing of coal by-products, building materials and hotels. China Kingho purchased Carabella Resources in 2014, gaining ownership of the Bluff thermal coal project. Guangdong Rising Asset Management purchased Caledon Coal in 2011, with the Cook Colliery, a share in the Wiggins Island Coal Export Terminal and the Minyango project now Caledon's major assets. The new owners of Cook invested in a new longwall system in the mine together with upgraded mine infrastructure. Cook ceased production in March 2017, and Caledon and associated companies are in the hands of an administrator. The Taroborah coal project, near Emerald, is owned by Shenhuo International, part of the Henan Shenhuo Group Co Ltd, a large Chinese company involved in coal exploration, electricity generation and aluminium production.

CITIC is a major Chinese company listed on the Hong Kong stock exchange; it holds a 14% interest in the Coppabella and Moorvale projects in Queensland. CITIC was also a shareholder with Peabody Resources in the Olive Downs project in Queensland until 2016 when the project was sold to Pembroke Resources. In 2003 China Huaneng Group purchased half of OzGen's 54% interest in the Millmerran power station and coal mine and half of its 50% interest in the Callide C power station, both in Queensland. In 2011 China Huaneng bought 50% of InterGen. The Millmerran power station and coal mine are now owned by InterGen (which in turn is owned by the Ontario Teachers' Pension Plan and China Huaneng Group/ Guangdong Yudean Group). The Callide C power station is now jointly owned by CS Energy and InterGen.

Other Chinese interests in Queensland include the China Stone project which is owned by Mac Mines P/L, a subsidiary of Meijin Energy, a privately owned Chinese company. This project is now in doubt as the company has terminated its application for the necessary mining leases.

India now a major investor and importer

India is now the world's second largest coal producer and has become a major coal importer in recent years. India's rapid economic growth and difficulties in expanding local coal production quickly enough have led to shortages of coal, leading to some Indian companies looking to invest in coal mining in Australia and other countries. In 2018-19 India produced an estimated 730 million tonnes of black coal, 94% of which was thermal coal; it also produces around 47 million tonnes per year of lignite.[553] The coal industry is dominated by State-owned companies, with national Government owned Coal India Limited accounting for over 80% of production. Much of its black coal is of relatively poor quality, high in ash and with a low energy content. The ash content of Indian coal typically ranges between around 25% and 35%, with some coal having a content of up to 50%[554]; this compares with the ash content of typical Australian export thermal coal between around 8% and 16%, with Australian coal having gone through modern coal washing and preparation processes. This comparison shows the tremendous challenge faced by India to lift the quality of the coal which goes into its power stations, and the long term benefits of the use of higher quality imported coal where the design of the boilers of power stations is able to cater for imported coal.[555]

Adani is the most recognised Indian coal investor in Australia with its Carmichael mine in the Galilee Basin in Queensland. The Adani Group was founded in 1988 by Gautam Adani and over the years has developed into a major company with interests in electricity generation, gas, coal mining, ports and logistics. It claims to be the largest importer of coal into India and also owns coal mines in Indonesia. Adani made its first major purchase in Australia in 2010, buying large coal tenements in the north of the Galilee Basin held by Linc Energy. The deal involved an upfront payment of $500 million to Linc plus a royalty of $2 per tonne on coal produced in the first 20 years of production. The tenements are around 100 kilometres north of the projects proposed by GVK Hancock and around 160 kilometres north west of the town of Clermont. In 2011 Adani purchased a 99 year lease on the coal terminal at Abbot Point owned by the North Queensland Bulk Ports Corporation, a deal worth $1.8 billion. The terminal is used by a number of major exporters including BHP and Glencore. Adani's initial plan for the Carmichael mine complex was to develop six open cut mines and five underground mines which at full capacity would produce around 60 million tonnes per year of thermal coal for export to India and other markets. It has revised this plan and is now looking at a much more modest operation producing around 10 million tonnes per year.[556] The company has also changed its plans to build its own dedicated rail link to the port in favour of a shorter line which will connect with Aurizon's existing rail network near Moranbah. Adani's Carmichael mine has been mired in controversy and legal challenges by environment groups and Aboriginal groups, but finally won the development approvals necessary for it to proceed. The company's plans involve minimal washing of the coal, with a target ash content of the coal of 25%.[557]

Another Indian company, GVK Power & Infrastructure, paid around $1.3 billion in 2011 to buy a 79% stake in Hancock Coal's thermal coal assets in the Galilee Basin. The GVK group's operations include power generation, airports, transportation and hospitality. The Australian joint venture, GVK Hancock Coal, has plans to develop thermal coal mines for export, with its Alpha Coal Project planned to be a massive mine producing around 30 million tonnes per year. GVK Hancock also has plans to develop two other major projects in the Basin, Kevin's Corner and Alpha West, to construct a 500 kilometre rail line to Abbot Point in a joint venture with Aurizon, and to build a new coal terminal at Abbot Point.

Other Indian companies which have invested in Australian coal include Gujarat NRE and Lanco Infratech. Gujarat NRE Coke Ltd purchased the Bellpac No.1 mine around 3 kilometres north of Wollongong in 2004 and re-opened the mine in March 2005. The company ran into major financial difficulties, and Jindal Steel bought a controlling interest in 2013, changing its name to Wollongong Coal in 2014. Wollongong Coal now operates the Wongawilli and Russell Vale mines which produce coking coal for the export market. Griffin Coal, which operates an open cut mine in Collie, WA, was purchased in 2011 by Lanco Infratech, an Indian company involved in a range of industries including electricity generation and infrastructure projects.

Japan still a dominant overseas investor

From the time of the development of the Moura/ Kianga mines in which Mitsui was a major investor, and Mitsubishi's involvement in the Utah/ CQCA developments, Japanese companies have been major investors in the Australian coal industry. In recent years, investment from China and India has tended to generate the most media attention, but Japan has continued to be a major investor, with Mitsui and Mitsubishi leading players, together with a range of other companies including Nippon

Steel and Sumitomo (both now part of Nippon Steel & Sumitomo Metal Corporation), Itochu, Sojitz, JFE, Marubeni, Idemitsu, J-Power and Japan Coal Development. Japan's early involvement was of course related to coking coal, with its steel mills looking for alternative sources of supply at a time when the USA dominated the coking coal export trade to Japan. When Japan's power industry began to look to thermal coal from the mid to late 1970s, the power utilities and related companies also looked to invest in coal projects in Australia, and the major trading companies including Mitsubishi and Mitsui also became involved.

In 2001 BHP and Mitsubishi Development established the BHP Mitsubishi Alliance (BMA) which saw the two companies become 50/50 owners of the group of mines which included Blackwater, Goonyella, Norwich Park, Peak Downs, Saraji and Crinum. Mitsubishi's other investments have included interests in Ulan (sold to Glencore in 2019), and in Warkworth (sold to Yancoal in 2018). Mitsubishi had been involved in Warkworth since the joint venture was established in the 1970s, with the company holding a 15% stake in 1978 when the project won its first export contract to supply thermal coal to Japan. Mitsubishi was also a long term investor in the Ulan coal mine, initially taking a 10% interest in White Industries in 1976 and then acquiring a 40% interest in Ulan in 1978 from White Industries. White Industries subsequently sold its interest in Ulan to Exxon which in turn sold to Glencore. Another investment was the company's purchase of 40% of the Howick mine in the Hunter Valley from CRA in 1989. Mitsubishi was also an investor in the Camberwell mine in the Hunter Valley through Mitsubishi Materials. Following the sale of its interests in Ulan and Warkworth and the sale of its 31.4% interest in Clermont in Queensland (to Glencore and Sumitomo), Mitsubishi is no longer an investor in Australia's thermal coal mines.

Mitsui has a 49% share of the Dawson mining complex in Queensland which began as the Moura mine, and has been the only company continuously involved in Moura since it began. The company acquired a 10% interest in the Curragh project in the 1980s and following the BHP purchase of Utah's coal assets in Australia, Mitsui's investments then also included 13.3% of the Moura, Nebo and Riverside mines. Through its shareholding in BHP Mitsui Coal, it now has a 20% stake in BHP's Poitrel and South Walker Creek mines in the Bowen Basin. Mitsui's other interests include 30% of Anglo's CapCoal mines and 20% of the Kestrel mine in Queensland, and 10% of the Bengalla mine and 32.5% of the Liddell mine in NSW. It also continues to be heavily involved in the Japanese Australia coal trade, marketing to the major steel mills, power utilities and to other industrial users.

In 1989 Nippon Steel bought into the Warkworth joint venture, taking a 7.5% interest in the mine with which in 1980 it and Sumitomo had contracted to buy 500,000 tonnes per year of coal on behalf of the Japanese steel mills. Another major mine in which Nippon Steel is involved is Hail Creek which was a potential project in the 1970s, but with development not proceeding until 2001-2002; by the time of development Nippon Steel's interest was a modest 5.33%.

Sumitomo was an early investor in Australia, setting up its Australian office in Sydney in 1972 to look for trade and investment opportunities in resources and other sectors. In the 1970s it held a 10% interest in the Hail Creek project. In 1978 it bought 15% of Lithgow Valley colliery from Oakbridge; this was followed by its purchase in 1979 of 20% of Wallerawang Collieries also from Oakbridge, with Sumitomo Metal Industries taking 15% and Sumitomo Corporation 5%. During the 1990s Sumitomo was the owner and operator of the Wambo mine in the Hunter Valley which was later purchased by Excel Coal and is now owned by Peabody.

In 2010 Nippon Steel acquired Itochu's 10% interest in Anglo's Foxleigh PCI mine in Queensland. Nippon Steel merged with Sumitomo in 2012 to form Nippon Steel & Sumitomo Metal Corporation and is Japan's largest steel producer. The new company inherited Sumitomo's investments in Australia which included interests in several Glencore assets; these were purchased by Sumitomo and Itochu from Xstrata in 2003 and included 20% of the NCA joint venture, 20% of Oaky Creek and 25% of four coal projects (Rolleston, Wandoan, Red Rock and Pentland). Sumitomo's other investments included 5% of Baal Bone in

NSW and 25% of Clermont in Queensland. Mining ceased at Baal Bone in 2011 and the mine is now used by Glencore for training purposes. Sumitomo also bought 10% of the North Goonyella mine project in 1989. Sumitomo has also been a long term investor in Hail Creek and joined with Oakbridge to develop the Saxonvale-Bulga mine in 1993.

JFE was formed in 2002 from the merger between NKK Corporation and Kawasaki Steel and is Japan's second largest steel producer. JFE Steel and JFE Shoji Trade Corporation bought into two AMCI coking coal joint ventures in 2005 - Carborough Downs in Queensland and Glennies Creek in NSW. JFE also has small interests in the Oakbridge mines which are majority owned by Glencore. The Byerwen coking coal mine near Glenden in Queensland's Bowen Basin is being developed by QCoal and JFE Steel; JFE acquired a 20% interest in this project in 2009.

Itochu Corporation is another major Japanese trading company that is less well known to the general public but which has been involved in the Australian coal export trade since the 1970s; it was known as C Itoh until 1992. In 1994 it began to invest in the industry, buying into the Stratford project near Gloucester in NSW. It then formed Itochu Coal Resources Australia Pty Ltd to invest in MIM's NCA project, and now also has interests in the Oaky Creek mines (20%), the Rolleston mine (12.5%), the Wandoan project (12.5%), the Maules Creek mine (15%) and the Ravensworth North mine (10%).

Another large trading company, Sojitz Corporation, formed in 2004 with the merger of Nissho Iwai and Nichimen Corporation, has made some significant investments in Australia. Nissho Iwai's involvement in the coal industry dates back to 1974 when it negotiated the first export contract for steaming coal to Japan, a deal for the supply of 500,000 tonnes per year which was increased in 1979 to 2.5 Mt per year. Sojitz is unique in being the only Japanese trading company which owns and is responsible for the operation of a coal mine in this country – the Minerva mine in Queensland which produces thermal coal for export to Japan and Korea. Sojitz owns 96% of Minerva, with Korea Mining Promotion Company owning the other 4%. The company has also developed the Meteor Downs South mine in Queensland in partnership with U&D Coal; it also operates that mine. Sojitz's other investments have included the Moolarben mine in NSW, and Jellinbah, Lake Vermont and Moorvale mines in Queensland. The company also purchased the Crinum Gregory mine in Queensland from BHP, a deal which was announced in May 2018 and completed in March 2019. Sojitz was also a long term shareholder in NSW coal company Coal & Allied. Sojitz is another large Japanese company which states that it is moving away from thermal coal, announcing the sale of its 30% interest in an Indonesian mine in March 2019, although also saying it would continue to be the sales agent for the mine in Japan.[558]

Japanese investments in steaming coal projects in Australia have also been made by J-Power, formerly EPDC (Electric Power Development Corporation), and by Japan Coal Development (JCD). J-Power is a large Japanese company whose assets include seven coal fired power stations in Japan with a total capacity of 8.4 GW. One of its first investments was in the Blair Athol mine in Queensland in 1982. In 1980 the Federal Government rejected the application by CRA and ARCO to proceed with the development of the mine as it did not at that stage meet the Government's requirement for at least 50% local equity. With CRA and ARCO agreeing to increased local equity, in 1981, Blair Athol was approved with EPDC and Japan Coal Development (JCD) holding a combined 10% interest. J-Power has other investments in Australia including 15% of Clermont, 7.5% of Narrabri and 10% of Maules Creek. EPDC was also an investor in Idemitsu's Ensham mine in the 1990s.

JCD is owned by the major Japanese power utilities; its major investments in Australia include 3.5% of Glencore's Clermont mine in Queensland and a small interest in Port Waratah Coal Services, the operator of the Kooragang and Carrington export coal terminals in Newcastle. JCD and EPDC initially took a 7% interest in the new Kooragang Coal Terminal in 1982, along with a consortium of 20 other Japanese companies which took 3% (the consortium included Mitsubishi and Mitsui, other trading companies and industrial companies). KCL was merged into PWCS in 1990 with the Japanese shareholdings now in the enlarged PWCS.

Idemitsu is part of the Idemitsu Kosan group, a major Japanese company involved in petroleum refining, petrochemicals, coal and mineral resource development and a wide range of other industries. Idemitsu owns 85% of the Ensham mine in Queensland and the Muswellbrook and Boggabri mines in NSW. Other major Japanese companies which have interests in Australia include Marubeni which has investments in Hail Creek, Coppabella/ Moorvale, Jellinbah East and Lake Vermont in Queensland. Marubeni was also a shareholder in Glencore's West Wallsend mine in NSW which ceased mining in 2016.

Investment in the Australian coal industry has also come from a number of other countries, notably the USA, South Korea and Thailand (through Banpu's ownership of Centennial Coal). South Korean investors include its major steel company POSCO and power utility KEPCO. POSCO began to source coal from Australia in the 1970s and its investments now include Mt Thorley, the Hume Coal project in the NSW southern highlands, Ravensworth Underground and Carborough Downs in Queensland. KEPCO is a shareholder in Moolarben and owns the Bylong Coal project north of Mudgee. Korean investment is also significant in several of Centennial's mines, with the SK group having interests in Angus Place (50%), Clarence (15%), Charbon (5%) and Springvale (25%). Korea Resources Corporation has an 82.25% interest in the Wyong Areas project in NSW, and a 25% interest in Springvale. SK Energy also has an interest in the Wyong Areas project.

Brazilian iron ore giant Vale, the largest iron ore miner in the world, is another major company which has had significant investments in Australia, including Carborough Downs and Isaac Plains, a 50/50 joint venture with Sumitomo. Carborough Downs was sold to AMCI subsidiary Fitzroy Australia Resources in 2016. Isaac Plains was purchased by Stanmore Coal in 2015. Vale was also the major shareholder in the Integra mine in the Hunter Valley which was placed on care and maintenance in 2014. Integra's other shareholders include Nippon Steel, JFE Shoji, JFE Steel, Posco, Toyota Tsusho Australia and Chubu Electric Power Company. Integra was purchased by Glencore in 2015.

The USA, UK and other European countries have also been a major source of investment in the industry, with Peabody currently the leading US coal miner operating in Australia. A number of other US companies have been major investors but are no longer involved; these include Utah International which sold out to BHP in 1984, Cyprus Mining which was part of the Phelps Dodge group, ARCO and Exxon. UK investment includes Anglo American, which is listed on the London Stock Exchange, and BHP which is dual listed (London and ASX). The Swiss-based Glencore, which is listed on the London stock exchange, is one of the leading investors in the Australian industry.

Coal pricing is no longer controversial

As we saw in earlier chapters, the export coal market and the issues of prices paid by the Japanese steel mills and power utilities, foreign investment in coal mining and the role of government in regulating coal prices and coal production have been extremely contentious issues in Australia. The CFMEU and its predecessor the Miners' Federation pushed extremely hard over many years for the creation of a national government authority to control the marketing of export coal, and at times for such an authority to also control the rate of production. That push effectively ended in the 1990s with the release of the Taylor review of the industry, although concerns about coal prices and over production continued during the 1990s.

With the growth of the Chinese economy and its impact on coal demand from around 2004, and with the growth in other major coal markets, particularly India, the dominance of the Japanese steel companies and power utilities in the metallurgical and steaming coal markets has declined significantly in recent years, and with that decline has come an acceptance that the market is operating effectively. The Japanese, however, still play a major role in the setting of prices in the thermal coal market, with one of the major power utilities each year negotiating what is effectively a benchmark price with Glencore, the major exporter, with that price applicable to coal sold under longer term contracts to the JPU. That price for the next Japanese financial year acts as the benchmark for contract prices for the Asian coal market, although significant coal continues to be sold on a spot basis, with the Newcastle spot price for 6000 kcal coal one important benchmark for spot sales. Japan's is now the number three importer of thermal coal behind China and India, and only accounts for around 12% of global thermal coal trade; it no longer has the market power that it wielded so effectively in the period from the 1960s to the 1990s.

The price formation process for metallurgical coal has evolved over the last decade to one which is quite different to that for thermal coal. In 2010, BHP as the dominant supplier of metallurgical coal to Japan pressured the Japanese steel mills to accept quarterly contract prices rather than annual prices. The BHP push for quarterly prices was reported as coming after China agreed to this change with one supplier. In March 2010, BHP struck what was heralded as a landmark deal, when it negotiated with JFE Steel, then the number two steel producer in Japan, to supply coal on an agreed quarterly price. The price agreed was believed to more closely reflect the prevailing spot price. That new contract was the first time a Japanese steel producer had agreed to move away from annually negotiated prices, and was seen as the beginning of the end of annual prices.

While the JSM were not happy with the move away from annual prices, by 2011 BHP appeared to have won over its customers, or at least had twisted their arms to agree, as it reported that in the year to June most of its contracts were "annual or long term volume contracts with prices largely negotiated on a quarterly or monthly basis."[559] The JSM's concerns had included the fact that sales to their local automobile manufacturer customers, one of their major markets, were based on annual contracts. While contracts then continued to be based on annual tonnages, the quarterly price system evolved to be based on price indexes and spot sales, although negotiations were still necessary to take account of differences in the quality of different coals. The quarterly pricing system, however, began to break down around 2016, as constraints on supply sent spot prices soaring. When spot prices are rising rapidly, exporters relying on fixed quarterly prices are disadvantaged; conversely, when spot prices are falling, the advantage tends to be with the exporter holding those fixed price contracts. And for the Japanese customers, the reverse is true.

Those customers began to seriously question being tied to fixed quarterly prices in 2016 and 2017, with Nippon Steel in May 2017 reported as trialling a new system which would see prices based on average spot prices over the previous three months. By 2018, BHP publicly acknowledged that the quarterly contract pricing system for coking coal was "dead and buried"[560]. The reality was that since around mid-2017, for the markets in northeast Asia, while quarterly prices continued to be applied to many sales, the basis was now spot prices or indexes.

Queensland privatises QR – but not without a fight

As we have outlined, major changes in the ownership of coal mines in Australia have occurred since the start of the 2000s, with major players exiting the industry, new players emerging and overseas investors continuing to play a changing, but dominant, role. The coal rail freight systems in both Queensland and NSW have also seen fundamental changes in this period, with the NSW Government privatising its freight rail operator FreightCorp in 2002 and agreeing to a long term lease of its Hunter Valley coal rail system to the Australian Rail Track Corporation in 2004. Queensland also moved to privatise the Central Queensland coal network of Queensland Rail (QR), but that took a little longer and involved a major battle with the Queensland producers.

In June 2009 Premier Anna Bligh announced a program of sales of government assets, including the coal rail network, one of the jewels in the crown of the Queensland Government. Other assets to be sold included the Abbot Point Coal Terminal, the Port of Brisbane Corporation, Queensland Motorways Corporation and Forest Plantations Queensland. This privatisation program would prove to be so unpopular with Queensland voters that the Bligh Government would lose office in the 2012 election. The Queensland Government had invested huge amounts in coal related infrastructure since the industry began to expand in the 1960s to supply the booming coal markets. QR was a major beneficiary of this growth, and as we have seen, its rail assets were the basis for the Government earning lucrative de facto tax revenues from the 1970s to the 1990s.

In parallel with the industry's export tonnages, QR's coal freight tonnages grew dramatically to over 150 million tonnes by 2004-05 and to around 176 million tonnes by 2006-07. However, the booming demand for coal from around 2004 saw the industry and infrastructure providers struggling to meet the demand. Coal producers were running into capacity constraints in key rail and port systems that are crucial to getting the coal to the customers overseas. The producers were not happy, as QR admitted in its annual report for 2006-07: "It is also clear our coal customers have been concerned about our performance in Central Queensland, and we need to do more to meet their expectations."[561] QR said that "The pressures associated with the coal boom resulted in the appointment of an independent expert, Mr Stephen O'Donnell, to review the performance of the Goonyella coal supply chain. QR welcomed his report not only because it will result in better outcomes but also because, for the first time, it provided a clear and independent picture of the complexity of this issue." O'Donnell was a former CEO of Pacific National and was reported to have estimated that the coal industry had lost export sales of $900 million over a period of 11 months (July 2006 to May 2007) as a result of lack of rail capacity, and also incurred demurrage of $300 million due to rail and port capacity constraints.[562] O'Donnell made a range of recommendations, including that QR urgently invest in new rolling stock.

QR's annual report for 2006-07 claimed that the organisation had made good progress with the review's recommendations, saying "We're working hard to move more coal for our customers and deliver long-term investment programs in new infrastructure. We are more than halfway through a $1.4 billion program for 120 new or upgraded locomotives and 1,500 new coal wagons. We're also well advanced in procuring additional locomotives and wagons. A business improvement program has also been established to deliver an extra five million tonnes per annum out of the Goonyella system. A similar process has been implemented in the Blackwater system to deliver more tonnages for customers."[563]

The capacity of the rail system was not the only challenge facing QR. By 2009, the threat of competition for the rail freight contracts with Queensland coal producers had materialised. The other major rail freight company, Pacific National (a division of Asciano), the company which acquired the above rail assets in the Hunter Valley of the NSW Government's FreightCorp, secured its first freight contract in Queensland in 2009; Pacific National won a contract to haul 5.75 million tonnes of coal per year from Anglo's Moranbah North mine. In 2010 the company also won a contract to haul almost 11 million tonnes per year from Anglo's German Creek mine. Competition for coal freight contracts, among the most lucrative in the country, was heating up. QR would now be less of a risk for Queensland taxpayers if it was privatised, rather than remaining as a government owned corporation. And with the Queensland Government keen to reduce its high debt levels, QR was an attractive entity to privatise.

The decision by the Bligh Government to sell QR's coal assets quickly stimulated interest from the major Queensland coal producers, with BHP lodging a statement of intent with the Government in December 2009, registering its interest in buying the coal rail network. Queensland Treasurer Andrew Fraser quickly rejected the BHP bid, saying that the Government had no intention of selling the rail tracks to one company, and that it had placed a 15% limit on ownership of the new company that would be floated on the ASX later in 2010.[564] Shortly after BHP was rebuffed, a group of coal producers, known as the Queensland Coal Industry Rail Group (QCIRG), met with senior Queensland Treasury officials to confirm that the industry intended to submit a proposal to buy the network.

The Queensland Government's plan to sell both the track and the above-rail assets (locomotives, wagons etc) as one parcel incurred the ire of the producers. Michael Roche, QRC's CEO, said that the industry had to change the Government's mind: "It is a must for the industry that this integrated business, initial public offering, does not proceed. We believe the Government has not properly assessed the risk to competition for coal haulage, nor have they properly assessed the risk to timely investment by this new company which would have monopoly control over the tracks."[565] The coal producers' proposal at that stage, once the Group had purchased all the assets, was to lease the track infrastructure to the Australian Rail Track Corporation, the Federal Government body which had earlier purchased the track infrastructure in the NSW Hunter Valley. The producers were wary of QR being sold to a private operator which, they believed, would not have the necessary incentive to invest in the track and associated infrastructure, but rather find it more convenient to force the coal producers to pay higher rates by restricting the capacity of the infrastructure.[566]

The Federal Government and the ACCC were also reported to have been supportive of the industry's bid to purchase the assets and lease the track infrastructure. The ACCC was reported as having warned that it "had great difficulty in the privatisation of a natural monopoly like the 'under rail' rail network and that those concerns were only exacerbated by the idea that a service provider would own that network."[567] Asciano was also worried about the QR privatisation, and in a submission to the ACCC contended that the sale of QR as an integrated track infrastructure and freight service provider would be detrimental as it would to "sharpen incentives to discriminate against above-rail competitors". Asciano argued that up to that time QR had had "purely commercial motives" but had "acted partly as an instrument of government policy. A privatised QR would have a fiduciary duty to maximise its value for shareholders by, amongst other things, minimising above-rail competition provided it remained within the law."[568]

The Queensland Government was determined to float the new company, believing that that option would give the state the best return. The coal producers employed Nick Greiner, the ex NSW Premier, to assist them to win the right to buy the assets. Greiner was also the chairman of Citibank Australia which was given responsibility for managing the financial aspects of the Group's bid. In March 2010 the producers were reported to have offered a substantial premium to the $3 billion the Government was expecting to receive from the sale. The approach from the producers was also dismissed by Premier Bligh and Treasurer Fraser, with Fraser saying "What has arrived today is not an offer but a letter to sit down and talk about a different process… But there is no offer on the table from the coal companies and, if you look to the industry's performance over Dalrymple Bay where they talked big about an offer for what was a troubled asset and that offer never eventuated, then you know it is London to a brick that there will not be a serious offer…When, if, there is an unconditional offer, then we will have to look at it. Right now, there is no offer…But what we have right now is what we think is the best model for QR National, for the Queensland economy and the workers of QR. What the coal companies have is a media strategy, not a strategy for QR. If the nature of the bidders was serious, they would have arrived at this position a lot earlier than this."[569]

One of the smaller coal producers, QCoal, withdrew from the consortium in early 2010, with its CEO, Chris Wallin, critical of the proposal being put forward by the consortium. Wallin said that QCoal would have had only a 1% stake in the consortium's company and noted that: "If you have ever owned 1 per cent of the shares in the company you will find out you do not get much say."[570] The withdrawal of a small producer was not a major setback for QCIRG, but with a determined Queensland Government, the industry was facing a major challenge to get its hands on the rail assets. But the Federal Government's support was encouraging, with Resources Minister Martin Ferguson saying that the Queensland Government's proposed model to float QR was "a recipe for disaster." Ferguson was not the sole Federal Government Minister voicing concern; Finance Minister Lindsay Tanner and Infrastructure Minister Anthony Albanese were also critical of the Queensland Government proposal. But Fraser countered, saying that in 2007 Ferguson had praised the efficiency of the vertically integrated WA iron ore supply chain and he was therefore "very perplexed about why Mr Ferguson now has a different view about our proposal for a vertically integrated QR."[571]

Of course, the WA iron ore industry, with BHP and Rio Tinto running their own rail systems from their mines to their own port facilities, was arguably not comparable to the QR situation. QR was a government-owned corporation, with ownership of the track, holding the great bulk of the freight contracts to haul the coal, and servicing a large number of mines owned by different coal companies, and the Government was planning to transfer ownership of that system to private owners.

The QCIRG pushed ahead and by May 2010 the Group was able to announce that it had secured funding for their bid. The 11 companies which had signed equity subscription agreements were Anglo American Metallurgical Coal, BHP Billiton, Ensham Resources, Felix Resources, Jellinbah Resources, Macarthur Coal, Peabody Energy, Rio Tinto Coal, Vale Australia, Wesfarmers Resources and Xstrata Coal. The other two members of the consortium, New Hope and Aquila Resources, were supporting members who may have been prepared to contribute equity at a later date.[572] Near the end of the month the consortium offered $4.85 billion, with the offer only for the track and associated infrastructure; it did not include QR's locomotives and wagons. Greiner said that the Government could sell QR's trains and other parts of the business, and achieve an overall $7 billion for the whole business. According to Greiner, the consortium's proposal was a no-brainer: "We think Blind Freddy can see a more attractive option for the government in terms of its asset sale program," he told reporters in Brisbane. "We think it's transparently obvious that if you put our $4.85 billion and the value of above-rail sold whenever they best think... they get a better outcome in terms of cash now and cash at the end of the day under this proposal than they do under the IPO."

Lance Hockridge, the CEO of QR, who was to become CEO of the new company, was dismissive of the industry group's offer. He claimed that the offer was not in the interests of the future expansion of the coal rail network and he also was dismissive of the call by the producers to gain access to QR's books to undertake detailed financial analysis of its operations. With QR having spent around $5 billion on the rail network in the preceding few years, there was also scepticism about the adequacy of the QCIRG plan to spend only $2 billion on the network in the following few years. [573] Fraser, however, said that the Government would consider the offer, but downplayed its chances of success: "We think that we've taken a decision for the structure of QR that's best for QR and best for the Queensland economy for the long-term future, but we've a responsibility to look at the offer and we will."[574] The Government agreed to open QR's books for the consortium to carry out due diligence and in August the Group submitted a higher bid, reported at over $5.2 billion, and gave the Government a deadline to take the bid seriously and negotiate. Failing that, the Group threatened to walk away.

The earlier bid had been insufficient to induce the Government to change its plans for a float of 75% of QR.[575] However, the new bid was soon to fall apart. "What a humiliation" were the first words in an article by Matthew Stevens in The Australian on 10 September. The QCIRG bid had been effectively sunk by one of the major companies. Anglo Coal had advised the other members of the Group that it was not prepared to contribute 13.2% of the capital funding for the bid (based on its share of the total tonnage on the rail system), but only 1%. Anglo would have contributed around $460 million of the $3.5 billion in equity for the bid; instead its contribution would fall to around $35 million. Stephens wrote that Anglo had emailed the other members of the QCIRG with its decision, "a decision that apparently snowballed through the consortium with two other of the tier-two players saying they would have to review their business case. Anglo's decision stunned the other QCIRG members and not just because it was delivered by email."[576] Stephens said that the companies claimed that the Government's tight deadlines for submitting the bid were a factor in the failure, together with the Government's demand that the bid be unconditional. Anglo's position was that, given the timing of the bid and the fact that due diligence on QR was delayed by the Government, it was not able to submit a firm proposal to its investment committee.

Keith De Lacy, chairman of Macarthur Coal and one of the members of the Group, later admitted that the consortium's proposal had some fundamental problems: "It wasn't efficient – you had half a dozen mining companies, all with different approaches…So it was fraught with its own problems…" However De Lacy also said that he thought that despite these problems the consortium would be successful. De Lacy also complimented the senior management of QR who worked to support the Government's proposal, admitting he thought that he had "underestimated the commercial capability of the QR team….I take my hat off to them."[577] Nick Greiner was reported to have claimed to QR CEO Lance Hockridge at one stage that "we will destroy your IPO".[578] Greiner had obviously been confident in his ability to lead the QCIRG in waging a successful campaign to buy the assets, but the QR management and the Queensland Government merged as winners, with the vertically integrated company now free to transition to a listed company.

The Queensland Government now lost no time and QR was floated on the ASX on 22 November 2010. The Government sold around 66% of its stake in QR and received $4.6 billion; it later sold down its stake, and in 2016 had reduced its equity to 2.68% of the company. QR became Aurizon on 1 December 2012; it continues to be the dominant rail freight provider for the Queensland coal industry, carrying 152 million tonnes in 2018.19, with 62 Mt carried in south east Queensland and NSW.[579] Aurizon also continues to own and maintain the central Queensland coal rail infrastructure.

NSW rail privatisation goes smoothly

In contrast to the drama surrounding the privatisation of Queensland Rail, the exit of the NSW Government from the heavy freight business was relatively smooth and uneventful. While the NSW Labor Government under Premier Bob Carr and Treasurer Michael Egan had been unsuccessful in winning support from the Party and key unions for the privatisation of the electricity generators and retailers, there was no significant opposition for the sale of the Government's freight haulage company, FreightCorp, or the rail system in the Hunter Valley which was largely dedicated to coal freight. FreightCorp was formed in the 1990s, one of the four units which were created out of the old State

Rail Authority. FreightCorp's business was highly dependent on its coal haulage; in fact it was believed that its Hunter Valley coal business was its only profitable business. With the tax element in coal freight rates having disappeared, it made eminent sense for the Government to sell off that business to the private sector which could be expected to make the operation even more efficient and which could provide the necessary capital for ongoing expansion of the rolling stock.

Michael Egan announced in 2000 that the NSW Government intended to privatise FreightCorp as part of a coordinated sale process also involving the National Rail Corporation (NRC). NRC had been established in 1992 following an agreement between the Federal, NSW and Victorian Governments and progressively took over responsibility for the interstate rail network between Brisbane, Sydney, Melbourne, Adelaide, Perth and Alice Springs. In January 2002, following a competitive tender process, the Federal, NSW and Victorian Governments announced the sale of FreightCorp and National Rail Corporation (NRC) to National Rail Consortium Pty Ltd for around $1.2 billion. National Rail Consortium was jointly owned by Toll Holdings and Lang Corporation, two public transport companies, which the Governments said would "bring significant large-scale national transport and logistics experience to the combined FreightCorp/NRC operation, together with additional financial and managerial capability." The benefits of the sale were expected to include "a significant injection of private sector expertise into the rail industry; a vigorous approach to increasing the rail share of the national freight task; (and) the integration of rail into the existing regional freight services operated by Toll and Lang."[580]

Following the sale, Pacific National Pty Ltd was established by Toll and Lang on 1 July 2002 to be the coal haulage operator in NSW, responsible at that time around 93% of the tonnage being hauled to the port of Newcastle and to Port Kembla. The sale of FreightCorp was well timed, as Pacific National's first major competitor for its NSW business emerged in 2004 when Queensland Rail won a 10 Mt per year contract to haul coal from BHP's Mt Arthur North mine. Pacific National was part of Asciano Limited when Asciano was demerged from

Toll Holdings in 2007; in 2016 it was purchased by a consortium of Brookfield Infrastructure Partners and other overseas investors. Asciano and QR have been actively competing for business in both Queensland and NSW and have made significant inroads into each other's home State. The other major rail haulage company is US operator Genesee Wyoming, which purchased Glencore's G'Rail business in 2016. G'Rail is the other major competitor in the NSW Hunter Valley, hauling around 40 million tonnes of Glencore coal per year. The sale came with a 20 year haulage contract for Glencore's coal.

In 2004, under a 60 year lease with the NSW Government, the Federal Government's Australian Rail Track Corporation (ARTC) acquired most of the Hunter Valley rail network and the NSW interstate mainline track. The ARTC now had responsibility for management and ongoing investment in these systems. ARTC had been established in 1998 following an agreement between the Federal Government and the States (not including Tasmania) for the ARTC to provide "one stop shop" access to the interstate rail network. In its 2018.19 financial year ARTC's Hunter Valley network carried 175 Mt of coal, of which 161 Mt was destined for the port of Newcastle. The growth of the system in the Hunter Valley is clear from the fact that total exports through the port in 2005-06 (just after the ARTC assumed control of the network) were around 80 million tonnes. Major investments over the intervening years by ARTC, Asciano, QR and G'Rail/ Genesee & Wyoming in rolling stock (as well as the investments in port capacity by PWCS and NCIG) have made that surge in exports possible.

Carbon taxes and emissions trading

As we saw in chapter 3, policy changes in the 1990s had major impacts on the coal industry, with the end of the Queensland Coal Board, the Joint Coal Board and the Coal Industry Tribunal among the most significant. The 1996-1997 industrial relations legislation, the end of export controls on coal, the end of the coal export duty, and the new national competition initiatives also had major impacts on the industry. Governments took a back seat in the first few years of the 2000s which were dominated by company mergers, takeovers and huge upheavals in export markets as Chinese growth impacted commodities markets. But with the election of the Rudd Labor Government in 2007, the coal industry became one of the industries which were the target of new Labor government policies; an emissions trading scheme and two mining profits taxes would have major implications for the coal industry.

When Labor won the 2007 election, one of Kevin Rudd's first acts as Prime Minister was to ratify the Kyoto Protocol which committed signatories to action to reduce emissions by 2012. Rudd then released a white paper in 2008 which proposed the introduction of an emissions trading scheme, this scheme gaining the support of Opposition leader Malcolm Turnbull and creating a fierce backlash from many of Turnbull's colleagues in the Liberal and National Parties. The ETS was to prove the undoing of Turnbull's leadership of the Liberal Party, when he lost support of his colleagues and was replaced by Tony Abbott in December 2009, just days before the fateful climate change conference in Copenhagen. That conference proved to be a watershed for policy in Australia. Rudd had gone to Denmark confident in obtaining an agreement on emission reductions, but returned home devastated that the conference was seen as a failure. China, with support from India, had played a clever game which effectively resulted in a non-outcome. Rudd had been damaged by the result, and would go on to lose the Prime Ministership to Julia Gillard in June 2010. After its leadership change, the Liberal Party was now firmly set against an ETS, a secret ballot held immediately after Abbott won the leadership ballot resulting in 54 votes rejecting an ETS, and only 29 in favour. As the new Opposition leader, Abbott now had a strong foundation to attack the Government on its policy, a position which was reinforced by the failure of Copenhagen. After Copenhagen, many Australians were more sceptical of their Government committing the country to potentially costly policies in the absence of strong international action. As Paul Kelly noted: "The viability of Australia's ETS depended on the extent to which other nations also priced carbon and traded permits. The global talks were about deals, trade-offs and compromise. Developing nations refused binding targets, making clear that any serious action they took required a financial bribe from rich nations. Despite Obama's efforts, the US Congress would not legislate 'cap and trade' without, at the least, binding commitments from China."[581] The agreement that did emerge from Copenhagen did not see the developing countries sign up to any binding commitments to reduce emissions, although Penny Wong, then Australia's environment minister, later stated that she believed that without Australia there would not have been the result that did emerge. Howard Bamsey, Australia's special envoy on climate change, said that there would not have been a Copenhagen accord without Rudd.[582]

In the August 2010 election Labor failed to win a majority of seats but was able to retain power with the support of the Greens and two independents. Prime Minister Julia Gillard then did a deal with the Greens in 2011 to introduce a carbon tax, albeit a temporary one. Gillard released her *Securing a clean energy future* report on July 10, with the centrepiece of the package of measures being the carbon tax which would be in place for three years before a transition to an ETS. Under the ETS there would be a variable price for carbon as reflected in the price of tradable emission permits. The carbon tax was to start from July 2012 at $23 per tonne of carbon or equivalent, and would apply to the 500 largest emitters, or to use the Government's term, polluters; the tax would rise by around 5% a year until 2015 when the ETS would take over.

The Government's package was based on the Treasury's analysis, also released in July 2011, *Strong Growth, Low Pollution – Modelling a Carbon Price.*[583] A critical assumption of the economic modelling by Treasury was that the carbon price in Australia would commence at $23 per tonne in 2012-13, rising to a world price of $29 per tonne in 2015-16. That assumption proved to be wide of the mark, as the European price, to which the Government later committed Australia to be tied, was actually on the way down. Australia's ETS was set on a path to involve a much higher carbon price than was in existence in other countries, thereby imposing higher costs on business. And with developing countries in many cases Australia's major competitors in terms of trade and investment, the new carbon regime was putting business at a competitive disadvantage. Carbon pollution and carbon polluters now became part of the commonly used terms by those supporting the Government's policies and by those keen to denigrate industries which used significant quantities of fossil fuels or which produced those fuels.

This carbon tax/ ETS policy gave Opposition leader Tony Abbott the ammunition he needed to fight Labor all the way to the 2013 election. Abbott's campaign to defeat the carbon tax took him to a range of industrial sites in the lead up to the July 10 announcement, and on the day after the announcement he visited Peabody's Wambo mine in the Hunter Valley with the media in tow. Abbott stressed the importance of the coal industry to Australia and warned of the damage the carbon tax would cause it.[584] He also thanked Peabody for the opportunity to visit the mine. However, on the same day, in a piece of exquisite timing Peabody (already holding 19.9% equity in Macarthur Coal), together with ArcelorMittal, one of the world's largest steel producers, announced a $4.7 billion takeover bid for Queensland coal producer Macarthur Coal. That announcement delighted the Government and Gillard seized on the Peabody bid, saying that "We are seeing the biggest takeover bid in Australian history for a coal company" adding that "You couldn't get a better indication that business people see a good future in coal mining in this country. There's more certainty now than there was before Sunday."[585] The reality was that Peabody saw Macarthur as a strategic asset, unlikely to be badly impacted by the carbon tax. Peabody had had Macarthur in its sights for some time, and was spurned by the company when it made an earlier bid. Macarthur was a major producer of PCI coal used in steelmaking and accounted for a significant share of Australia's PCI exports and a large percentage of internationally traded PCI coal. Peabody had paid a high price for Macarthur Coal, however, its takeover of the company at the time of such a critical policy announcement by the Government, did little for the campaign by the coal industry to defend the industry from the new tax and the ETS.

Coal mines were now exposed to a tax on the fugitive methane emissions which, based on the carbon price of $23 per tonne, the

Government said would cost a gassy mine an average of $7.40 per tonne of coal, with non-gassy mines predicted to see an average cost of $1.40 per tonne of coal. The gassiest mines were predicted to face a cost of around $25 per tonne of coal.[586] The Government also announced a Coal Sector Jobs Package with a budget of around $1.3 billion over 6 years. The package aimed at assisting the mines with the highest methane emissions, but specifically excluded any mine expansions or new mines which might be developed, with the Government saying that the carbon tax would be an incentive for expanded coal production to come from mines with lower methane emissions. Assistance under the package could be used to purchase carbon credits to offset carbon price liabilities, to undertake activities to reduce fugitive emissions, or for other activities to support the ongoing mining operations.[587] The Government was intent on discouraging any new gassy mines and its policy also implied little, if any, investment in new coal-fired electricity generation.

The coal industry had become an even more prominent target but the industry took up the fight to the Federal Government through the Australian Coal Association (ACA). In June 2011 the ACA released the results of a study by ACIL Tasman which forecast mine closures and job losses of over 4000 in the industry as a result of the carbon tax. Climate Change Minister Greg Combet hit back at the ACA, criticising the study for its assumption on carbon prices rising to $55 per tonne by 2020 and for not taking into account the Government's package of measures to assist the coal industry. Combet assured the industry that it had a bright future under a carbon price, pointing to Treasury modelling which showed the industry's production more than doubling by 2050.[588] Combet also pointed to the $4.7 billion bid one week earlier by Peabody for Macarthur Coal, which he said was "a clear vote of confidence in the future of the industry with a carbon price." June also saw the release of a report by the Centre for International Economics, commissioned by the ACA, which found that no other major coal exporting country imposed a tax on fugitive emissions from coal mines, the report also noting that the European Union did not include such emissions in its emissions trading scheme.[589]

The ACA was particularly critical of the Government's scheme, with Executive Director Ralph Hillman saying that the legislation contained a "poison pill" that would forever exclude coal mining from being treated the same way as every other trade exposed industry. Hillman pointed out that the legislation to enact the scheme specifically excluded coal mining from the definition of trade exposed industries, despite meeting the criteria.[590] In October the ACA released the results of further research by ACIL Tasman which found that around 27% of employment in coal mining projects would be under threat with a carbon tax.[591] ACA chairman John Pegler said that international competition for investment in new coalmine projects was intense and any increase in costs would "very quickly render them uncompetitive and drive this investment offshore to our competitor countries". Pegler said the analysis had also examined the effect of the government's coal job package and found it would have little impact, delaying the closure of four mines by one year.

The Rudd and Gillard super profits taxes

International markets for minerals tend to run in cycles, with relatively short periods of strong demand and high prices, and longer periods of weaker demand and lower prices. But when prices for commodities including gold, coal and iron ore are high and mining companies are earning what are seen as excessive profits, politicians tend to see the industry as a target from which they can extract additional revenue in the form of royalties, super profits taxes or high charges for transport or other services which are supplied by government businesses. In the mid-1980s and the 1990s, the Australian mining industry was generally not a target for governments seeking to extract greater returns from our minerals. The reason was simple — mining profits were generally poor to modest (and negative in the case of some companies), and there was little or no super profit element to catch the eye of money hungry treasury officials or politicians.

When the world economy started to pick up speed in the early 2000s, the fortunes of the minerals industry also started to change. Following the Asian financial crisis of the late 1990s and the 'tech wreck' in the early 2000s, when the bubble in the prices in Internet and technology stocks burst, global growth was poor. Global GDP growth in 2000 had been strong at 4.3%, but the tech wreck hit the US economy hard and global GDP rose by only 1.9% in 2001. Growth was only marginally better in 2002 at 2.15%, but picked up in 2003 to 2.9%, and it was in 2004 that growth really began to surge, with global GDP up by 4.5%.

The Chinese economy, which began to open up to the world in 1978, had been growing at an average rate of around 10% a year for two decades. By 2000, the Chinese economy had grown to account for almost 4.5% of global GDP. By 2004, China accounted for around 5.7% of global GDP, and its phenomenal growth in the next few years would see this figure rise to almost 8% in 2008, the year which the global financial crisis (GFC) commenced. It was around 2004 when we started to see commodity prices begin to surge. Commodity prices were depressed throughout the 1990s, with the RBA's US dollar index of non-rural commodity prices at 25.0 in mid-2000. Prices then began to improve a little, with the index creeping up to almost 34 by mid-2004. But by October 2008, the peak of the index before the GFC hit, the index had surged to 100.1. China was accounting for a huge share of the growth in the global economy and also for a major share of global imports of key industrial raw materials, including oil, coal, iron ore and copper.

One of the new Prime Minister Rudd's early decisions in 2007 was to hold the 2020 Summit, a two day talk fest in Canberra which produced a grab-bag of recommendations. One key recommendation from the Summit was picked up, and the Government gave the task of a detailed review of our taxation system to a committee headed by Treasury Secretary, Ken Henry. The Henry Review reported in December 2009, detailing a list of 138 recommendations to change our tax arrangements. The Government was nervous about many of the Henry Review's recommendations, but Treasurer Wayne Swan zeroed in on recommendation 45, the key elements of which were as follows: "The current resource charging arrangements imposed on non-renewable resources by the Australian and State governments should be replaced by a uniform resource rent tax imposed and administered by the Australian government that: (a) is levied at a rate of 40 per cent, with that rate adjusted to offset any future change in the company income tax rate from 25 per cent, to achieve a combined statutory tax rate of 55 per cent; (b) applies to non-renewable resource (oil, gas and minerals) projects, except for lower value minerals for which it can be expected to generate no net benefits."[592] A resource rent tax was now to be implemented.

The commodities which would be subject to the new resource rent or super profit tax were to include petroleum and natural gas, uranium, black coal, iron ore, gold, silver, the base metals (copper, lead, nickel, tin, zinc, bauxite), diamonds and other precious stones, and mineral sands. The Budget papers for the 2009-10 financial year noted that Australia's terms of trade had been hard hit by the global recession, with prices for thermal coal down by 44% in US dollar terms, metallurgical coal prices down by around 60%, and iron ore prices expected to see large falls when the contracts were renewed. However, the Treasury took a very positive view on the long term outlook for Australia's commodity prices, saying that even after the large fall in the terms of trade expected in 2009-10, our terms of trade would remain around 45% above the average in the decade prior to the commodity boom.[593] Picking up the optimistic Treasury line, the Henry Review concluded that Australia's terms of trade would remain well above average for decades to come.[594] The minerals industry was expected to be earning super profits for the foreseeable future, and the Government wanted its share.

Paul Kelly has written in detail about the way in which the proposed tax was handled by Treasurer Swan and Prime Minister Rudd.[595] His conclusion, that "As a policy mistake perpetrated by the Rudd Government, nothing matches the mining tax for its scale of failure", is hard to dispute.[596] The mining industry was in fact not opposed in principle to a resource rent tax. In its submission to the Henry Review, the Minerals Council of Australia, the peak industry body whose members include the major mining companies, accepted the principle of a profits-base tax, although it also understood the difficulties in being able to devise and introduce such a tax, as well as the need to work with the State governments who were the ones who owned the mineral resources and managed the systems of royalties.[597]

Resources Minister Martin Ferguson was briefed by the Treasurer in January 2010 on the Henry Review recommendation, but was given no details of how the tax would be designed. "Swan asked Ferguson to reassure the mining industry that it would be fully consulted on the tax... He asked (Ferguson) .... to talk to the mining companies and get their agreement. The message was that we want to develop a profits-based tax but give the industry an undertaking on behalf of the government there will be full and proper consultation."[598] Ferguson said that he had discussions "at the most senior levels" with BHP, Rio and Xstrata and also the Minerals Council. "I said this will be the agreement: you give the government time to think this through and nothing will occur without full and proper consultation."[599] BHP Billiton's CEO, Marius Kloppers, Rio Tinto Australia's CEO, David Peever, and Fortescue Metals CEO, Andrew Forest, also were also involved in discussions with the Treasurer or Prime Minister. Peever, who met Swan in April 2010, told Kelly: "We were assured by Wayne Swan and his chief of staff that whatever tax policy was settled upon, it would not be in the budget's forward estimates... It meant there would still be time after the tax announcement and the budget to work through the proposed policy because the numbers were not settled upon."[600]

Swan released his tax package, *Stronger, Fairer, Simpler*, on 2 May 2010, the centrepiece of which was the new mining super profits tax (MSPT). All hell then broke loose. The tax was to apply from mid-2012 and would raise $12 billion in the first two years, with the revenue to fund a reduction in the rate of company tax from 30% to 28%, as well as a new infrastructure fund, assistance for small business and superannuation concessions for people on low incomes. The mining industry accused the Government of an ambush, and this was also the view of Martin Ferguson, who believed that he had been deceived by the Treasurer and had in turn misled the mining industry. The mining companies saw the new tax as threatening the long term viability of the industry and agreed to fund an advertising campaign to be run through the Minerals Council of Australia.[601] The campaign kicked off on 7 May, and ran for over two months at a reputed cost of $25 million. Paul Kelly concluded that "It ranks with the ACTU's anti-Work Choices paid campaign as the best special interest campaign in recent decades."[602] The advertising campaign included full page ads in major newspapers and prime time television ads. While there was support for the MSPT from some elements of the media, including some of the journalists writing in The Age and Sydney Morning Herald, the Minerals Council's campaign was very effective in mobilising public opinion against the tax. Some of the ads featured real mining industry employees, others focussed on the contribution of the industry, and in particular the income tax and corporate tax the industry paid.

The golden years for mining profits

The few years up to the global financial crisis did see major mining companies making eye-watering profits and these profits continued for several years after the GFC unfolded in 2008 and 2009. The Treasury and Wayne Swan no doubt were well aware of the levels of profit and wanted the Commonwealth Government to cream off a greater share. For example BHP Billiton Ltd's accounts for its global operations showed that pre-tax profits hit $US23.5 billion in 2008, up from $US19.2 billion in 2007. The company's pre-tax profit then halved in 2009 to $US11.6 billion as commodity prices fell, before surging once again to reach a peak of $US31.3 billion in 2011 as commodity markets benefited from China's policies to stimulate investment. After 2011, profits fell and hit a low of $US 8.1 billion in 2015. BHP, of course, was also paying substantial taxes during this period, averaging around $US6 to $US7 billion per year, much of that to Australian Governments.

Beneficiaries of the high commodity prices during this period also included the major Japanese companies involved in Australia's minerals industry, including Mitsubishi and Mitsui. These companies tend to adopt a low profile and so do not attract the same media attention as the major miners such as BHP and Rio Tinto. However, their substantial investments in Australia paid off handsomely in the early 2000s and in subsequent years. For example, BRW magazine's financial data for the top 1000 companies in Australia for 2009 reported Mitsubishi Development recording net profit after tax of \$2.4 billion, which gave a return on shareholders' funds of over 37%.[603]

Swan and Rudd attempted to portray the mining industry as paying little tax. In his *Treasurer's Economic Note* issued in May 2010 the week after his tax package, Swan stated: "In releasing *Stronger, Fairer, Simpler* the Government made the point that the amount the Australian community charges mining companies for our non-renewable resources has fallen from one dollar in three for the first half of the decade to one dollar in seven today". In response to criticism that this did not include company tax, Swan agreed and added that "...even if we include company tax, the point holds. The amount the Australian community charges for its non-renewable resources has halved, as a share of profits, compared to about ten years ago. And the amount the Australian community receives in both taxes and charges for our non-renewable resources has also halved." [604] The Government's figures had royalties plus company tax for the mining industry at 55% profits in the period 1999-2000 to 2003-04, falling to only 27% in 2008-09. In other words, Swan was accusing the mining industry of not paying its way. Not only was the industry making super profits, Swan claimed it was paying a smaller share of those profits in taxes and royalties. This would be unfortunate for the mining industry as it would leave the impression in the minds of many people that the industry was not contributing to the nation the way it should.

The mining industry's major concern with the MSPT was the 40% rate of tax on super profits, or profits earned above the long term bond rate. As Rio Tinto's David Peever noted: "We are concerned about the inclusion of existing operations and the apparently arbitrary way the new resources tax was set at 40 per cent. Taxing 40 per cent of profits over the long-term bond rate, together with corporation tax, would make the Australian minerals sector the highest taxed in the world, seriously eroding competitiveness."[605] As the details of the new tax began to be analysed, the theoretical basis for the tax also came to be increasingly questioned. Ross Garnaut, who had been involved in the 1980s in the development of the Hawke Government's petroleum resource rent tax, concluded that the tax policy was not "world's best practice."[606] The "silent partner" aspect of the tax was one of its key attributes, but doubts were growing about its real world acceptance. Being a silent partner meant that the Government in effect would become a 40% equity owner in existing and new projects, taking 40% of the super profits (plus its normal share of company tax), but also being liable for 40% of any accumulated losses. The Government assumed that its 40% share would mean that companies would be able to finance 40% of their projects at the long term bond rate.

Paul Kelly quotes some of the key players he interviewed in the 2012 to 2014 period, people who, with the benefit of time, saw the flaws in the design of the tax. Chris Barrett, one of Rudd's advisors, said: "Having the government take a 40 per cent share was an elegant position but the 40 per cent rebate was not bankable. It was not valued by the market, so it didn't work as intended." Wayne Swan admitted that: "We didn't necessarily get the design right. It became clear from the initial contacts with the industry that refundability was never going to fly." Martin Ferguson, Labor's then Resources Minister, was "appalled by the Treasury design" and called it "a textbook proposal you would apply to an immature mining industry such as Nigeria or Mozambique." David Peever explained that "you cannot go to your board saying we have this new project but don't worry about it because if we lose money the taxpayer will look after it."[607]

Of course the silent partner issue was just one of the key concerns of the industry which also included the provision for the rate of tax to be based on earnings over and above the long term bond rate, with the tax also to apply to existing projects, and the whole question of sovereign risk and the damage to Australia's reputation as an attractive destination for investment. There was also the fact that while the RSPT was supposed to have the Government and the mining companies sharing the risk, mines which were operating at the time could be argued to have been generally the more viable operations. Unprofitable mines, and there had been many over the years, particularly in the coal industry, had often been closed or had been sold on to new owners at significant loss to the original owners. "The government would have been creaming off its share of the successes while avoiding its share of past losses."[608]

It was not long before the controversy over the tax, combined with a range of other concerns about his personality, management style and policies, led to Prime Minister Rudd being deposed by his deputy, Julia Gillard on 24 June, 2010. As the new Prime Minister, Gillard moved swiftly to work through a revised mining tax with the industry. Gillard invited BHP, Rio Tinto and Xstrata Coal to meet to thrash out the details, and on 2 July, 2010 a new tax was announced. The new tax, called the Minerals Resource Rent Tax (MRRT), had been agreed. It would apply only to iron ore and black coal mining operations with annual profits above $50 million. It would be levied at a rate of 30%, but, with an extraction allowance, this was reduced to an effective 22.5%. It was still a tax on above-normal or super profits. And rather than the historical cost of investments being the base for depreciation calculations, companies were now allowed to revalue their investments to a current market value. The three major mining companies had clearly been able to force very favourable design features on a new Prime Minister keen to put this issue behind her and move on to fighting an election to give her some legitimacy. Miraculously, the new tax was still going to raise billions for the Government - $10.5 billion over two years – and only a little less than the revenue which was supposed to be generated by the ill-fated Resource Super Profit Tax. The Treasury had revised up its estimates of iron ore and coal prices, and bingo, the revenue to the Government now looked respectable. And early in 2011, the revenue stream forecast by the Treasury over the next nine years, was revealed as a huge $38 billion.[609]

Forecasting commodity prices of course is a difficult and, for Treasuries reliant on tax revenues, a practice full of potential traps. The reality soon became apparent when, in the first half year of its application (July to December 2012), the tax raised a mere $126 million. Treasury Secretary Martin Parkinson appeared before the Senate Economics Committee in February 2013 and the Treasury's problems in being able to estimate the revenue from the tax became much clearer. In the May 2012 Budget, the Treasury's forecast revenue for the 2012-13 year was around $3 billion. That figure was reduced to $2 billion in October 2012 to take account of lower commodity prices, the higher Australian dollar and increases in State royalties. But the Treasury was in the dark about the values companies had put on their assets, and also on the share of profits which related to operations not covered by the mining tax.[610]

The revenue from the new tax, as with the RSPT, was also dependent on what happened with mining royalties, the bulk of which are paid to State Governments as owners of the minerals. The RSPT allowed royalties to be deducted from revenue prior to calculation of the tax payable to the Commonwealth. But as the Commonwealth had not negotiated with the States to hold royalties at the rates applying when the tax was agreed with the big three miners, it was now at the mercy of the State Governments. The Governments of WA, Queensland and NSW had all raised their royalty rates on iron ore (WA) and coal (Queensland and NSW) in the period 2008 to 2012, further reducing the revenue to the Commonwealth. Western Australia in its 2011-12 Budget pushed up royalty rates on iron ore fines, with the new rates expected to reap an additional $1.9 billion over the four years 2011-12 to 2014-15.

In Queensland, Treasurer Andrew Fraser announced a new two tier royalty for coal in June 2008: coal sold for less than $100 per tonne would retain the old 7% ad valorem rate, while coal sold for over $100 would pay the 7% rate up to $100, plus 10% on each dollar over $100. The Queensland Government said it expected to receive an additional $578 million in coal royalties in the 2008-09 year, but only $183 million in 2011-12, as coal prices eased. Queensland Treasury's guesses at the additional royalty revenue proved to be way off the mark, as revenue jumped by over $2 billion to $3.1 billion in 2008-09 (helped by higher production and high prices); and by 2011-12 coal royalties were still generating a windfall for Queensland, generating revenue of almost $2.4 billion.

In NSW, the Government brought down a mini budget in October

2008 which included provision for higher coal royalty rates: 8.2 % for open cut mines (up from 7%), 7.2% for underground mines (up from 6%), and 6.2% for deep underground mines (up from 5%). The effective rates also rose as transport costs from mine to port were no longer deductible in the royalty calculation. With the debate on the MSPT still raging, the NSW Government introduced a supplementary royalty in its 2011-12 Budget, which was to apply to companies subject to the new Commonwealth MRRT. NSW was quite candid about its reasons for the higher rates, which were "intended to protect NSW revenue from Australian Government changes, while minimising financial impact on NSW coal mining."[611] The NSW Budget papers had the supplementary rate raising $235 million in the 2011-12 financial year, rising to $465 million in 2014-15. The supplementary rate was to be determined by December 2012, however in the Budget papers for 2013-14 the Government revealed that with weaker coal prices, the higher Australian dollar and low revenue paid by companies under the MRRT scheme, it had set the supplementary royalty rate at zero over the forward estimates (the period to 2015-16). The NSW Government had made sure it did not suffer from the MRRT, but ended up collecting no revenue from its supplementary royalty.

Treasury gets it wrong

The Liberal-National Coalition won the 2013 election, with two of new Prime Minister Tony Abbott's priorities being the abolition of the mining tax and the abolition of the carbon tax. These taxes were abolished in 2014, and the mining tax saga in particular has become a lesson in how not to conduct the development of a major public policy. In March 2015, the Minerals Council of Australia (MCA) released a report by Chris Richardson of Deloitte Access Economics which threw some light on how the Treasury and the Treasurer got it wrong in relation to the mining industry's tax contribution. However, by this time a lot of damage had been done, with many people assuming the mining industry was effectively ripping off the nation. The Treasury had forecast that the ratio of royalties and company tax paid to profit would continue to fall. Richardson found it difficult to replicate the Treasury data but concluded that it was clear that, contrary to the Treasury's calculations, the ratio had risen. Richardson's report also contained data from Deloitte's annual survey for the MCA of the Council's member companies which showed that by 2013-14 the ratio of royalties and company tax to profit had risen to 47.1%, the highest on record.[612] The Deloitte report for 2014-15 saw the ratio rising to a peak of 54%, with the 2015-16 year seeing a fall to 51%.[613]

As noted by the MCA, the Australian minerals industry (not including petroleum and gas) "paid $225 billion in company tax and royalties in the 11 years to 2017-18. This includes $18.6 billion in company tax during the financial year 2017-18, accounting for 22 per cent of all company tax paid, despite the industry comprising less than 1 per cent of all companies... Of the $225 billion, $94 billion was directed to state governments in royalties...In addition, the sector paid over $9 billion in payroll taxes to state governments over the same period[614]." It would appear that the mining sector has been paying its way after all.

A welcome trend in recent years has seen some major mining companies publishing data on the taxes they pay in Australia and on their economic contribution through payments to suppliers, employees etc. BHP published its first economic contribution report for its 2014-15 financial year, and this continues to be a feature of its ongoing annual reporting. Over the five years 2014-15 to 2018-18, BHP paid $US22.7 billion to Australian governments in corporate income tax ($US11.6 billion), royalties ($US9.4 billion) and other taxes and payments ($US1.7 billion). BHP's royalty payments to the Queensland Government over that period totalled $US3 billion, most of which was for its coal mining operations; its royalty payments to the NSW Government totalled $US469 million.[615]

Comparable data published by Rio Tinto began in 2011. In that year, when the MSPT was introduced by the Rudd Government, and thanks largely to very high iron ore prices, Rio Tinto paid $US4.9 billion in corporate income tax and $US2 billion in royalties to Australian Governments; including what it describes as employer payroll tax, Rio's total payments to Australian governments totalled $US7.5 billion. Rio's company tax paid in Australia rose to $US5.25 billion in 2012, but as commodity prices fell, these tax payments also fell, reaching only $US1.3 billion in 2016. Improved prices have seen Rio's corporate income tax paid to the Australian government rise to $US1.9 billion in 2017, $US3.2 billion in 2018 and $US4.2 billion in 2019. Royalties paid to governments in these last three years have totalled $US1.6 billion in 2017, $US1.15 in 2018 and $US1.7 billion in 2019. Total payments to governments in Australia by Rio in 2019 were $US6.2 billion.

Glenore was the subject of a scathing article in the Sydney Morning Herald in June 2014 which accused Glencore Coal International Australia Pty Limited of paying no tax in Australia over the previous three years and of avoiding tax through its exposure to "large, unnecessarily expensive loans from its associates overseas", with interest rates on those loans up to 9%. The article also pointed to a significant increase in the company's sales to related companies, which it said was "indicative of transfer pricing – also known as profit shifting.[616] Peter Freyberg, its Global Coal CEO, wrote to the Herald saying that Glencore had paid $3.4 billion in "royalties and tax" over the previous three years, and that it complied with all the tax rules and regulations. He said that Glencore's economic contribution was far broader than taxes or royalties and that it had operated in Australia for more than 15 years, employing 15,000 people in its coal, copper, grain, nickel and zinc operations. In the previous 5 years, Freyberg said that Glencore's coal business had invested heavily in sustaining, expanding and building new coal mines, with the additional $8 billion investment having created thousands of jobs that the investment would serve Australia for decades.[617] At the time of the Herald article, Glencore's annual report gave no information on its corporate tax or other payments the company made in various countries, but Glencore has published a tax payment statement for recent years, with its first statement for the year 2013 showing that the company paid Australian company tax of $88 million and royalties of

$501 million. For 2013 Glencore noted that it had acquired Xstrata in 2013 and inherited tax losses including from Xstrata's $3 billion Jubilee nickel acquisition in Western Australia."[618] The Jubilee mine proved to be uneconomic and was subsequently closed. In comparison with BHP and Rio Tinto, company tax paid in Australia by Glencore has continued to be modest, with $77 million paid in 2014, $73 million in 2015 and a negative $37 million in 2016. Royalties for the years 2014 to 2016 averaged around $500 million per year, rising to $747 million in 2017. For the 2016 year, Glencore explained that it was not liable for corporate tax, with a subsidiary actually receiving a tax refund; the major reasons for not being liable for tax were the prices for key commodities which it described at being at multi year lows, and the fact that it was continuing to utilise the inherited tax losses from the Xstrata acquisition. Glencore said that with higher prices in 2017, it expected to pay more than $400 million in Australian corporate income tax for the 2017 tax year.[619] The company's Voluntary Tax Transparency Australia Report for the 2018 year showed actual corporate income tax paid totalling $588 million and government royalties $941 million. That report also showed that for the period 2008 to 2018, corporate income tax totalled $2498 million and government royalties $6506 million.[620]

In recent years, BHP and Rio Tinto have been subject to investigation by the Australian Taxation Office (ATO) in relation to transfer pricing. The BHP dispute with the ATO centred around the pricing of commodities purchased by its Singapore marketing operations and also whether the profits earned by Singapore should be subject to additional tax in Australia under the Controlled Foreign Companies rules. The dispute was finally resolved in November 2018, with BHP and the ATO agreeing on the company paying a not insubstantial $529 million relating to earnings for the period from 2003 to 2018 (of which BHP had already paid $328 million to the ATO). BHP also agreed to change the ownership structure of its Singapore operations which from July 2017 became fully owned by the ASX listed BHP Limited. Previously, BHP Singapore was 42% owned by the London listed arm of the company, a structure which allowed 42% of the profits of the Singapore operations to avoid paying tax in Australia. BHP says it is planning to continue to base its marketing operations in Singapore and all profits from these operations will be subject to Australian tax. BHP states that being located in Singapore provides the company with access to a highly skilled global trading centre for its commodities in close proximity to its customer base.[621]

In Rio Tinto's 2016 annual report, auditors PricewaterhouseCoopers (PwC) noted the uncertain tax position relating to the transfer pricing of certain transactions between Rio Tinto's Australian entities and its commercial centre in Singapore. PwC said that the company was in discussion with the ATO, and that it had provided for an increase in its tax provision of $US380 million for the 2016 year.[622] The company's 2018 annual report referred to ongoing discussions with the ATO relating to transfer pricing of certain transactions with the group's Singapore centre.[623] In its 2019 annual report, the Audit Committee noted that it had received updates from management "on the status of ongoing discussions with the Australian Tax Office relating to the transfer pricing of certain transactions with the Group's commercial centre in Singapore and considered the appropriateness of provisions for uncertain tax positions." Rio's battle with the ATO appears to be continuing.

The productivity puzzle

The boom years for commodity prices from around 2004 produced high prices and healthy profits for resource exporting companies as major economies recovered from the "tech wreck" and as China's growth began to impact much more seriously on commodity markets. However, while shareholders were happy and governments were looking at ways of extracting a greater share, mining companies throughout the world were experiencing a rapid deterioration in their levels of productivity. In their efforts to rapidly increase production to meet the growing demand, many companies were sacrificing the hard won gains in efficiency that they had made in the 1990s and early 2000s. McKinsey & Company, for example, noted that worldwide mining operations in 2013 were as much as 28% less productive than a decade earlier, even after allowing for declining ore grades and mine cost inflation, with the decline evident across different commodities, companies and regions.[624]

Inefficiencies in the coal industry in Australia were highlighted by the Productivity Commission inquiry in 1997-98, with a study by Tasman Asia Pacific finding a major gap between best practice in Australia and the USA, and between the best mines and other mines in Australia in areas including truck and shovel operations and draglines. The Tasman report noted a "very mixed productivity performance" by Australian black coal mines, while also finding that Australia also had a number of mines which were at or close to world best practice performance levels. The report concluded that the major problem for the industry in Australia was the large number of poorly performing mines.[625] Productivity in the Australian coal industry surged from the mid-1990s as competitive pressures forced producers to reduce employee numbers, overhaul work arrangements and look for ways to extract greater efficiency out of expensive capital equipment.

In the decades up to the 1980s, output per man shift (OMS) was the standard productivity indicator in the industry, measuring the number of tonnes of coal produced divided by the number of shifts worked in a day at each mine. A mine employing 100 workers all working a standard shift length and producing 1000 tonnes of coal per day would have had an OMS of 10. The OMS, however, was a very crude metric, and as working arrangements changed and shift lengths became longer at some mines, output per man hour became another common measure of productivity. In the 1990s the limitations of these measures of labour productivity were more widely recognised and the focus broadened to include measures of the efficiency with which capital equipment was being utilised and to measures of the efficiency with which all inputs were utilised (the total factor productivity approach). The Tasman Asia Pacific study mentioned above used this total factor productivity approach.

While it is important to recognise the limitations in productivity measures which have been commonly used since the 1980s and 1990s, such as saleable production per employee, these measures can still be useful guides to trends in productivity over time. In NSW saleable production per employee rose by around 64% from 1995 to 2000, by which time it averaged 10580 tonnes per year. The measure then averaged around 11250 tonnes for the next six years, before it began a steep decline, bottoming out in 20011-12 at only 7000 tonnes before rising again, reaching almost 10000 tonnes in 2016-17. In Queensland, saleable coal production per employee peaked in 2001 at around 16000 tonnes per year, but by 2010 was down to around 8500 tonnes per year. For the 2016-17 year, it appears that Queensland's productivity was still low at around 7700 tonnes.[626]

There are many factors driving trends such as these, some of which relate to geology (for example the need to mine at deeper levels or the need to adopt underground methods rather than open cut methods). However, when productivity falls, the cost per tonne tends to rise, and if prices for the coal or mineral start to fall, the consequences for the viability of mines can be dramatic. The drop in productivity in Australian mining was not a serious problem until around 2010 to 2012 when commodity prices began to fall and the high cost structures of mining companies were exposed to the spotlight. Shareholders did not like what they saw and senior managers were forced to institute in some cases drastic programs to lift productivity by retrenching workers, looking for better ways of operating and reducing the cost of work done by contracting companies. Many contractors had also enjoyed high profits when the market conditions allowed but were hit hard and were forced to lower prices as contracts came up for renewal or as producers took back some work in-house.

BHP Coal is one example of what happened in the Australian coal industry following the GFC. The company reported in 2016 that it had slashed its cash costs per tonne of coal over the previous four years by 56% in its Queensland metallurgical coal mines, and by 29% in its NSW Mt Arthur mine. A further 20% reduction in its Queensland coal costs was expected in its 2017 financial year, and a 9% reduction in NSW energy coal, bringing costs per tonne down to $US52 and $US38 per tonne for the two parts of the business.[627] Data in BHP's 2017 Annual

Report for actual costs per tonne in 2016 and 2017 differ slightly from the above costs, but on the basis of both sets of numbers, it appears that in 2012, BHP Coal's cost per tonne in 2012 was around $US150 for its Queensland metallurgical coal business and around $US60 for its NSW energy coal. The dramatic reductions in costs achieved by BHP were, of course, not unique, as coal companies in Australia and around the world were working hard to restructure their businesses in the face of declining coal prices and strong competition. The rapid escalation in costs up to 2012 is also evident from BHP's earlier annual reports, with its metallurgical coal business having operating costs per tonne of $US54.5 in 2006, rising to $US158.6 in 2012 (in those years the business included the higher cost Illawarra coal mines which are now part of South32).[628]

The reductions in the cost structure of coal producers over the period from around 2012 were associated with major falls in employment in the Australian industry, cutbacks which hit many local communities in Queensland and NSW very hard. Job numbers in both States peaked in 2012 after seeing rapid growth since the early 2000s, Queensland at just over 41000 and NSW at just under 25000. By 2017 Queensland's employment was down to just under 31000, having reached a low of 28359 in 2016. In NSW the decline has been less severe, with job numbers falling to 19388 in 2016, before recovering to 20538 in 2017 and 22308 in 2019.[629] The Bowen Basin towns were hit particularly hard, and there were also significant effects on centres such as Mackay, Rockhampton and Townsville from which many coal mine workers commute to their minesites.

From an industry in turmoil to a world leader

The Australian black coal mining industry is now totally unrecognisable from the industry that emerged at the end of World War Two, an industry in a poor state, unprepared to meet the needs of a country seeking to grow and develop. It is now a world leading industry, technologically advanced, with a good safety performance, producing around 450 Mt of saleable coal per year. Queensland has been the largest producing state for some years, with output in 2019 of 250 million tonnes, with NSW producing 197 Mt, WA around 7 Mt and Tasmania less than 1 Mt. Australia accounts for around 6% of world black coal production, but is the dominant supplier of coking coal to the international market, with a share of 54% in 2019, and is the second largest exporter of steaming coal, with around 19% of the export market. Our coking coal exports in 2018-19 totalled 184 Mt, with China, India, Japan, South Korea and the EU our major markets. Australian coking coal underpins the steel industry in Japan and Korea which produce little of their own coal, and plays a major role in China and India which also have large domestic coal industries. Australia exported 210 Mt of steaming coal in 2018-19, well behind Indonesia's 466 Mt. Our steaming coal is a critical source of energy for power generation in a number of countries, including Japan, Korea and Taiwan; China also imports significant quantities of steaming coal from Australia. Our steaming coal is also finding growing markets in Asian countries which are investing heavily in new, more efficient coal-fired power stations, and is also a key input into the cement industries in a number of Asian countries.

In the 1980s, 1990s and early 2000s black coal was Australia's major export commodity, but was overtaken by iron ore in 2010-11 as iron ore prices surged. In the financial year 2018-19, Australia's metallurgical coal exports totalled $44 billion and thermal coal $26 billion.[630] The export values for both categories of coal will be significantly lower in the 2019-20 year due to lower prices. The black coal mining industry employs around 53000 people, with thousands more employed in companies which provide services to coal mines, in the transport of coal to export ports and local power stations, steel mills and cement plants, and in the export terminals in both NSW and Queensland. Coal mining employs 21393 in NSW, 30925 in Queensland, around 700 in WA and around 70 in Tasmania.[631] The NSW Government's recent Strategic Statement on Coal Exploration and Mining quotes coal mining employment in the State as 22000, with the industry generating a further 89000 indirect jobs. Black coalmining ceased in South Australian in 2015 with the closure of the Leigh Creek mines and following the closure of the power stations in Port Augusta which were the only market for Leigh Creek coal. The Bowen Basin in central Queensland and the major coastal towns and cities which service the industry (including Mackay, Gladstone, Rockhampton, Townsville and Bowen) are the areas most directly impacted by coal mining in Queensland. In NSW the key areas are the Hunter Valley, Lake Macquarie area, the Illawarra region, and the Lithgow, Mudgee and Gunnedah areas.

The industry is now dominated by several large and medium sized companies, including BHP, Glencore, Anglo American, Peabody, New Hope, Centennial, Yancoal, Whitehaven and South32. Among the Japanese companies, Mitsubishi and Mitsui continue to play the most important role in the industry. The use of contracting companies to operate coal mines or supply specialised services is common throughout the industry and sees major companies such as Thiess, Leighton and Downer EDI playing a significant role, along with many smaller contractors, and employing many of the workers in the industry.

Black coal was the energy source for 46.6% of Australia's electricity in 2017-18 and provides a much higher percentage in NSW and Queensland. Victorian brown coal fuelled another 14% of national electricity generation, with gas 21%, hydro 6% and other renewables 11%.[632] The coal industry is also a significant sector within the broader Australian mining industry, accounting for around 20% of the value added in 2014-15.[633] The value added by the mining industry and the related mining equipment, technology and services industries is estimated at $133 billion in 2015-16, with 484,000 jobs directly supported by these sectors. Allowing for the indirect impact on other areas of the economy, the total economic contribution of these industries was $237 billion, and the jobs supported across Australia were over 1.1 million (or around 10% of all employment).[634] The black coal industry is also a significant contributor to government revenues through royalty payments which totalled $4363 million in Queensland in 2018-19 and around $2000 million in NSW.[635]

Cattle grazing on rehabilitated mining land at New Hope's Oakleigh West.
Courtesy New Hope Group.

Kogan Creek power station, Qld; one of Australia's HELE power stations.
Courtesy CS Energy.

6

The Future for Australian Coal

The domestic market is declining

The domestic market for thermal coal in Australia has been declining in recent years with the closure of a number of coal-fired power stations. In Queensland, the 190 MW Collinsville power station closed in 2012 and the four 120 MW units at Swanbank B were decommissioned between 2010 and 2012. In NSW the 1400 MW Munmorah station closed in 2012, the Wallerawang 1000 MW station was decommissioned in 2014 and the 150 MW Redbank station closed in 2014. In Victoria the 165 MW Energy Brix power station closed in 2014; the 150 MW Anglesea power station and the dedicated coal mine closed in 2015; and the 1600 MW Hazelwood station in the Latrobe Valley closed in 2017. In South Australia, the 240 MW Playford and 520 MW Northern power stations in Port Augusta and the Leigh Creek coal mine closed in 2016.

AGL, which owns the 2640 MW Bayswater power station and the 1680 MW Liddell station in NSW and the 2225 MW Loy Yang A power station in Victoria, has stated that it will cease using coal for power generation by 2050. This policy reflects the company's plans to transition into other forms of energy for power generation, but also the dates by which its power stations will reach the end of their economic life. Liddell will be around 50 years old by 2022; AGL plans to close the first of four units in April 2022, with the other three units to close in April 2023. Loy Yang A's four units were commissioned between 1984 and 1988, and Bayswater's units were commissioned in 1985-86. AGL says that these two stations can be expected to cease operating in 2030s and 2040s. AGL says it is committed to an "orderly, respectful, and smooth transition for the Latrobe Valley away from coal-fired power generation". According to AGL's CEO, Brett Redman, that transition could occur more rapidly depending on customer demand for low-carbon power, lower costs of new technologies or changing community demands.[636] Origin Energy is the owner of the 2880 MW Eraring power station located on the edge of Lake Macquarie in NSW. Eraring, Australia's largest power station, was commissioned in the early 1980s, but each of its 660 MW turbines was upgraded to 720 MW between 2011 and 2012; Eraring is planned for closure in 2035.

Another ageing station in Victoria is the 1480 MW Yallourn power station, owned by Energy Australia, which was developed in the 1970s; Yallourn may be nearing the end of its life by the late 2020s. The other major coal fired station in Victoria is Loy Yang B, an 1100 MW station commissioned between 1993 and 1996 and owned by Hong Kong-based Chow Tai Fook Enterprises (which also owns Alinta Energy).The other ageing power station in NSW is the 1320 MW Vales Point B station. The Vales Point A station had four generating units which came online in the 1960s but were decommissioned in 1989. Two 660 MW units were added to Vales Point in 1978 and became the B station which is operated by Sunset Power International (a company owned by Trevor St Baker and Brian Flannery). The other major coal-fired power station in NSW is Energy Australia's 1400 MW Mt Piper station commissioned in 1992-1993 and which has a potential life into the 2040s.

In Queensland there is over 8000 MW of coal fired capacity, much of it less than 20 years old. The 1680 MW Gladstone power station is the oldest currently operating station, having been commissioned between 1976 and 1982. Tarong's 1400 MW of capacity came on stream between 1984 and 1986, the 700 MW Callide B power station was commissioned in 1989 and the 1460 MW Stanwell station was commissioned between 1993 and 1996. Other coal fired power stations in Queensland are relatively new, with the 810 MW Callide C, the 443 MW Tarong North and the 850 MW Millmerran stations not commissioned until between 2001 and 2003. The 750 MW Kogan Creek station is younger again and was only commissioned in 2007. Kogan Creek, Callide C and Tarong

North are super critical plants. Given its age, Gladstone is expected to be the next to close of the current stations in Queensland, potentially later in the 2020s. With the Queensland Government's CS Energy and Stanwell Corporation owning most of the coal fired generating capacity in the State, the timing of closure of each station will be a sensitive political decision, potentially influenced by community concerns, not only about CO_2 emissions, but also the availability of dispatchable power. And how the ongoing role of the State's relatively young coal fired stations meshes with the Government's emission policies remains to be seen.

At the national government level, the Federal Government's Technology Investment Roadmap Discussion Paper issued in May 2020 provides an insight into Australia's future energy policies. The goal of the Roadmap, in the words of the Minister for Energy and Emissions Reduction, is "to bring a strategic and system-wide view to future investments in low emissions technology."[637] The paper acknowledges the future role of gas: "Domestically gas will play an important role in balancing renewable energy, ramping up and down to match supply and demand. Gas is already playing an increasingly important role in South Australia to balance intermittent renewable electricity...." The Paper, however, does not refer specifically to coal, although technologies "that increase the efficiency of the existing thermal generation fleet" are included on the "indicative list of priority technologies" for the electricity sector, together with a number of other technologies including pumped hydro, large-scale batteries, ultra-low cost transmission, gas generation to firm variable renewables, solar thermal and wind. In essence, the Paper's neglect of coal means it accepts its phasing out as a source of power generation in Australia in the coming decades.

The State Governments in Australia have their own climate change policies which give a clear indication of the future for electricity generated from coal. The NSW policy commits the government to "the aspirational objectives of achieving net zero emissions by 2050 and helping NSW to become more resilient to a changing climate." The Queensland policy has three key objectives: 50% renewable energy by 2030, achieving zero net emissions by 2050 and an interim emissions reduction target of at least 30% below 2005 levels by 2030. In Victoria, under the Climate Change Act 2017, the state is committed to a "long term emissions reduction target of net zero by 2050" and the State's climate change policies acknowledge that achieving net zero emissions by 2050 will require "a transition from brown coal-fired electricity to clean energy sources."

None of the major private sector companies currently operating coal fired power stations (AGL, Origin and Energy Australia) has any plans for new coal fired stations; their plans revolve around investment in renewables (mainly solar, wind and pumped hydro) and batteries, with gas fired power stations also playing a major role, at least in the short to medium term.

With the debate over climate change intensifying in Australia, and the fierce opposition to the use of coal in power generation from some sections of the community, it is unlikely that we will see another coal fired power station constructed in Australia unless the investment is underwritten by, for example, the Federal Government. The likelihood, therefore, is that the use of coal for local power generation will cease by 2050, and possibly by 2040.

The international coal market outlook

Global coal consumption peaked in 2013 and 2014 and has declined slightly since, with the 2019 level about 2.5% below the peak. Rises in south and southeast Asian countries have tended to offset much of the decline in consumption in Europe and the USA. Consumption in China has also declined marginally, with the 2019 level down around 1% on its 2013 peak. The international outlook for coal is a mixed one, and varies by region and whether we are looking at thermal coal or metallurgical coal. The outlook also varies widely depending on the assumptions made about future policies adopted by the various governments in Europe, Asia and North America.

The International Energy Agency (IEA) publishes its authoritative annual World Energy Outlook report each year, attracting the attention of the media, governments, analysts and investors around the world. The 2019 WEO published in late 2019 stated that "Coal is ... being steadily squeezed out of the energy mix in many advanced economies by a mix of environmental policies and competitive pressures from increasingly cost-competitive renewables and, in some markets, from natural gas." The WEO also noted that in Asia, "demand for electricity has continued to grow and coal remains the largest source of generated electricity."

The 2019 WEO looked at the impact of different scenarios on energy demand to the year 2040, including its Stated Policies Scenario (SPS - which reflects current government policies and targets) and its Sustainable Development Scenario (SDS - which maps out a way to meet sustainable energy goals in full, requiring rapid and widespread changes across all parts of the energy system). There is a dramatic variation in terms of coal demand between the two scenarios. While global energy use continues to grow under both scenarios, under the Stated Policies Scenario global coal demand is essentially flat, while under the other scenario it falls rapidly. The IEA also notes that "What happens in Asia will be pivotal, given the region's large coal supply industry and the young average age of the coal-fired (power station) fleet." It also points to the potential for coal capture, use and storage (CCUS) to play an important role, with large scale deployment of CCUS technologies potentially allowing a distinction to be made between coal use and the emissions from its combustion.[638] Under the SPS, global coal demand is almost steady over the period from 2018 to 2040, with only a 1% decline in that period. Under the SDS, global demand falls by over 60% between 2018 and 2040, with dramatic falls in each region of the world.

The SPS scenario sees coal demand for power generation in deep structural decline in many advanced economies, with coal demand also down in China due partly to moves to improve air quality. However, in other developing countries in Asia, coal demand will increase to satisfy fast-rising demand for electricity and industrial development. The SDS scenario sees coal's share in the global energy mix falling dramatically towards 10% by 2040.

Another organisation which publishes long term forecast for the world energy markets is the US Department of Energy's Energy Information Administration (EIA). The EIA's latest forecasts, which are its "reference case" based on current trends and some planned changes to infrastructure,[639] see world coal consumption remaining roughly steady at around 7.25 billion tonnes per year until 2040, after which it rises to around 8.16 billion tonnes by 2050, the increase driven by increases in India and Asian countries (not including Japan, China and Korea). China remains the largest coal producer and consumer, its total consumption in 2050 around 3175 Mt compared with the peak of around 4260 Mt in 2013.[640] China's coal imports decline slowly under this EIA scenario, from around 253 Mt in 2020 to 222 Mt in 2050.

India is the standout case in the EIA's scenario, with its coal production forecast to grow by an average of 2.7% per year over this period, from around 770 Mt in 2018 to just over 1800 Mt in 2050. However, consumption is forecast to grow more rapidly, by 3.1% per year, from around 1000 Mt in 2018 to around 2600 Mt in 2050. Under this scenario, India will be the world's largest coal importer by 2030 with almost 360 Mt of imports, with the annual tonnage rising to around 780 Mt by 2050. The growth in India's power generation sector will see coal consumption growing by 2.5% per year over the period to 2050. However, with renewable generation growing by just over 8% per year, coal's share in power generation will fall significantly. There is also a strong growth in coal consumption and imports in other non OECD Asian countries in the EIA's scenario, where imports grow from around 158 Mt in 2018 to 416 Mt in 2050; this region includes countries such as Indonesia, Thailand, The Philippines, Malaysia and Vietnam. The EIA sees Australia and Indonesia remaining the world's largest coal exporters, with Australia having 37% of total exports by 2040 and Indonesia 28%.

Outlook for metallurgical coal

In looking at what this mixed outlook for coal may mean for Australia's coal exports, we need to look separately at the metallurgical coal and thermal coal markets. In the 1970s, 1980s and 1990s the international coal market tended to be dominated by Japan as the largest importer of both thermal coal and metallurgical coal. With the emergence of China and India as major coal importers, the relative stagnation of the Japanese economy from the 1990s, and the growth of other coal importing countries, Japan has ceased to dominate the market, although it continues to be a significant importer and influence on the market. With Japan now only accounting for around 16% of the international metallurgical coal trade and 13% of the thermal coal trade, coal exporters in Australia and other countries now have more options, and the controversies which characterised the market in the earlier decades have largely died away. Australia has dominated the metallurgical coal market since it overtook the USA as the major exporter in 1984 and that dominance is expected to continue. Australia's coking coal exports in 2018-19 represented around 54% of world trade, with Canada, USA and Russia the other major exporters.

The growth in the metallurgical coal market is very largely dependent on the growth in steel production from blast furnaces, with every tonne of crude steel requiring around 780 kilograms of coking coal. However, a significant proportion of steel is produced from scrap which is recycled through electric arc furnaces, with around 517 million tonnes of steel produced by this method in 2019 out of total global steel production of 1.869 billion tonnes. China is a significant and growing producer of electric arc furnace steel (with 10.4% of crude steel production totalling 103 Mt in 2019). India's electric arc steel production (62 Mt) is already relatively large for the size of its steel industry which produced a total of 111 Mt of crude steel in 2019.

China already accounts for 53% of world steel production and so the trend in its production is crucial to estimating the likely trend in metallurgical coal demand. India's 2017 National Steel Policy "sets an ambitious steel capacity target of 300 million tonnes, and a consumption and production target of 255 million tonnes by 2030."[641] India has the potential to see significant growth in its steel industry, and with limited domestic metallurgical coal resources, the country should also continue to see rapid growth in metallurgical coal imports.

Growth in the global steel industry is likely to depend largely on emerging economies including India which have the potential to grow at a relatively rapid rate over the next few decades. However, irrespective of the rate of growth of global steel over the long term, it is expected that the market for metallurgical coal will continue to support Australia's high quality metallurgical coal producers provided they remain competitive. Competition may well intensify, with traditional exporters in Canada and the USA likely to continue to compete with Australia, with Russia looking to compete more for Asian markets and production in countries such as Mongolia and Mozambique also expected to grow.

At present it is generally accepted that metallurgical coal will continue to be an essential input into the production of steel, a process which involves the conversion of the coal to coke and the use of coke in the blast furnace, together with iron ore, limestone and scrap steel. Coke burns to produce carbon monoxide which in turn reduces the iron ore to liquid iron metal, with the metal absorbing carbon from the carbon monoxide. Carbon and small amounts of manganese are what gives steel its strength. However, on the horizon as a potential replacement longer term for metallurgical coal is hydrogen. While currently "There is limited or no use of hydrogen in steel making and the process is not being widely pursued in research and development"[642], with increasing pressure on steel producers to reduce carbon emissions, the coal industry will be closely watching developments in relation to hydrogen.

Outlook for thermal coal

Indonesia has been the major steaming coal exporter since 2005, when it took the number one position from Australia. Indonesia and Australia continue to be the leading exporters into the Asian market for steaming coal, being closer to major coal importing countries. Russia, Colombia and South Africa are other major exporters, with Russia making increasing moves into Asian markets. Colombia's natural markets are in

Europe while South Africa exports to both European and Asian markets.

Indonesia has some significant question marks over its long term ability to continue to export at the current rate, given that country's own expected growth in demand for coal to supply its growing electricity generation sector. The Australian Department of Industry notes that the Indonesian government has previously flagged plans to limit annual production in order to safeguard coal reserves for future domestic use. The Indonesian government is targeting an output cap of 550 million tonnes for 2020. However, this is below the 610 million tonnes produced in 2019 and, given past performance, the Government's ability to restrict the industry to its target is uncertain.[643] Indonesia may also face pressure from coal importing countries because of the quality of its coal, much of which has a low energy content (between 5100 and 6000 kcal/kg), and with a proportion of its coal below 5100 kcal/kg. While Asian demand has been strong for lower quality coal up until now, as we outline below, Australia's higher quality steaming coals are likely to be favoured by the new generation of higher efficiency power stations under construction and planned in the Asian region.

Russia is now the third largest thermal coal exporter, exporting around 181 Mt in 2019, and "has been investing heavily in transport infrastructure to its eastern ports – targeting the Asian premium market, where Japan's utilities are diversifying their supply sources, and South Korea's new regulations are lifting demand for Russia's low sulphur coal." Further growth in Russia's coal exports "will be supported by ongoing government plans to invest in the coal industry and in associated rail and port infrastructure."[644]

South Africa's higher quality coal gives it an advantage over Indonesia in selling into countries such as India which are seeking to use better quality coals. However, there are question marks over its ability to attract investment from major mining companies, with transportation infrastructure also remaining a problem.[645] The constraints on the rail system include the main Transnet line to the Richards Bay export coal terminal which has a capacity of around 76 million tonnes, while the export terminal's capacity is much higher at 91 million tonnes.[646] In October 2019, the South African government approved the National Development Plan, which sees coal-fired power generation capacity falling from 37 GW at present to 33 GW by 2030. This reduction in generating capacity if achieved will free up the export capacity of the thermal coal producers to some extent.

In looking to the future demand for thermal coal, the IEA's South East Asia Energy Outlook 2019 report provides some valuable insights for this key region which includes Indonesia, Vietnam, Thailand, Malaysia, The Philippines, Singapore, Lao, Cambodia, Brunei and Myanmar. In the Stated Policies Scenario, overall energy demand grows by 60% between 2018 and 2040, with the region's economy more than doubling over this period.[647] While the lower rate of growth in energy demand reflects a structural shift towards less energy intensive economies and greater efficiency, all fuels are expected to play a role in meeting the growth in demand, including coal. Coal demand is forecast to grow by 3% per year, doubling by 2040, with thermal coal accounting for over 90% of the increase, mainly due to new coal fired power stations. The report also notes that almost 100 GW of coal fired capacity is planned to come online in the region, with most of this in Indonesia, Vietnam and Malaysia (of which 30 GW is already under construction). With Indonesia's coal exports forecast to decline significantly over this period as its domestic market absorbs greater coal production, the IEA sees Australia and Russia as the countries well positioned to offset the decline. Russian export volumes in fact are seen as overtaking Indonesia's exports by 2040.

Under the Sustainable Development Scenario, the southeast Asian region's energy demand increases only marginally and there are major shifts in the pattern on energy use and power generation. Coal demand peaks after 2020 and is 80% below the SPS level by 2040, with coal demand declining by 4.3% per year. Around 90% of the decline in coal use by 2040 is in power generation, where coal's share falls from 40% today to only 4%.[648] The report does not show the impact on coal imports, but by 2040, there would be a dramatic reduction in imports of thermal coal compared with the more positive Stated Policies Scenario.

One developed country not planning to phase out coal fired electricity is Japan. Under its current policy, coal remains an important part of the energy mix, although its share in power generation declines from 32% in 2018 to 26% in 2030, with the share of renewables rising from 22% to 24%, nuclear up from 6% to 22%, and gas accounting for the remaining share. Japan's Minister for Economy, Trade, and Industry (METI) Hiroshi Kajiyama has indicated that the Government will look to close around 100 of the country's old coal fired power stations by 2030. Media reports in July 2020 said, however, that under the Government's plans coal would still remain a key part of the energy mix.[649] In January 2020, Japan had 9.3 GW of coal fired capacity under construction, with a further 2.6 GW planned.[650] For Japan to close many old power stations and meet its targeted energy mix, more new coal fired stations are likely to be needed.

For the Australian thermal coal producers, the future trend in demand from the current major consuming countries (Japan, China and South Korea), from India and other south Asian countries, and from the developing countries in southeast Asia will be critical. Australia will continue to be a significant thermal coal exporter for the foreseeable future. What the longer term future holds remains to be seen and will become clearer as the current decade unfolds. If the future corresponds to the IEA's Stated Policies Scenario, the outlook for our producers is positive. However, under the Sustainable Development Scenario the outlook would be quite grim. But whether the SDS is a realistic scenario, particularly for the developing countries of southeast and south Asia, is debatable given the drastic changes which governments and investors in those countries would have to accept and given the fact that under the 2015 Paris Agreement on climate change, developing countries did not accept any strong commitments to reduce emissions over the period to 2030.

New generation high energy low emission (HELE) power stations

One of the major developments in recent years has been the widespread incorporation of more efficient technology into many of the new power stations being constructed or planned, particularly in the Asia Pacific. The term high efficiency low emission (HELE) is now often used to describe these new power stations. HELE power stations using coal to generate power operate at higher temperatures and pressures than older power stations, meaning that they are able to generate electricity with significantly lower CO_2 emissions per unit of electricity.

In 2017, the Minerals Council of Australia noted that there were more than 725 HELE coal power plants in operation in east Asian countries, with another 110 plants under construction or planned.[651] In China there were 579 operating plants, with 575 new plants planned; in India the numbers were 49 operating and 395 planned; in Japan 44 operating and 17 planned; in South Korea 38 operating and 18 planned; Taiwan 7 operating and 9 planned; in Indonesia 3 operating and 32 planned; in Vietnam 2 operating and 57 planned. A report issued by the IEA Clean Coal Centre in June 2018 provides another perspective on the potential for coal demand in several countries in Asia – Indonesia, Bangladesh, Malaysia, The Philippines, Thailand and Vietnam – which have a combined population of over 700 million.[652] This report looked at the possible trends in demand for coal-fired electricity in these countries in the period from 2015 to 2040, and forecast demand rising from 845 TWh in 2015 to 1231 TWh in 2040. The report does not forecast coal consumption in these countries, but while consumption will not increase as rapidly as coal-fired electricity demand, it will still grow substantially, as will coal imports. The report sees many of the new coal-fired power stations being HELE plants using the more efficient super critical (SC) and ultrasuper critical (USC) technologies. And towards the latter part of the period, it sees even more efficient advanced ultrasuper critical (ASC) plants being constructed in some of these countries. These SC, USC and ASC plants are significantly more efficient than older subcritical plants, with, for example, the USC plants emitting around 19% less CO2 than a new subcritical plant, and up to 40% less if the new HELE plants are replacing older plants.

It therefore seems that while there are uncertainties regarding the long term future for Australia's steaming coal exports, there will continue to be a major global trade in the decades ahead. And to the extent that coal consuming industries, particularly power generators in Asia, turn to higher quality coal, Australia should be well placed to capture a significant share of the market.

Carbon capture and storage

Carbon capture and storage (CCS) is a process that captures CO2 "from industrial processes, energy generation, or the atmosphere and stores it underground. (It may) also involve capture, utilisation and storage (CCUS) where CO_2 is used, for example, to manufacture biofuels, to boost plant growth, or in food and drink production."[653] CCS has existed for many years and, according to the Global CCS Institute, there were 18 large-scale CCS facilities in commercial operation in 2018, with 5 under construction; these facilities were in industries including natural gas processing, electricity generation, steel, plastics and chemicals. There were an additional 28 pilot and demonstration scale facilities in operation or under construction in 2018. One of the electricity plants operating with CCS technology is NRG's plant in Texas where the Petra Nova Carbon Capture joint venture project commenced in December 2016 and by September 2018 had captured 2 Mt per year of CO_2.[654] This project uses a process developed by Mitsubishi and the Kansai Electric Power Company and the captured CO_2 is used to improve oil recovery at an NRG oil field located around 130 kilometres away, with the CO_2 being sequestered underground.

The IEA notes that prior to the Covid-19 crisis, CCUS was gaining momentum, with 5 new CCUS power plants announced in 2019, mainly stimulated by investment incentives in the USA, and bringing the total number of plants under development around the world to 14. Under its Sustainable Development Scenario, the IEA is looking for CCUS power plants to play an important role in reducing global CO_2 emissions, with a target to reach 310 Mt per year of CO_2 capture by 2030. But the IEA notes that CCUS is at an early stage of commercialisation, and new investments will require "targeted policy measures such as tax credits and grant funding."[655]

One of the major CCS projects to date in Australia has been the Callide Oxyfuel project at CS Energy's old Callide A power station in Queensland. This project was a joint venture involving CS Energy, ACA Low Emissions Technologies (ACALET), Glencore, Schlumberger, J-Power, Mitsui and IHI Corporation; it received funding from the Commonwealth Government's Low Emissions Technology Demonstration Fund. It began as an initiative in 2003 of the Australian Coal Association's Coal21 and ran through to 2015, with financial support also provided by ACALET, the Japanese and Queensland Governments and technical support from JCoal. Oxyfuel technology allows coal to be burnt more efficiently in the boiler of a power station and produces a smaller stream of waste gas that contains CO_2 in a more concentrated form. More than 70% of the CO_2 is then extracted from the waste gas using established cryogenic technology. According to the project website, the project confirmed that CCS could be applied to a coal-fired power station to generate electricity with almost no emissions, and it represented a first for Australia as the world's first industrial scale demonstration of oxyfuel combustion and carbon capture technology.[656] The Callide plant was retrofitted with the oxyfuel carbon capture technology (also a world first) and saw the first injection underground of CO_2 from an Australian power station. The major current CCS project in Australia is at Chevron's Gorgon facility which became operational in 2019 and is expected to reduce greenhouse emissions by 40% over the life of the project. The Gorgon project injects CO_2 into a deep reservoir more than 2 kilometres beneath Barrow Island in WA.

As the IEA has noted, "CCS technologies can play a critical role in the sustainable transformation of the global energy system. They offer a solution to some of the most vexing energy and climate challenges we face, including the need to significantly reduce emissions from industrial processes and from a large and relatively young global fleet of coal and gas-fired generation units. CCS also provides the means to deliver 'negative emissions' to offset emissions from sectors where direct abatement is not economically or technically feasible."[657] The IEA has also emphasised that there could be a growing role for CCS with "greater climate ambition" where countries commit to achieve greater reductions in CO_2 emissions.[658] How practical and economic the widespread use of CCS by power stations will be is difficult to predict and will depend on many factors, not the least of which is the additional cost of CCS compared with non CCS stations and other power generation technologies (gas, nuclear and renewables). While there is significant potential worldwide for CCS, it is unlikely that CCS will have any significant impact on the global coal market in the short to medium term.

The reality of coal

Despite the political and community pressures on the coal industry in Australia and in many other countries, the reality is that coal mining will continue to be an important industry for many years to come. While coal consumption for use in electricity generation will continue to fall in Europe and the USA, and very likely in Australia, this is not the case in many rapidly developing countries in Asia or in Japan where new coal-fired power stations are being built, many of them with significantly higher efficiency levels and lower carbon emissions per unit of electricity. And, of course, with the world heavily reliant on steel for the building industry, for roads and bridges, for manufacturing and for many other uses, and with metallurgical coal an essential ingredient in blast furnaces for the production of steel, there will continue be a strong demand for the various types of metallurgical coal (including the high quality coking coals produced in the Bowen Basin and in the NSW Illawarra region, and the soft coking coals and PCI coals produced in both Queensland and NSW).

Finance for coal mining is becoming harder to obtain from banks and other traditional sources in Australia, with several of the major banks adopting policies in recent years which limit their ability to lend to the coal mining sector. The National Bank announced in December 2017 that while it would continue to support its existing customers in the mining and energy sectors, including those with coal assets, it would no longer finance new thermal coal mining projects. Westpac will now lend only to thermal coal projects in existing basins and where coal quality ranks in the top 15% globally. The Commonwealth Bank's policy is to continue to reduce its exposure to thermal coal mining and coal fired power generation, "with a view to exiting the sectors by 2030, subject to Australia having a secure energy platform." The National Australia Bank says it will not finance new or material expansions of coal fired generation unless there is technology in place to reduce emissions. NAB will also reduce thermal coal mining finance to effectively zero by 2035. A number of major banks in Europe have also moved to limit or ban lending for coal mining projects.

However, as the CEO of New Hope Coal has noted, while "Most of the European banks have turned away from financing thermal coal in particular", this is not the case in Asia, where the banks "are opening up because they can see the demand in their home markets for the resource." These Asian banks, he said, were "gaining the expertise to evaluate debt financing opportunities in Australia."[659] The amounts to be financed over the next few decades will be huge. Even with the constraints on the industry, finance will be needed for the development of new mines and the expansion of existing mines to cater for growth in demand and to replace mines which close due to the resource being depleted or if they are no longer financially viable. And the funding needed to sustain mines throughout their normal lives will also be significant to replace old machinery and equipment, for major maintenance and overhauls, and to introduce new technology into the operations. If Australian banks and other financiers turn their backs on thermal coal mining companies, it is likely that banks and financiers overseas will step in to ensure a continuing flow of coal to key consuming industries.

While trends in international markets point to a positive outlook for Australian coal exporters, major constraints on coal mining companies in Australia may not be just market conditions or the lack of finance, but rather obstacles due to planning decisions by governments, delaying tactics by governments and decisions by courts. Governments can claim to be amenable to new development but can impose conditions which deter investment from proceeding.

Australia has some major decisions to make in the coming years. Will we allow the premature closure of coal fired power stations which are providing much needed stability and dispatchable power to the grid? Who will fund the major investments in the electricity grid necessary to support the surge in renewable generation dispersed across large areas in Queensland, NSW and other States? Will Australia be prepared to look at nuclear energy as a realistic option for the future? Will governments and courts continue to allow new thermal coal mines to be developed for the export market? If we see new coal mine projects fail to proceed because of decisions by governments, courts or other stakeholders such as banks and insurers, will we be prepared to see demand for coal move to other supplying countries whose coal quality may be inferior? What will be our stance on the thermal coal produced by mines that are developed largely to produce metallurgical coal?

The coal industry will certainly live in interesting times in the years ahead.

Abbreviations

ACA	Australian Coal Association
AFR	Australian Financial Review
BC	Brisbane Courier
CT	Canberra Times
CM	The Courier-Mail
CPD	Commonwealth Parliamentary Debates
DT	Daily Telegraph
DMR	NSW Department of Mineral Resources
IEA	International Energy Agency
HRA	Historical Records of Australia
HRNSW	Historical Records of New South Wales
IM	Illawarra Mercury
JCB	Joint Coal Board
MM	Maitland Mercury
NAA	National Archives of Australia
NMH	Newcastle Morning Herald; also Newcastle Morning Herald and Miners' Advocate
NSW	New South Wales
NSWCA	New South Wales Coal Association
NSWCCPA	NSW Combined Colliery Proprietors' Association
NSWMC	NSW Minerals Council
NSWPD	NSW Parliamentary Debates
OECD	Organisation for Economic Cooperation and Development
QCB	Queensland Coal Board
QPD	Queensland Parliamentary Debates
QT	Queensland Times (Ipswich)
QGMJ	Queensland Government Mining Journal
QMC	Queensland Mining Council
QRC	Queensland Resources Council
RBA	Reserve Bank of Australia
REQ	Resources and Energy Quarterly, Department of Industry, Innovation and Science
SMH	Sydney Morning Herald
WA	Western Australia

Endnotes

Chapter 1

1   CT 19 August 1971, p.8
2   SMH 26 October 1970, p.19
3   SMH 24 November 1971, p.3
4   ibid
5   NSW Combined Colliery Proprietors' Association Mission to Japan, October 1972 (NSWCCPA 1972)
6   NSWCCPA 1972, October 2 section, p.5
7   Swartz, Ministerial Statement to House of Representatives: Australia's Natural Resources, 28 October 1972
8   NSWCCPA 1972, October 4 section, p.2
9   NSWCCPA 1972, October 4 section, p.8
10   ibid
11   SMH 6 October 1972, p.1
12   Sun Herald 22 October 1972, p.124
13   Swartz, October 1972
14   SMH 23 March 1971, p.7
15   Jay, p.179
16   JCB Annual Report 1972-73, pp.43-44
17   SMH 1 November 1972, p.3; 2 November 1972, p.2
18   JCB Annual Report 1972-73, p.237
19   ibid
20   JCB Annual Report 1974-75, pp.23,45
21   NAA, Series A5908 submission 630 Export of Black Coal, April 1972
22   Utah Development Company Submission to Senate Select Committee on Foreign Ownership and Control, 11 May 1972, pp.11-12
23   NAA, Submission 3 Export of Black Coal, 2 January 1973
24   JCB Annual Report 1972-73, p.97
25   Quoted in JCB Annual Report 1972-73, p.23
26   Hansard 23 October 1973
27   Department of Minerals and Energy 1975 p. 1
28   Bowen and Gooday, Economics of Coal Export Controls, p. 1
29   Bowen and Gooday, p. 1
30   Australian Dictionary of Biography online
31   Whitlam p. 254
32   Menadue p. 126-127
33   T Fitzgerald, The Contribution of the Mineral Industry to Australian

Welfare, 1974

34   SMH 19 April 1974, p.23

35   CT 26 June 1974, p.19

36   Fitzgerald, p.19

37   CT 17 April 1974, p.15

38   Jay, p.180

39   CT 17 July 1972, p.7

40   Foreign Investment – Statement by the Leader of the Opposition, 2 May 1975

41   SMH 26 October 1972, p.3

42   CT 30 October 1973, p.1

43   CT 30 October 1973, pp.1,12

44   CT 1 November 1973, p.2

45   Department of Minerals and Energy 1975 p. 5

46   CT 5 December 1973, p.23

47   CT 8 August 1974, p.1

48   There are differing versions of these projects to be funded. See Commonwealth Parliamentary Debates, 9 July 1975; G Whitlam, The Whitlam Government 1972-1975 p. 254; and B Hayden, Hayden An Autobiography, p.250

49   SMH 12 June 1975, p.9

50   CT 24 October 1974, p.11

51   F Crean, CPD, 30 October 1975

52   NAA, Decision 3096, Submission 1515, Newcastle Coal Loader, 7 January 1975

53   Menadue p. 127

54   CPD, 21 October 1976

55   SMH 21 August 1975, p.20

56   SMH 20 August 1975, p.13

57   SMH 21 August 1975, p.20

58   ibid

59   SMH 6 September 1975, p.55

60   SMH 10 September 1975, p.30; 12 September 1975, p.20

61   Media Release – Coal Export Tax, Rt Hon J D Anthony, 20 August 1975

62   CT 6 November 1971, p.19

63   CPD, 5 March 1975

64   Press Release, Deputy Prime Minister and Treasurer, 31 March 1976.

65   CT 13 October 1975, p.3

66   JCB Annual Report 1973-74, p.22

67   S Everett, Port-oriented coal transport infrastructure: an analysis of locational decision-making, University of Wollongong, 1984, p.66

68   D. Anthony, Foreign Investment in Australia's Resources, Speech to Chamber of Manufactures of W.A., 28 July 1975

69   CPD, 1 April 1976

70   CPD 1 April 1976

71   JCB Annual Report 1973-74, p.110

72   David Lee in The Second Rush (pp.239-244) goes into some detail on the 1974 and 1975 negotiations, including Connor's role

73   Lee, The Second Rush, p.240

74   CT 7 July 1975, p.3 and Lee, p.243

75   CT 7 July 1975 p.3

76   JCB Annual Report 1973-74, p.109

77   SMH 12 May 1975, p.19

78   SMH 13 December 1975, p.33

79   JCB annual reports

80   JCB Annual Report 1974-75, p.113

81   JCB Annual Report 1974-75, p.112

82   AFR 26 August 1976

83   CT 4 April 1978, p.15

84   ibid

85   CPD 5 April 1978

86   ibid

87   CT 14 October 1978, p.1

88   SMH 25 October 1978, pp1,6

89   Statement of Minister for Trade and Resources, 24 October 1978

90   CT 7 December 1978, p.39

91   JCB Annual Report 1979-80, p.14

92   Butlin Archives, Deposit Z224, Box 38, letter from ACA to Australian Mining Industry Council, 8 December 1975

93   ACA letter to TPC 11 March 1976, Butlin Archives, Deposit Z224, Box 39

94   ACED Minutes of Second Meeting 24 February 1976, Butlin Archives, Deposit Z224, Box 39

95   CT 2 July 1973, p.8

96   ibid

97   SMH 16 Oct 1973, p.19

98   Lee, p.240

99   AFR, 14 May 1974

100   Butlin Archives Deposit Z224, Box 40, Minutes of special meeting of ACA, 6 September, 1978

101   ACA Report on Export Committee, 25 October 1978, Butlin Archives, Deposit Z224 Box 40

102   ibid
103   Sun Herald 22 June 1975, p.103
104   ibid
105   SMH 16 October 1976, p.1
106   CPD 3 September 1975
107   Galligan, p.141
108   ibid
109   Cabinet Decision 1296 and attachments, 21 July 1976
110   Cabinet Decision 3489, 27.July 1977
111   This section is based largely on material in the QCB annual reports, for
       the years from 1974-75 to 1979-80.
112   Murray, p.303
113   CT 29 July 1975, p.3
114   QCB Annual Report 1975-76, p.21
115   Hansard 26 August 1976, p.52
116   QCB Annual Report 1976-77, p.20
117   QCB Annual Report 1976-77, p.18
118   The QEGB was established to merge the various regional authorities
       including the Southern Electricity Authority
119   Galligan, p.111
120   ibid
121   CT 7 August 1980, p.1
122   CT 2 August 1980, p.1
123   CT 3 September 1980, p.1
124   CT 5 September 1980, p.3
125   D Yergin, The Prize, p.5
126   BP Statistical Review of World Energy 2015
127   Yergin,p.598
128   Yergin p,599
129   SMH 4 June 1975, p.16
130   CT 10 Sept 1975 p.15
131   CT 11 February 1976, p.1
132   CT 10 September 1977, p.15
133   SMH 30 June 1977, p.15
134   SMH 3 July 1976, p.1
135   SMH 17 March 1977, p.19
136   CT 3 February 1979, p.20
137   Yergin, p.680
138   Yergin, p.688
139   IEA, The First Twenty Years, Volume Two, p.172
140   IEA, The First Twenty Years , Volume Two, p.171

141   IEA, The First Twenty Years, Volume Two, p.174

142   IEA, The First Twenty Years, Volume Two, p.175

143   IEA, The First Twenty Years, Volume Two, pp.176-177

144   Election speech, 30 September 1980

Chapter 2

145   E Kimura, Interest Returns to northeastern coalfields, Mineral Exploration, Spring 2011

146   Contract prices are from Coal Information 2000 edition, International Energy Agency, Part 1, p.73

147   SMH, 17 September 1983, p.27

148   www.northernminer.com/news/quintette-contract-talks-threaten-production/1000172066/ accessed 4 April 2017

149   SMH 12 May 1980, p.32

150   SMH 3 June 1980, p.13; CT 3 June 1980, p.1

151   SMH 3 September 1980, p.7

152   ibid

153   Joint Coal Board, Prospective Expansion - black coal industry in Australia, 1980. p.13

154   CT 11 November 1981, p.32

155   SMH 18 June 1983, p.26

156   SMH 18 Oct 1980 p.34

157   SMH 18 Oct 1980 p.34

158   SMH 27 Oct 1981 p.19

159   SMH 6 Nov 1982, p.26

160   SMH 23 August 1983, p.26

161   SMH 30 October 1984, p.14

162   Joint Coal Board, Prospective Expansion, p.14

163   QGMJ, December 1980, p.576

164   QCB Annual Report 1984-85, p.21

165   SMH 25 March 1987, p.1

166   Industry Commission, Energy Generation and Distribution, 1991,Vol.2 p. C6

167   SMH 23 May 1981, p.30

168   SMH 8 April 1981, pp.8-9

169   SMH 21 January 1981, p.12

170   SMH 18 April 1981, p.5

171   SMH 27 May 1981, p.3

172   SMH 4 November 1981, p.30

173   QCB Annual Reports: 1983-84, p.33; 1989-90, p.16

174 AFR 22 March 1983, pp.8,15
175 SMH 8 August 1985, p.3
176 CT 21 August 1985, p.7
177 Denney, p.89
178 Denney, p.89
179 ibid
180 CT 19 June 1984, p.12
181 R Stuart, Resources Development Policy: The Case of Queensland's Export Coal Industry, in A Patience (ed) The Bjelke-Petersen Premiership, 1985, p.61
182 Galligan, p.64
183 QPD, 3 December 1968, p.1979
184 QPD 10 September 1974, p.660
185 Galligan, p.209
186 Ibid
187 Galligan, p.206
188 QR submission to Productivity Commission Inquiry into the Australian Black Coal Industry, 1997, p.10
189 CT 17 May 1984, p.2
190 Ibid
191 CM 13 May 1984, p.38
192 CM 19 May 1984, p.1
193 CT 6 June 1984, p.24
194 CM 1 June 1984, p.1
195 CM 19 June 1984, p.12
196 Butlin Archives, Deposit Z224, Box 40, Letter from NSWCCPA to Premier, 6 September 1979
197 Productivity Commission, The Australian Black Coal Industry, 1998, Vol.1, p.282
198 SMH 8 September 1982, p.12
199 SMH 2 June 1982. P.7
200 SMH 2 June 1982, p.7
201 SMH 27 May 1982, p.2
202 SMH 1 June 1982, p.3
203 SMH 28 May 1982, p.10
204 SMH 25 March 1983, p.3
205 NSWPD, 23 March 1983, p.4959
206 SMH 15 June 1982, p.28
207 AFR 22 March 1982, p.3
208 SMH 5 May 1984; data prepared by Dr Ian Story, Meares and Philips
209 Industry Commission, Mining and Minerals Processing in Australia,

1991, Vol.3, pp.433,436

210   SMH 14 December 1982, p.28. Survey was by Coopers & Lybrand; coal results were extracted from the national mining industry survey results.

211   SMH 11 December 1985, p.40

212   SMH 16 December 1985, p.19 and 17 February 1987, p.22

213   SMH 22 July 1987, p.3; CT 22 July 1987, p.7

214   SMH 16 November 1987, p.7

215   SMH 21 June 1989, p.16

216   SMH 6 August 1989, p.157

217   SMH 30 July 1989, p.41

218   SMH 24 September, p.40

219   J Welsh, Jack Welsh: What I've learned leading a great company and great people

220   CT 30 January 1983, p.1

221   CT 29 January 1983, p.1

222   Thompson and Macklin, The Big Fella, p.160

223   Thompson and Macklin, p.166

224   Thompson and Macklin, p.153

225   SMH 4 April 1984, p.26

226   Galligan, p.42

227   CM 24 October 1979, p.1

228   SMH 25 October 1979, p.17

229   CT 12 January 1980, p.16

230   SMH 5 December 1992, p.40

231   CT 27 November 1980, p.25

232   ibid

233   CT 29 April 1987, p.20

234   CT 19 October 1988, p.42

235   CT 8 June 1988, p.28

236   SMH 25 October 1992, p.36

237   C. Copeman, Light on the Hill: Industrial Relations Reform in Australia: The Robe River Affair,   http://archive.hrnicholls.com.au/archives/vol3/vol3-8.php

238   SMH 4 July 1987 p.30

239   SMH 2 July 1987, p.13

240   SMH 25 November 1986, p.28; 24 June 1987, p.3

241   K Killin, Drovers Diggers and Draglines, A History of Blair Athol and Clermont, p.68

242   SMH 7 November 1974 p.23

243   CT 23 May 1975, p.11

244   CT 2 February 1978 p.11

245 NAA, Cabinet Decision 5536 May 1978 and Decision 5705 June 1978
246 Australia Japan Coal Conference papers, April 1978, p.22
247 Sun Herald 23 July 1978, p.106
248 SMH 16 April 1980, p.27
249 CT 18 October 1977, p.1
250 SMH 19 October 1977, p.23
251 CSR's submission to the Government is contained in NAA Cabinet Decision 5705 of June 1978
252 SMH 18 July 1978, p.28
253 CT 13 January 1979, p.21
254 SMH 22 October 1972, p.124
255 CT 19 July 1979, p.11
256 SMH 26 October 1982, p.28
257 SMH 30 March 1983, p.1
258 CT 29 March 1983, p.10
259 Ibid
260 CT 31 March 1988, p.3
261 NAA, Cabinet Decision 2434, 10 November 1983
262 SMH 3 May 1985, p. 18
263 SMH 31Oct1985 p.21
264 NAA, Decision 8491, I September 1986, Submission 4309
265 SMH 21 February 1986, p.20
266 SMH 18 June 1985, p.20
267 CT 18 September 1986, p.16
268 SMH 18 September 1986, p.1
269 ibid
270 SMH 26 September 1986, p.21
271 SMH 29 October 1986, p.2
272 SMH 16 March 1987, p.29
273 SMH 14 July 1987, p.2
274 CPD 25 November 1986
275 ibid
276 SMH 23 July 1987, p.3
277 SMH 21 August 1987 p.2
278 SMH 5 September 1987, p.3
279 SMH 4 April 1988, p.24
280 SMH 4 April 1988, p.4
281 AFR, 21 April 1988, p.6
282 CT 22 April 1988, p.15
283 SMH 22 April 1988, p.26
284 CT 6 May 1988, p.3

285   SMH 6 May 1988, p.4
286   SMH 9 August 1988, p.1
287   Media release, Minister for Primary Industries and Energy, 5 May 1988
288   Data from ABS Year Book Australia, various years
289   SMH 26 October 1982, p.28
290   SMH 26 October 1982, p.13
291   SMH 25 January 1983, p.2
292   CT 5 May 1983, p.7
293   SMH 11 May 1983, p.10
294   SMH 19 May 1983, p.3
295   CT 31 May 1983, p.8
296   AFR 1 July 1986, p.8
297   SMH 22 July 1986, p.5
298   ibid
299   SMH 5 March 1987, p.5
300   NAA, Cabinet Decision 1042, 5 July 1976
301   NAA, Cabinet Decision 2846, 3 May 1977
302   SMH 7 March 1987, p.7
303   SMH 7 March 1987, p.7
304   CPD 28 October 1987
305   CT 12 May 1986, p.1
306   P Costello, In Search of the Magic Pudding, The Dollar Sweets Story, http://archive.hrnicholls.com.au/archives/vol5/vol5-5.php
307   SMH 12 November 1986, p.2
308   CT 25 March 1987, p.16
309   CT 6 October 1989, p.5
310   SMH 14 September 1988, p.14
311   SMH 7 October 1988, p.4
312   CT 8 October 1988, p.2
313   IEA, Energy Policies and Programs of IEA Countries, 1980 Review, p.183
314   Ibid
315   CT 9 November 1986, p.7
316   S Culter, Managing Decline: Japan's coal industry restructuring and industry response, p.143
317   IEA, Energy Policies of IEA Countries: Japan 2003 Review, p.72
318   Media Release, Treasurer, 1 August 1990

Chapter 3

319   CT 28 May 1990, p.8

320   SMH 19 May 1990, p.39
321   ibid
322   SMH 22 February 1993, p.34
323   SMH 25 March 1993, p.22
324   SMH 11 May 1991, p.11
325   J Haywood, Dartbrook mine – a case study,
326   SMH 25 January 1995, p.7
327   Hansard 10 October 1995
328   Anglo Coal buys into Foxleigh coalmine, Courier Mail, 26 December 2007
329   CT 26 January 1994, p.17
330   CT 1 April 1994, p.13
331   IEA, International Coal Trade, The evolution of a Global Market, 1997, p.47
332   CT 2 March 1994, p.14
333   J McPhillips, Wharfies and Miners, p.45
334   CT 15 March 1994, p.17
335   McPhillips, p.46
336   SMH 18 April 1994, p.7
337   ACIL/ Australian Coal Association, Winning Coal: Securing a Viable and Expanding Future for Australia's Black Coal Industry, 1994, pp.24-25.
338   CFMEU United Mineworkers, Out of the Red into the Black, 1994,pp.i-ii
339   CFMEU 1994, p.iii
340   Study of the Australian Black Coal Industry, A Report to the Australian Coal Consultative Council, Canberra, November 1994 (Taylor Report), p.31
341   Taylor Report, pp.12-13
342   Taylor Report, p.13
343   CT 6 March 1995, p.2
344   Media Release, Minister for Resources, 1 August 1995
345   JCB Annual Report 1967-68, p.14
346   JCB Annual Report, 1971-72, p.122
347   JCB Annual Report, 1972-73, p.23
348   Cabinet Submission 630 (withdrawn), Export of Black Coal, 4 April 1972
349   M Byrnes, Australia and the Asia Game, Allen & Unwin, Sydney, 2006. This is the second edition of the book; the first edition was published in 1994.
350   Byrnes, 2006, pp.17-19
351   Byrnes, 2006, p.19

352   Byrnes, 2006, p.81

353   IEA, Coal Information 2000, p.173

354   Asia Pacific Strategy, Submission to Productivity Commission Inquiry into Australian Black Coal Industry, 5 February 1998, p.5

355   Asia Pacific Strategy, p.60

356   ABARE, Quality adjusted prices for Australia's black coal exports, 1997, p.55

357   Camberwell, pp.2-3

358   CPD, 26 November 1986

359   B Bowen & P Gooday, The Economics of Export Controls, ABARE Research Report 93.8, 1993, p.47

360   ABARE, 1993, p.47

361   ABARE, 1993, p.48

362   NAA, Cabinet Submission 844, Joint Coal Board (and Coal Industry Tribunal), 1976

363   ibid

364   Butlin Archives, Deposit Z224, Box 39, Minutes of meeting of ACAED, 15 September 1977

365   B Kelman, Review of the Joint Coal Board, February 1991, p.6

366   SMH 17 February 1984, p.4

367   SMH 17 February 1984, p.4

368   Kelman, p.11

369   Kelman p.17

370   Productivity Commission, The Australian Black Coal Industry, 1998, p.XLIV

371   Jeff Shaw had acted as a barrister for the CFMEU on a number of occasions before he entered Parliament as a member of the Legislative Council.

372   Coal Industry Bill 2001 debate, 6 December 2001

373   Industry Commission, Mining and Minerals Processing in Australia 1991, Vol.3, p.554

374   ibid

375   QPD 10 April 1991, p.7013

376   Industry Commission 1991, Vol.3, p 555

377   QCB Annual Report 1989-90, p.4

378   QPD 30 November 1992

379   QCB Annual Report 1991-92, p.12

380   www.queenslandspeaks.com.au/peter-ellis - accessed 15 March 2016

381   Hansard 29 October 1997, pages 4037 – 4040.

382   QPD 29 October 1997 p.4039

383   SMH 22 April 1993, p.1

384   CT 17 April 1991, p.1

385   SMH 29 April 1988, p.2

386   Industry Commission, 1991, Vol.3, p.545

387   National Archives, Cabinet Minute 15066 of 26 March 1991

388   Australian Coal Association, Industrial Relations Reform for the Australian coal industry, August 1993, p.3 quoted in Sheldon and Thornthwaite, Employer Associations and Industrial Relations Change, p.133

389   Industry Commission 1991, Vol.3, p.547

390   Industry Commission 1991, Vol.3, pp.545-546

391   Common Cause, July 1994, p.6

392   ABC Radio PM Program, 19 April 1994

393   D Watson, Recollections of a Bleeding Heart, pp.361-362

394   CT 21 April 1994, p.1

395   Common Cause, July 1994, p.6

396   ibid

397   AFR 28 July 1994, p.8

398   Common Cause, November 1994, p.6

399   SMH 18 October 1994, p.2

400   SMH 21 October 1994, p.8

401   CPD Second Reading Speech 2 November 1994

402   Media Release, 9 December 1994

403   Common Cause, December 1994/ January 1995, p.6

404   ibid

405   Graham Gillespie, QMC industrial relations manager, quoted in Sheldon and Thornthwaite, Employer associations and industrial relations change, p.133

406   SMH 7 December 2013 Business p.10

407   M Wooden and F Robertson, Employee Relations Indicators: Coal Mining and Other Industries Compared, NILS, Flinders University of South Australia, June 1997, p.i

408   Wooden & Robertson, p.28

409   CPD 23 May 1996 page 1295

410   CT 5 September 1993, p.5

411   M Lee, Griffith Law Review 2002, Vol 11 No 1, p.171

412   Coal War Flares Up: Miners At BHP Strike, SMH 19 September 1997

413   SMH 30 September 1997, p.1

414   Lee p.182

415   ABC PM program, 25 June 1999

416   ibid

417   SMH 16 February 1996, p.5

418 A Davies, Coal Reform: The Hunter Valley No. 1 Story. Presentation to HR Nichols Society, 2001
419 SMH 3 April 1997, p.3
420 SMH 3 April 1997, p.4
421 SMH 28 May 1997, p37
422 ibid
423 SMH 23 October 1997, p.6
424 NH 21 October 1998, p.1
425 Davies, 2001
426 Productivity Commission, 1998 p.87
427 ABS data from Productivity Commission 1998, p.117
428 Ann Hawke and Frances Robertson, Enterprise Bargaining in the Coal Industry: Implications for Work Practices, National Institute of Labour Studies, 1996. Quoted in Rio Tinto submission to Productivity Commission Inquiry, 1998.
429 Productivity Commission, 1998, Vol.1, p.109
430 CT 9 April 1992, p.2
431 The following quotes are from BHP's appearance at the Productivity Commission inquiry's hearings in Brisbane, 24 November 1997, pp.175-199
432 J Hayward, Dartbrook Mine – a case study, Coal Operators' Conference 1998, University of Wollongong Research Online, p.225
433 ibid
434 Camberwell Coal Submission to Productivity Commission Inquiry into Australian Black Coal Industry, 1997, p.6
435 BHP Billiton submission to Productivity Commission Inquiry into Workplace Relations Framework, March 2015, p.9
436 BHP Submission to Productivity Commission, p.12
437 SMH 13 November 1981, p.35
438 SMH 1 February 1983, p.2
439 SMH 9 Dec 1982 p.9 and SMH 15 Sept 1982 p.1
440 SMH 10 February 1989, p.27
441 SMH 2 May 1997, p.36
442 SMH 29 April 1997, p.1
443 Industry Commission, Minerals and Minerals Processing in Australia, 1991, Vol.1,p.104
444 Productivity Commission 1968, Vol. 1 p.281; based on submissions by QR and the Queensland Government
445 Queensland Mining Council submission to Productivity Commission inquiry into the Australian Black Coal Industry, November 1997. This section of chapter is based on the details of that submission.

446   Productivity Commission 1968, Vol. 1 p.281

447   Industry Commission 1991, Vol.3, p.431

448   Productivity Commission, Micro Economic Reforms and Australian Productivity: Exploring the links; 1999, Vol. 2 Case Studies, p.111

449   PC Micro Economic Reforms 1999, p.114

450   IPART, Aspects of the NSW Rail Access Regime – Final Report, April 1999, p.11

451   Productivity Commission 1998, p.210

452   IPART defined efficient costs as an "estimate of total costs assuming best practice industry benchmarks are achieved across all activities" IPART 1999, p.11

453   NH 2 August 1946, p.5

454   Australian Academy of Technological Sciences and Engineering ,Technology in Australia 1788-1988, Online edition 2000, p.755

455   People Passion and the Pursuit of Excellence, 20 years of ACARP, The Edge Public Relations, 2012, p.4

456   Coal Industry Policy Statement, Department of Primary Industries and Energy, January 2002

457   20 years of ACARP, p.6

458   20 years of ACARP, p. 22

459   ACARP Annual Report 2019, p.8

460   Australian Academy of Technological Sciences and Engineering, Technology in Australia 1788-1988, p.755

461   NSW Mine Safety Performance Reports, NSW Resources Regulator, various years.

Chapter 4

462   Queensland Mines and Quarries - Safety and Health Performance Reports, various years

463   Queensland Dept Mines Annual Report 1972, p.82

464   Explosion at Appin Colliery 24 July 1979, Report of His Honour Judge A J Goran, p.169

465   Goran Report, p.170

466   Goran Report, p.173

467   Howard Jones, Spontaneous Combustion is Deadly in Underground Mines, Queensland Dept Mines in conjunction with Queensland Coal Owners Association and Queensland Colliery Employees Union, 1976

468   Moura No. 2 Warden's Inquiry Report p.60

469   Moura No. 2 Warden's Inquiry Report p.46
470   A Hopkins, Managing major hazards: the lessons of the Moura Mine Disaster,p.4
471   SMH 18 January 1996, p.1
472   Gretley Inquiry report p. 692, 694
473   Gretley report p. 687
474   Gretley Inquiry report p.688
475   Gretley Inquiry report p.746
476   Gretley Inquiry report p.748
477   ABC *PM*, 12 November 1999. See www.abc.net.au/pm/stories/s66107. htm. From Ken Phillips, The Politics of a Tragedy, Institute of Public Affairs, October 2006 p.27
478   Managers guilty over mine deaths, SMH 10 August 1994
479   Hopkins, Gretley, p.15
480   S Johnston paper, Queensland Mining Industry Health and Safety Conference Proceedings, 1997
481   SMH 9 April 1997, p.3
482   SMH 11 April 1997, p.8
483   Johnston, 1997
484   Harrison et al in J Brune (ed) Extracting the Science: A Century of Mining Research, p.289
485   Queensland Mines and Quarries Safety Performance and Health Report 2014-15, p.1
486   Coal Workers' Pneumoconiosis Select Committee, Inquiry into the Re-identification of Coal Workers' Pneumoconiosis in Queensland – Interim Report, March 2017, p.1
487   Coal Workers' Pneumoconiosis Select Committee, Final Report, May 2017, p.18
488   SMH 23 November 2000, p.30

Chapter 5

489   The Age, 23 May 2001, p.3
490   BHP Billiton Annual Report 2001, p.3
491   BHP Billiton Annual Report 2002, Part C, p.29
492   Rio Tinto Annual Report 2008, p.14
493   The Age, 5 June 2009, p.1
494   The Age, 6 June 2009, p.1
495   The Age 11 December 2010, p.1
496   The Age 20 August 2010, p.1
497   CT 8 September 1994, p.15

498   SMH 23 November 1984, p.4

499   The Age 29 August 2000, p.2

500   ibid

501   The Age 29 August 2000, p.2

502   SMH 7 June 2003, p.45

503   GlencoreXstrata CEO hails 'managers as owners' model, The Australian, 20 May 2013

504   ibid

505   Glencore boss berates BHP, Rio CEOs for new mines, SMH 26 February 2013

506   REQ December 2013, p.16

507   Glencore media statement 8 August 2017

508   Glasenberg shows his distrust of greenfields coal, AFR 14 May 2013

509   ibid

510   Glencore media release 25 October 2013

511   Glencore, Rio Tinto, BHP Billiton defy coal supply glut, The Australian, 13 September 2014

512   SMH 17 August 2015, p.8

513   Rio confirms rejection of $182b Glencore merger bid, AFR 8 October 2014

514   Glencore Media statement, Furthering our commitment to the transition to a low-carbon economy, 20 February 2019

515   Glencore Annual Report 2018, p.21

516   Glencore's coal cap a comfortably loose fit, AFR 21 February 2019

517   ibid

518   Macquarie research report, quoted by Tim Treadgold in www.forbes.com 12 October 2018

519   Rio Tinto Annual Report 2018, p.47

520   Peabody Energy, New York, News Release 17 June 2010

521   ibid

522   Coal titan Peabody Energy files for bankruptcy, Washington Post, 13 April 2016

523   G Boyce, Energising the world one BTU at a time, October 2012.

524   ibid

525   ibid

526   Peabody sees role for coal in energy policy 17 April 2018, The Australian 17 April 2018

527   SMH 25 Sep 1989, p.1

528   SMH 25 Sep 1989, p.1

529   SMH 20 Dec 1989 p.28

530   SMH 1 Dec 1990 p.3

531   SMH 30 Oct 1992 p.5
532   SMH 9 Dec 1992 p.9
533   SMH 24 December 1992, p.6
534   NSWPD 29 April 1993
535   Questions on Notice, 23 May 1993.
536   Powercoal CEO Phillip McCarthy evidence to Productivity Commission,
        17 November 1997, pp.17-18
537   NH 14 August 2001, p.5
538   NH 13 August 2001, p.12
539   NH 16 August 2001, p.13
540   Final report of the Special Commission of Inquiry into Electricity Trans-
        actions, Hon B J Tamberlin QC, October 2011, p.120 (Tamberlin Report)
541   NSW Govt sells Eraring cancels Cobbora contract, AFR 2 July 2013
542   Tamberlin report, pp126-127
543   Tamberlin Report, p.125
544   Source: World Bank. These figures are in current prices.
545   BP Statistical Reviews of World Energy, various years
546   World Coal, October 2018, p.12
547   ibid
548   J Melanie, The World Coal Market: Prospects to 2010, ABARE Con-
        ference Paper 02.6, Joint 8th APEC Coal Flow Seminar, Kuala Lumpur,
        March 2002.
549   Rio Tinto China, China's Coal Export Potential, September 2001.
550   Department of Energy, Energy Information Administration, Interna-
        tional Energy Outlook 2002, Table A.2
551   SMH 12 August 2016, p.3
552   www.australianmining.com.au 12 August 2016
553   Coal in India, Department of Industry, Science and Technology, 2019
554   Coal in India, Department of Industry, Science and Technology
555   Provisional Coal Statistics 2015-16, p.24
556   Adani downsizes coal ambitions, The Australian, 1 November 2018
557   Carmichael Coal Mine and Rail Project, Environmental Impact State-
        ment, Volume 2 Section 2, p.2-9
558   Media Release 11 March 2019, Sojitz.com
559   BHP Billiton Annual Report 2011, p.34
560   BHP says no return for quarterly coking coal contract, AFR, 7 May 2018
561   Queensland Rail Annual Report 2006-07, p.12
562   S Baines, Steel on Steel, pp.17-18
563   QR Annual Report 2006-07, p.12
564   AFR, 18 December 2009, p.5
565   AFR 18 December 2009, p.5 and Baines, p.81

566   Miners bid for slice of Queensland Rail sale, The Australian, 12 December 2009

567   ibid

568   BHP behind bid to buy Queensland Rail, The Australian, 16 December 16, 2009

569   Coal giants eye Queensland Rail bids, The Australian, 13 March 2010

570   Queensland Rail chairman rails against float detractors, The Australian, 28 April 2010

571   Fraser hits back over Ferguson's QR privatisation 'disaster' claims, The Australian, 25 March 2010

572   Mining firms offer $4.85b for QR assets, SMH, 26 May, 2010

573   AFR 2 June 2010, p.3

574   Coal miners offer $4.85 billion for Queensland's coal track network, The Courier-Mail, 26 May,2010

575   Greiner's miners boost their bid for Queensland rail infrastructure, The Australian, 10 August 2010

576   QR bid withdrawal an embarrassment for coalminers, The Australian, 10 September 2010; Anglo shaky from the start of QR consortium, The Australian, 11 September, 2010

577   Baines, p.111

578   Baines, p.86

579   Aurizon Investor Presentation FY2019 Results, p.45

580   Joint Media Release, Combined Sale of Freightcorp and National Rail, 31 January 2002

581   P Kelly, Triumph and Demise - The Broken Promise of a Labor Generation, p.275

582   Kelly, p.275

583   The Treasury, Strong Growth, Low Pollution – Modelling a Carbon Price, 2011, p.139

584   See transcript of media conference www.liberal.org.au

585   www.theaustralian.com.au 12 July 2011

586   Commonwealth of Australia, Securing a Clean Energy Future, the Australian Government's Climate Change Plan, 2011, p.134

587   Department of Resources Energy and Tourism, Coal Sector Jobs Package, Frequently Asked Questions, 2012

588   101 Government's climate change economist strikes out at coal claims, The Australian, 22 June 2011

589   102 Centre for International Economics, Coverage of coal mine fugitive emissions in climate policies of major coal exporting countries, June

590   103 Carbon legislation "special clause" targets coal for punishment, ACA

media release 22 August, 2011

591   More fuel on carbon fire, Sydney Morning Herald, 10 October 2011,

592   Henry K et al, Australia's future tax system, Report to the Treasurer, Part One, December 2009, Commonwealth of Australia, 2010, p.89 (Henry Review)

593   Budget Papers 2009-10: Budget Paper 1 Statement 2 Economic Outlook p.2-31

594   Henry Review Part Two Vol.1 p.218

595   P Kelly, Triumph and Demise - The Broken Promise of a Labor Generation

596   Kelly, p.295

597   Kelly, p.298

598   Kelly, pp.297-298

599   Kelly. P.298

600   Kelly. P.302

601   Kelly, p.303

602   Kelly, p.308

603   BRW, Vol 32 Number 3, January 2010

604   Swan W, Treasurer of Australia, Treasurer's Economic Note, 9 May 2010

605   Rio Tinto Media Release 2 May 2010

606   Sydney Morning Herald, Weekend Business, 22 May 2010, p.6

607   Kelly, pp.310-311

608   C Richardson, Mining tax ratios revisited, Minerals Council of Australia, March 2015, p.28

609   Kelly, p.344

610   The Australian, Treasury exposes mining tax flaw, 15 February 2013

611   Budget Statement 2011-12, Chapter 5, p.3

612   Richardson, p.15

613   Deloitte Access Economics, Minerals Industry Tax Survey, Minerals Council of Australia, December 2017, p.vi

614   Tax Facts, October 2019

615   BHP Economic Contribution Reports 2015, 2016, 2017,2018 and 2019

616   SMH 27 June 2014, p.1

617   SMH 30 June 2014, p.11

618   Glencore Australia Tax Fact Sheet (2013), p.2

619   Glenore Australia Tax Fact Sheet (2016), p.1

620   Voluntary Tax Transparency Australia Report 2018, Glencore, p.2

621   BHP settles longstanding transfer pricing dispute, BHP media release, 19 November 2018; BHP Annual Report 2019, p.296

622   Rio Tinto Annual Report 2016, p.210

623   Rio Tinto Annual Report 2018, p.91

624   Productivity in mining operations: Reversing the downward trend, McKinsey & Company metal and Mining Practice, May 2015, p.1

625   Tasman Asia Pacific, Benchmarking the Productivity of Australia's Black Coal Mines, 1998, p.xv

626   Data for employment in Queensland coal mines appears to be subject to major variations and its accuracy from year to year is questionable. The 2016-17 figure is based on total saleable production of 238 million tonnes and employment at June 2017 of 30925(Department of Natural Resources and Mines).

627   Coal: The path to improve results, Presentation by Mike Henry, BHP Coal, 21 June 2016

628   Calculated from BHP accounts. Operating costs - revenue minus EBITDA – divided by production.

629   Queensland Mines and Quarries, Safety Performance and Health 2015-16 and 2016-17; Coal Services P/L annual reports

630   REQ March 2019

631   NSW number refers to production employment as at June 2018 - from Coal Services P/L Annual Report   2017-18, p.49. Queensland number is from www.dnrm.qld.gov.au as at June 2017. WA number is an estimate based on Premier Coal's employment of around 400 and production of around 4 million tonnes – www.premiercoal.com.au. Griffin Coal produces over 3 million tonnes – www.griffincoal.com.au. Tasmania number is from Cement Australia website www.cementaustralia.com.au which says employment at its subsidiary Cornwall Coal is more than 70.

632   Department of Environment and Energy, Australian Energy Update 2019, p.29

633   Deloitte Access Economics, Mining and MTS, 2017, p.6

634   Deloitte Access Economics, p.ii

635   This is estimated actual from Qld Budget and Strategy 2019.20. NSW number is from Budget Statement 2019.20, Revenue, p.4.4 which has total mining royalty revenue in 2018.19 as $2074 million. Coal accounts for the great proportion of mining royalties in NSW.

Chapter 6

636   AGL outlines early-exit pathways from coal power, The Australian, 30 June 2020

637   Department of Industry, Science and Resources, Technology Investment Roadmap Discussion Paper, May 2020. P.3

638   IEA World Energy Outlook Executive Summary and Outlook by Scenario – Coal. From www.iea.org

639 EIA, International Energy Outlook 2019; these forecasts quoted are the EIA's "reference case"

640 The EIA's coal numbers are expressed in terms of short tons; the data has been converted here to tonnes. The EIA data for coal differ from, for example, data published by BP.

641 Department of Industry, Science & Innovation, Coal in India 2019, p.52

642 COAG Energy Council, Australia's National Hydrogen Strategy 2019, p.68

643 REQ June 2020, p.60

644 REQ, June 2020 p.61

645 ibid

646 www.businesslive.co.za 9 February 2018

647 IEA, South East Asia Energy Outlook, October 2019, p.99

648 IEA, South East Asia Energy Outlook, p.109

649 uk.reuters.com 3 July 2020

650 Arrow Research Ltd report in The Coal Hub Issue 189, 17 July 2020

651 New Generation Coal: Reliable, Secure and Low Emissions, Minerals Council of Australia media release, 17 January 2017

652 Dr Ian Barnes, HELE Perspectives for Selected Asian Countries, IEA Clean Coal Centre, June 2018

653 Technology Investment Roadmap Discussion Paper, p.45

654 Global CCS Institute, The Global Status of CCS: 2018, p.17

655 See Tracking Power 2020 at iea.org/tcep

656 www.callideoxyfuel.com

657 D Turk, International Energy Agency, in The Global Status of CCS: 2017, Global CCS Institute, p.19

658 ibid

659 The Australian, Australian Coal Special Report, 23 March 2019, p.1

Glossary of terms

Anthracite
    A hard variety of coal, having a very high carbon content (over 90%) and low moisture content; mainly used for residential and commercial heating. In geological terms, the oldest form of coal.

Ash
    The material remaining after coal is burnt eg in a boiler. Typically composed mainly of iron, silica and alumina.

Australian workplace agreement (AWA)
    An individual, legally enforceable agreement between an employer and employee about the employee's terms and conditions of employment.

Base load
    The component of electricity generated by power stations which is continuously in operation. In Australia the base load is typically fired by coal in NSW and Queensland and by brown coal in Victoria.

Benchmark pricing
    The system which applied until the 1990s/ early 2000s whereby the Japanese steel mills would agree to negotiate collectively through one of the mills as the lead negotiator. The negotiated price became the benchmark accepted by the other mills, with variations in price due to quality factors. The Japanese power utilities (JPUs) adopted a similar system, with one the utilities leading the negotiations and the others adopting the price negotiated, again subject to quality variations. While the benchmark system no longer operates in a formal sense, the JPUs still nominate a lead negotiator each year.

Bituminous coal
    The category or rank of coal that includes most thermal coal and all metallurgical coal. There are sub-bituminous coals which are or were used for power generation (eg from Leigh Creek in South Australia, Callide in Queensland and Collie in Western Australia).

Blast furnace
    A furnace in a steel mill in which coke reacts with iron ore to reduce it to iron.

Bord and pillar
    The dominant type of underground coal mining in Australia until the 1980s and 1990s when longwall mining became more common. A bord is a "heading" or tunnel or road. The coal is removed as the bords are developed, leaving the pillars of coal to support the roof of the area being mined. Pillars can also be extracted under certain conditions.

Brown coal.

Also called lignite. Contains a lower percentage of carbon than bituminous coals and has a high moisture content. Used in Victoria for power generation. Because of its high moisture content it is unsuitable to transport long distances and so is essentially a domestic fuel in Australia.

Carbon capture and storage (CCS)

Carbon capture and storage (CCS) is "A process that captures ...(CO_2) from industrial processes, energy generation, or the atmosphere and stores it underground. (It may) also involve capture, utilisation and storage (CCUS) where CO_2 is used for example to manufacture biofuels, to boost plant growth, or in food and drink production. (Dept Industry, Science & Resources, Technology Investment Roadmap Discussion Paper 2020)

CCS involves capturing carbon dioxide from sources such as power stations, oil wells and LNG plants, compressing it, transporting it to a suitable storage site, and injecting it into deep geological formations where it can be safely and permanently stored.

Carbon tax

A tax based on the emission of carbon dioxide or certain other gases (including methane) into the atmosphere.

Cents per tonne kilometre

A measure used to compare rail or other transport costs. It is the cost to transport one tonne of a commodity one kilometre.

CEO

Chief Executive Officer

Certified agreement

A legally enforceable collective agreement about terms and conditions of employment between an employer and a group of employees or between an employer and one or more unions. Such agreements were certified by the Australian Industrial Relations Commission.

CIF

Cost, insurance and freight. A term used in marketing and transport. When coal is purchased on a CIF basis, the supplier is contracted to arrange and pay for it to be shipped to the customer, with the price including all the costs to the import port, not including any import duties.

Coal

Vegetable matter which has been changed through a process of heat, chemical action and pressure over long periods of time. See also anthracite, bituminous coal, thermal coal, metallurgical (coking) coal, PCI coal and brown coal (lignite).

Coalification

The process whereby heat and pressure turn decomposing plant material to coal over millions of years. Peat is the first stage in this process, but is not regarded as coal. See also Rank.

Coal dust

Fine powdery dust produced by mining which can be explosive under certain conditions. Also dangerous to humans, causing lung disease.

Coal face

The working location where coal is being cut from the seam.

Coalfield

An area or district containing coal. In NSW the major coalfields are the Newcastle, Hunter, Gunnedah, Western and Southern coalfields. In Queensland the major coalfields are commonly referred to as basins, with the North Bowen Basin, South West Bowen Basin, South East Bowen Basin, Surat Basin, Callide Basin and Galilee Basin, plus the Clarence Moreton Basin as the major areas.

Coal handling and preparation plant.

A plant which washes coal to remove impurities and carries out other processes such as screening to produce coal of a certain size. Often simply called a washery.

Coal seam gas

Also called coal seam methane and coal bed methane. The gas which is predominantly methane is contained in seams of coal and is released during mining or by the "fracking" process now common in the USA and in Queensland.

Coal terminal (also called export coal terminal)

The facility at a port where coal is delivered from the mine, stockpiled and finally loaded onto the ship.

Clean coal technologies

Technologies and processes which improve the environmental performance of coal, including by removing impurities, increasing the efficiency with which coal is burnt, and decreasing carbon dioxide, sulphur or other emissions.

CFMMEU

The Construction Forestry Maritime Mining and Energy Union; its members in the coal industry belong to the Mining and Energy division of the union.

Coking coal

See Metallurgical coal

Collier

A term now rarely used. Can refer to someone who mined coal or worked in a coal mine or to a ship used to transport coal.

Colliery

A coal mine, including its infrastructure (roads, plant, machinery, equipment and coal washery).

Commonwealth Government

The national government of Australia; also called the Federal Government.

Continuous miner

An electrically powered machine mounted on tracks which has a cutting arm which rips the coal from the face. The coal is gathered by mechanical arms and loaded onto a conveyor. The machine is also used to drive roadways or headings.

Crib room

An underground area where miners eat their meals.

Custom and practice

The range of informal rules, procedures and standards applying in the coal industry or at different minesites at different times and which at one time were recognised in industry awards and by the Coal Industry Tribunal.

Decline

A sloping tunnel big enough for vehicles, loaders and trucks to drive up and down to get to the coal or ore and take it out.

Demurrage

The cost typically paid by the exporter as a result of port delays. Typical contracts allow for a certain time for a ship to enter a port, load the coal and depart and for payment of a certain amount of demurrage per day if that time is exceeded.

Deputy

A coal industry work classification in NSW. Deputies are underground supervisors.

Dragline

A large machine operated by one person which removes overburden in open cut mines. Draglines have a long arm or boom from which hangs a bucket, with the largest buckets capable of holding over 100 cubic metres of rock and earth.

Drift

An inclined haulage road to the surface. Also a heading used for exploration or ventilation.

Emissions trading scheme (ETS)

These schemes exist in a number of countries and regions and typically place a limit on the volume of carbon dioxide or other gases which can be emitted. Emitters may have a level of emissions which they must not exceed and may be able to buy or sell permits depending on whether they have too few or too many permits.

Enterprise agreement (EA or EBA)

Enforceable documents that cover the employment conditions of a group of employees and their employer. Enterprise agreements or enterprise bargaining agreements can include single or multi-enterprise agreements and must meet a number of tests before they can be approved by Fair Work Australia.

Face

Also coal face. The working area where coal is being extracted.

Federal Government

The national government of Australia; also called the Commonwealth Government.

Fracking

Short for "fracturing". The process involving drilling into seams of coal or shale and injecting a combination of water, sand and chemicals to fracture and open the coal seams and release the gas or oil.

Free on board (FOB)

Most Australian coal is sold on an FOB basis where the customer pays for the costs of shipping, insurance, unloading and other costs. The supplier is responsible for costs up to and including the loading of the coal onto the ship at the export port.

Goaf

That area in a mine from which coal has been extracted. The goaf will generally have been filled up by the collapse of the roof of the mine or can be expected to filled once the roof collapses.

Greenhouse gases

The "greenhouse effect" is a natural phenomenon caused by gases emitted into the atmosphere which trap some of the infrared heat from the sun's rays which are reflected from the earth's surface. Greenhouse gases include water vapour, carbon dioxide (the major man-made greenhouse gas), methane, nitrous oxide and chlorofluorocarbons. Human activities which emit these gases include industrial processes (eg power stations, cement works, steelworks), transport (eg use of petroleum in cars) and agriculture (eg rice farming and livestock).

Hard coal

A term used for example by the International Energy Agency to include thermal coals, metallurgical coals and anthracite. Also a term for black coal (as distinct from brown coal).

HELE technologies

High energy low emission technologies in power stations, including supercritical and ultrasupercritical technologies, allow those stations to operate at higher temperatures and greater pressures and mean that CO_2 emissions per unit of electricity are significantly lower than in older power stations.

Lead negotiator

The major Japanese power utilities nominate one of the utilities each year to be the lead negotiator ie the first to negotiate the next year's price with one of the Australian suppliers. That price tends to be influential in negotiations involving suppliers and other consumers in countries such as Korea and China. See also Benchmark pricing.

Kg

Kilogram

Lodge

The unit or local branch of the CFMMEU at individual mines.

Longwall mining

The predominant form of underground coal mining in Australia which is also very safe and highly efficient. Parallel headings or tunnels are driven up to several hundred metres apart and then connected to form the longwall face. A shearer travels along the coal face cutting the coal. The roof is supported by a series of hydraulic supports. As the coal is cut, the shearer and the supports can all move. As the system "retreats" the roof behind the hydraulic supports caves in.

Metallurgical coal

The type of coal used in the manufacture of steel and other metals; includes hard coking coal, soft coking coal and PCI coal. See also Steelmaking

Mine subsidence

The sinking or drop in the level of the land surface due to underground mining.

Miners' Federation

The major union in the coal industry, now the CFMMEU.

Mt

Million tonnes

Mtpa

Million tonnes per annum

MWh

One megawatt hour equals 1,000 kilowatt hours. A company consuming one megawatt hour of electricity is using 1,000 kilowatts continuously for one hour.

Oil shale

A sedimentary rock which may contain hydrocarbons which can be extracted to produce petroleum products. Oil shale was mined near Lithgow during World War 2 and processed to produce petroleum. It was also mined from the 1850s to produce kerosene for domestic use.

Open cut mining

Mining from an open hole in the ground. Also called open cast mining.

Overburden

The volume of material (rock, dirt etc) removed to uncover the seam of coal (or other mineral) in an open cut mine. Usually expressed in cubic metres. Also known as spoil.

PCI coal

PCI – pulverised coal injection. PCI coals are used in steel making to replace a proportion of higher cost metallurgical coals.

Pit

Can refer to a colliery, or just to the area of an open cut mine being mined, or to the underground workings.

Pit pony

A horse used underground, mainly to haul coal skips.

Pit-top

The surface area at the top of the mine shaft.

Productivity

The productivity of a mine is a ratio of its output to one or more of its inputs. Common measures of labour productivity are tonnes produced per person per year and tonnes per person per hour. Other measures of productivity can include total factor productivity (ie including labour and capital equipment) and productivity measures relating to major items of equipment such as draglines and longwall systems.

Rank

A method of classifying coal based on the amounts of carbon and volatile matter it contains. It signifies the coalification of the organic material or the degree to which the coal has changed and matured. The lowest rank coal is lignite; next come the sub-bituminous coals; then the bituminous coals (thermal and metallurgical); with the highest rank coal being anthracite.

Rehabilitation

In the mining industry this term refers to the process whereby mined land is returned to a state in which it can be used for farming, livestock, forestry, or for other purposes.

Renewable energy

Renewable energy includes the energy derived from hydro power plants, wind turbines, solar thermal equipment, solar photo voltaic equipment, geothermal plants (using the energy from deep underground) and plants burning biomass (eg waste from sugar cane, timber and other crops).

Reserves

In lay terms, reserves of coal or minerals refer to the quantity of coal or ore which can be mined under certain assumptions. Under the Australian standard for reporting (the JORC code), reserves can be proven or probable. The code defines an 'ore reserve' as the economically mineable part of a measured and/or indicated mineral resource.

Roof bolt

Modern underground coal mining and metalliferous mining use roof bolts to strengthen the roof of the mine and minimise the likelihood of the roof collapsing. Roof bolts are drilled into the roof and contain chemicals which solidify and anchor the bolt. Roof bolts have replaced props in modern mines throughout the world.

Raw coal

The coal produced from a mine before being washed.

Run of mine coal

Coal as it comes from the mine prior to any screening or other treatment.

Saleable coal

The coal which is in a form suitable for selling and transporting to the user. Most Australian coal is washed prior to sale (and so changes from raw to saleable coal), but some coal is sold without being washed.

Sales contracts

Most coal is sold under long term contracts between the producer and the customer which cover one or more years, with prices typically re-negotiated annually. Significant quantities are also sold on a spot basis involving one cargo or a series of cargos usually to be delivered within a period of a few months. Tenders have also become a popular method of purchasing coal and involve the customer issuing a tender document which contains details of the required quality, quantity etc.

Seam

The layer of coal or mineral. Coal seams can range in thickness from a few centimetres to 30 or more metres.

Shaft

A vertical hole or tunnel which gives access to the mine and can be used for transporting workers and equipment and for ventilation and other services.

Shale

A fine-grained, sedimentary rock composed of mud from flakes of clay minerals and fragments of other materials. Can contain methane gas. See also Oil Shale.

Shearer

A rotating cutting device used in underground coal mining.

Skip

A container or wagon used to transport coal from the face to the unloading point. In the early days of mining skips were wheeled by men or boys or horses to the bottom of the shaft or to the mine exit.

Soft coking coal

A lower quality of coking coal used in steelmaking. In Australia most soft coking coal comes from the Hunter Valley or Gunnedah coalfields.

Steaming coal

See Thermal coal.

Steelmaking

Steel is made through one of two major processes – the blast furnace/ basic oxygen furnace process (BF/BOF) or the electric arc furnace (EAF) process.

In the BF/BOF process coking coal is heated in large ovens to produce coke (which is composed almost entirely of carbon). The coke is then used in the making of iron, with a mixture of coke, iron ore and limestone poured into a blast furnace. The coke burns to produce carbon monoxide which in turn reduces the ore to liquid iron (or hot metal) which has absorbed the carbon from the carbon monoxide. The liquid iron is then passed through a basic oxygen furnace to produce crude steel. On average, the BF/ BOF process uses 1,400 kg of iron ore, 770 kg of coal, 150 kg of limestone, and 120 kg of recycled steel to produce a tonne of crude steel. (World Steel Association).

The EAF process uses primarily recycled steels and/or direct reduced iron and electricity. On average, the recycled steel-EAF route uses 880 kg of recycled steel, 150 kg of coal and 43 kg of limestone to produce a tonne of crude steel. (World Steel Association)

Strip (or stripping) ratio

The relationship between the volume of overburden which has to be removed for each cubic metre of coal. Can also be expressed in other ways eg the ratio of the thickness of overburden to the thickness of the coal seam. A mine with a low strip ratio will typically be able to produce coal more cheaply than a similar mine with a higher ratio.

Stone dusting

The use of dry stone dust (calcium carbonate) to prevent the spread of coal dust in underground mines and so reduce the risk of explosions.

Tailings

The waste material which is left after coal or minerals are washed or otherwise treated prior to sale.

Thermal coal

Also known as steaming coal and energy coal. Refers to coal which is suitable for use in power stations or other plants to generate steam; the steam then drives turbines which generate electricity. Thermal coal is also used in kilns in the manufacture of cement.

Washery

A plant to wash impurities from coal before it is shipped to users. See also Coal handling and preparation plant.

Bibliography

ACIL Economics & Policy P/L, *Review of Mine Safety in NSW*, ACIL 1997.

Australasian Coal & Oil Shale Employees' Federation, *Australia undermined: coal in crisis*, ACSEF, Sydney, 1972.

Australian Coal Association, *Coal Industry Taxation Policy*, ACA, Sydney, 1982.

Australian Coal Association, *Submission to the Minister for National Resources for repeal of the coal export duty*, ACA Sydney, 1983.

Australian Industrial Relations Commission Review Committee, *Review Report*, AGPS, Canberra, 1991.

Baines, Stephen, *Steel on steel: inside the battle for the future of Australia's biggest railroad*, University of Queensland Press, St Lucia, Qld, 2014.

Bowden, Bradley (ed), *Work and Strife in Paradise: the History of Labour Relations in Queensland 1959-2010*, Federation Press, Annandale, NSW, 2009.

Bowden, Bradley, *A study of resource dependency: the coal supply strategy of the Japanese steel mills 1960-2010*, Journal of Management History, Vol. 19 Issue 1, 2013.

Bowden, Bradley, *A History of the Pan-Pacific Coal Trade from the 1950s to 2011: Exploring the Long-term Effects of a Buying Cartel*, Australian Economic History Review, Vol. 52 (1).

Bowen, Bruce and Gooday, Peter, *The Economics of Export Controls, ABARE Research Report 93.8*, Australian Bureau of Agricultural and Resource Economics, Canberra, 1993.

Brady, Frank, *Electricity Supply in NSW – The First Century and Beyond*, Institution of Engineers Australia, 1994.

Branagan, David, Geology and Coal Mining in the Hunter Valley, Newcastle Public Library in assoc with Newcastle and Hunter District Historical Society, 1972.

Branagan, David (ed), *Coal in Australia: the Third Edgeworth David Symposium*, University of Sydney, September 1990.

Broomham, R., *First Light: 150 years of Gas*, Hale & Ironmonger, Sydney, c1987.

Brune, J. (ed), *Extracting the Science: A Century of Mining Research*, Society for Mining, Metallurgy and Exploration, 2010.

Bunker, S. and Ciccantell, P., *East Asia and the Global Economy: Japan's Ascent, with Implications for China's Future*, John Hopkins University Press, 2007.

Byrnes, Michael, Australia and the *Asia game: the politics of business and economics in Asia*, Allen & Unwin, 2nd ed., Crows Nest, NSW, 2006.

Christison, Ray, *A light in the Vale: Development of the Lithgow District Miners' Mutual Protective Association*, City of Greater Lithgow Mining Museum, Lithgow c2011.

Committee of Review into Australian Industrial Relations Law and Systems (Hancock Committee), *Report of the Committee of Review*, AGPS, Canberra, 1985.

Coopers & Lybrand, *Australian Coal Association Industry Report 1987: coal in crisis*, Sydney, 1988.

Culter, Suzanne, *Managing decline: Japan's coal industry restructuring and community response*, University of Hawaii Press, Honolulu, c1999.

Dabscheck, Braham, *The Struggle for Australian Industrial Relations*, Oxford University Press, Melbourne, 1995.

Daly, Fred, *From Curtin to Kerr*, Sun Books, South Melbourne, 1977.

Denney, Bruce, *From Lamplight to Daylight – A History of the New Hope Group*, Impact Press, Edgecliff, NSW, 2019.

Department of Resources and Energy, *Initiatives in minerals and energy: report by Rex Connor, M.P., Minister for Minerals and Energy, on progress during 27 months of Labor Government*, Department of Minerals and energy, Canberra, 1975.

Dingsdag, Donald, *The restructuring of the NSW coal industry, 1903-1982*, Ph.D. Thesis, University of Wollongong, 1988.

Drysdale, P. and Tsukuda, C., *Long-run adjustment of the Japanese iron and steel industry and its implications for Australia: an overview*, Research School of Pacific studies, ANU, Canberra, 1989.

Dunn, Col, *A History of Electricity in Queensland*, Bundaberg, Queensland, 1985.

Employee Relations Study Commission, *Working Relations: A Fresh Start for Australian Enterprises*, Allen & Unwin in association with Business Council of Australia, North Sydney, c1991.

Fisher, Chris, *Coal and State*, Methuen, Sydney, 1987.

Fitzgerald, T.M., *The contribution of the mineral industry to Australian welfare: report to the Minister for Minerals and Energy the Hon. R. F. X. Connor M.P.*, AGPS, Canberra, 1974.

Freudenberg, Graham, *Cause for Power: The Official History of the New South Wales Branch of the Australian Labor Party*, Pluto Press, Sydney, 1991.

Galligan, Brian, *Utah and Queensland coal: a study in the micro political economy of modern capitalism and the state*, University of Queensland Press, Brisbane, 1989.

Thomas, Pete and Gorman, Paddy, *The coal mines that workers ran: from Nymboida to United a remarkable chapter in Australia's industrial history*, CFMEU Mining & Energy Division, Sydney, 2000.

Gunningham, N., *Mine Safety: law and regulation*, Federation Press, Annandale NSW, 2007.

Harris, Stuart and Ikuta, Toyaki (eds), *Australia and the energy coal trade*, Australia – Japan Research Centre, Canberra, 1982.

Hargraves, A.J. and Martin, C.H. (eds) *Australasian coal mining practice*, AusIMM, Parkville Vic., 1993.

Hawthorne, W.L., *Coal exploration by the Queensland Department of Mines during the period 1950 to 1984*, Queensland Geological Record, 2011/07.

Hawke, Anne and Robertson, Frances, *Enterprise Bargaining in the Coal Industry: Implications for Work Practices*, National Institute of Labour Studies, Monograph Series Number 5, Adelaide, 1999.

Hayden, Bill, *Hayden; an autobiography*, Harper Collins, Sydney, 1996.

Hearn McKinnon, Bruce, *The Weipa dispute: ramifications for the spread of Australian workplace agreements*, ACIRRT Working Paper 50, 1997.

Hilmer, Fred, *Avoiding Industrial Action: A Better Way of Working*, Business Council of Australia, Melbourne, 1991.

Hilmer, Fred and McLaughlin, Peter, *A Better Way of Working*, Economic Planning Advisory Council, Canberra, 1989.

Hince, Kevin, *Conflict and coal: a case study of industrial relations in the open-cut coal mining industry of central Queensland*, University of Queensland Press, Brisbane, 1992.

Hopkins, Andrew, *Lessons from Gretley: mindful leadership and the law*, CCH Australia, Sydney, 2007.

Hopkins, Andrew, *Managing major hazards: the lessons of the Moura disaster*, Allen & Unwin, Sydney, 1999.

Industrial Relations Study Commission, *Enterprise Bargaining units, a better way of working*, Business Council of Australia, Melbourne, 1989.

International Energy Agency, *Coal Information 2000*, OECD/IEA, Paris, 2000.

International Energy Agency, *Steam Coal Prospects to 2000*, OECD/IEA, Paris, 1978.

International Energy Agency, Coal prospects and policies in IEA countries, OECD/IEA, Paris, 1979.

International Energy Agency, *Coal prospects and policies in IEA countries – 1981 review*, OECD/IEA, Paris, 1982.

International Energy Agency, *World Energy Outlook*, OECD/IEA, Paris, 1982.

International Labour Office, *Study of the Embargo of Coal Exports from South Africa*, ILO, Geneva, 1992.

Jay, Christopher, *The Coal Masters: The History of Coal and Allied 1844-1994*, Focus Publishing, Sydney, 1994.

Joint Coal Board, *Prospective expansion –black coal industry in Australia*, JCB, Sydney, 1980.

Kelly, Paul, *The March of the Patriots*, Melbourne University Press, Carlton Vic., 2009.

Killin, Kerry, *Drovers, Diggers and Draglines: A History of Blair Athol and Clermont*, Pacific Coal, Brisbane, 1984.

Kinninmonth, R.J. and Baafi, E.Y. (eds), *Australasian Coal Mining Practice*, AusIMM, Carlton Vic., 2009.

Koerner, Richard, *Pacific Rim Coking Coal Trade*, Australia-Japan Research Centre, Canberra, 1992.

Koerner, Richard, *Behaviour of Pacific energy markets: the case of the coking coal trade with Japan*, Australia-Japan Research Centre, Canberra, 1996.

Kume, Ikuo, *Disparaged Success: Labour Politics in Postwar Japan*, Cornell University Press, New York, 1998.

Lee, David, *The Second Rush: Mining and the transformation of Australia*, Connor Court Publishing, Redlands Bay Qld, 2016.

Lee, M. and Draper, S., *The coal industry: the current crisis and the campaign for a national coal authority*, Journal of Australian Political Economy, No.23, 1988.

McColl, G.D., *A review of labour relations in the Australian black coal industry*, Centre for Applied Economic Research, UNSW, Working Paper No.42, 1982.

McKern R.B., *An overview of foreign participation in the Australian minerals industry*, ANU, Canberra, 1975.

McPhillips, Jack, *Wharfies and miners say "Enough is Enough": the story of the 1994 wharfies' and miners' disputes*, Socialist Party of Australia, Surry Hills NSW, 1994.

Menadue, John, *Things you learn along the way*, David Lovell Publishing, Ringwood Vic., 1999.

Martin, C.H. et. al., *History of Coal Mining in Australia*, AusIMM Monograph No.21, 1993.

Mumford, K., *Structure and Disputation in New South Wales coal industry, 1952-1987*, Australian Economic History Review, Vol.34 No.1, 1994.

Murray, Alan, *No Easy Field: Ipswich coal mining 1920 to 2000*, University of Queensland Press, St. Lucia Qld, 2010.

Murphy, D. (ed), *The Premiers of Queensland*, University of Queensland Press, St. Lucia Qld, 1990.

NSW Combined Colliery Proprietors' Association, *Coal Mission to Japan October 2-5, 1972*, NSWCCPA, 1972.

NSW Commission of Audit, *Focus on Reform: Report on the State's Finances*, Commission of Audit, Sydney, 1988.

NSW Government, *Coal Export Strategy Study: report of task force*, Sydney, 1979.

NSW Parliament Select Committee, *Report of the Select Committee upon the Administration of State Owned Coal Mines*, Parliament of NSW, Sydney, 1993.

Norington, Brad, *Jennie George*, Allen & Unwin, St. Leonards NSW, 1998.

Parker, Paul (ed), *Pacific coal trade in the 1990s: energy or environmental priorities*, ANU, Canberra, 1993.

Patience, Alan (ed), *The Bjelke Petersen Premiership 1968-1983: issues in public policy*, Longman Cheshire, Melbourne, 1985

Patience, Alan and Head, Brian (eds), *From Whitlam to Fraser*, Oxford University Press, Melbourne, 1989.

Patience, Alan and Head, Brian (eds), *From Fraser to Hawke*, Longman Cheshire, Melbourne, 1989.

Pearse, G., McKnight, D. and Burton, B., *Big Coal: Australia's Dirtiest Habit*, University of NSW Press, Sydney, 2013.

Penney, K. and Cronshaw, I., *Coal in India 2015*, Office of the Chief Economist, Department of Industry and Science, Canberra, 2015.

Philalay, M., Drahos, N., and Thurtell, D., *Coal in India 2019*, Office of the Chief Economist, Department of Industry, Innovation and Science, Canberra, 2019.

Porter, Denis, *Coal: the Australian Story, from convict mining to the birth of a world leader*, Connor Court Publishing, Redlands Bay, Qld, 2019.

Pragnell, Brad, *Mapping enterprise agreements in the NSW and Queensland coal industry*, ACIRRT Working Paper 35, 1995.

Price, John, *Japan Works: Power and Paradox in Postwar Industrial Relations*, ILR Press, Ithaca NY, 1997.

Priest, Joan, *The Thiess Story*, Booralong Publications, Ascot Qld, 1981.

Productivity Commission, *The Australian Black Coal Industry*, Ausinfo, Canberra, 1998.

Queensland Coal Board & State Electricity Commission of Queensland, *West Moreton Coalfield Joint Investigation: report and recommendations*, Queensland Coal Board, Brisbane, 1975.

Reeves, Andrew, *Up from the Underground: Coalminers and Community in Wonthaggi 1909 to 1968*, Monash University Publishing, Clayton Vic., 2011.

Ross, Edgar, *A History of the Miners' Federation of Australia*, Australasian Coal and Shale Employees' Federation, 1970.

Scott, Keith, *Gareth Evans*, Allen & Unwin, St. Leonards NSW, 1999.

State Pollution Control Commission, *Botany Bay Coal Loader: report and findings*, SPCC, Sydney, 1975.

Sheldon, Peter and Thornthwaite, Louise, (eds), *Employer Associations and Industrial Relations Change*, Allen & Unwin, St. Leonards NSW, 1999.

Spires, Robert, *History of Kemira Colliery 1857-1984*, Wollongong City Council Library.

Sumiya, Mikio, *A History of Japanese Trade and Industry Policy*, Oxford University Press, New York, 2000.

Taylor, Ross, *In the Black: The Auscoal Super Story*, Auscoal Super, Newcastle, 2011.

The Edge Public Relations, *People Passion and the Pursuit of Excellence – 20 years of ACARP*, Australian Coal Association Research Program, Brisbane, 2012.

Thomas, Pete and Gorman, Paddy, *The coal mines that workers ran: from Nymboida to United a remarkable chapter in Australia's industrial history*, CFMEU Mining & Energy Division, Sydney, 2000.

Thompson Peter, and Macklin, Bob, *The Big Fella: the rise and rise of BHP*, Random House, North Sydney NSW, 2009.

Trengove, Alan, *Discovery: Stories of Modern Mineral Exploration*, Dominion Press, North Blackburn Vic., 1979.

Trengove, Alan, *What's good for Australia..! the story of BHP*, Cassell Australia, Stanmore NSW, 1975.

Tsokhas, Kosmas, *Beyond dependence: companies, labour processes and Australian mining*, Oxford University Press, Melbourne, 1986.

Whitlam, Gough, *The Whitlam Government 1972-1975*, Penguin Books Australia, Ringwood Vic., 1985.

Whitlam, Gough, *The inaugural R.F.X. Connor memorial lecture: the Connor legacy*, University of Wollongong Historical Society's Historical Journal, Vol.3 No.1 November 1979.

Wilcox, J.M., *International trade in coal: paper presented at UN seminar on future prospects for coal, AGPS, Canberra, 1980.*

Wilcox, Jack, *Coalman*, self-published, Maleny, Qld, 2014.

Wilkenfeld, G. and Spearitt, P., *Electrifying Sydney – 100 years of Energy Australia*, Energy Australia, Sydney, 2004.

Wilson Carroll, L., *Coal – bridge to the future: a report of the World Coal Study*, Ballinger Publishing Company, Cambridge Ma, c1980.

Wooden, Mark, et. al., *Coal industry awards and agreements and the implications for work practices and working conditions*, National Institute of Labour Studies, Adelaide, 1996.

Wooden, Mark and Robertson, Frances, *Employee Relations Indicators: Coal Mining and other Industries Compared*, National Institute of Labour Studies, Working Paper No.143, Adelaide, 1997.

Wooden, Mark, et. al., *The transformation of Australian industrial relations*, Federation Press, Leichhardt NSW, 2000.

Wright, Michael, *Muted Sirens: Demarcation and Union Coverage in the Australian Coal Mining Industry*, ACIRRT Working Paper 27, May 1993.

Yergin, Daniel, *The Prize: the epic quest for oil, money and power*, Free Press, New York, 2008.

Index

9 781922 449313